T3-BOQ-267

Statistics for Public Policy and Management

William F. Matlack

Graduate School of Public and International Affairs
University of Pittsburgh

DUXBURY PRESS *North Scituate, Massachusetts*

Statistics for Public Policy and Management was edited and prepared for composition by *Mary N. Lewis.* Interior design was provided by *The Book Department.* The cover was designed by *Edward Aho.*

Duxbury Press
A Division of Wadsworth, Inc.

Library of Congress Cataloging in Publication Data

Matlack, William F 1928-
Statistics for public policy and management.
Includes index.
1. Statistics. 2. Management—Statistical methods.
I. Title.
HA29.M28 519.5 79-11886
ISBN 0-87872-226-2

Printed in the United States of America
1 2 3 4 5 6 7 8 9—84 83 82 81 80

Contents

Preface

Statistics for Public Policy and Management is written for students who are preparing either for careers in government or for managerial careers in private organizations that are involved in public service. There are no mathematical prerequisites beyond elementary algebra.

It has been my experience in teaching students in both business and public administration that many are looking for applications of statistics that are relevant to the management of nonprofit, educational, health-care, and other institutions of a quasi-public nature. Yet most textbooks in statistics for management seem to place heavy emphasis on production, finance, marketing, and other areas that are of concern chiefly to profit-making enterprises. This text is an effort to broaden the applications of statistics in recognition of this wider range of interest.

Teachers of basic statistics courses are well aware of the fear and trepidation that many students bring to the subject. Since this fear and trepidation can be a major obstacle to effective learning, I have attempted to present the material in a clear, nontechnical, and nonthreatening manner.

Each chapter starts with a statement of objectives, an introductory overview, and a chapter outline. The main body of the chapter consists of the presentation of material, examples, and exercises for the student. The chapter ends with a chapter summary and a set of exercises.

Chapter 1 introduces the student to the concept of statistics, particularly as it applies to public management. Chapters 2 and 3 introduce descriptive statistical measures that will be used in later chapters on inference making. Chapters 4 through 6 introduce probability and probability distributions. In addition to preparing the student for making statistical inferences, these chapters are intended to develop the ability to think in a probabilistic way. Chapters 7 through 11 cover statistical inference, including both parametric and nonparametric statistical tests. Chapter 12 discusses the application of statistical techniques in experimental design, particularly as it relates to the evaluation of public programs.

The twelve chapters are designed to be covered in a one-semester or one-quarter course. If the instructor wishes to reduce the quantity of material to be covered, she or he can eliminate those advanced sections and problems marked with an asterisk (*), and possibly one or more of the last four chapters. Chapters 1 and 8 are cumulative and are essential to understanding the subsequent chapters.

I would welcome the comments of instructors using the book, particularly about the quantity of material and the general receptivity of the students. Please send any suggestions to me at the Graduate School of Public and International Affairs, University of Pittsburgh, Pittsburgh, PA 15260.

Acknowledgements. My debt to my friends, family, colleagues, students, reviewers, and typists is so great that my own contribution seems small in comparison.

I have made use of all the books listed in the Suggested Readings and am indebted to their authors. I acknowledge with deep appreciation the wise and warm counsel of Morris Hamburg of the University of Pennsylvania, whose influence over many years is reflected not only in the writing of this book but also in my teaching and my statistical thinking.

I want to thank my colleagues in the University of Pittsburgh, particularly William Brinckloe and Michael Gold, for their encouragement and suggestions. I am indebted to the many students whom I have had the privilege of teaching in the Graduate School of Public and International Affairs and in the Wharton School. It is from them that I received the inspiration to write this book.

I want to acknowledge the support of Alexander Kugushev, Robert West, and the staff of Duxbury Press. In particular I want to thank Jerry Lyons for all his work in putting this book together, and Mary N. Lewis for her painstaking and able work in improving every chapter.

I am grateful to David Presser for his very thorough and constructive review. I also want to thank George I. Balch, University of Illinois, Chicago; and John P. Dirkse, III, George Washington University, who reviewed and commented on an earlier draft of the manuscript.

I want to thank Patti Kearney, Sandi Mason, Mary Ann Welsh, Joyce Dzugan, and Ruth Buncher for their many hours of typing when they were under pressure from other equally overbearing professors.

I want to thank my father, David Matlack for, among many other things, inspiring a reluctant schoolboy to take some math.

My deepest and warmest thanks go to Cynthia, Meg, Donald, Liza, and Amy.

Introduction

Ms. Able, a section chief in a state agency, is choosing a new supervisor for a unit in her section. In order to evaluate the candidates for the position, she collects such readily available information as age, performance evaluations, and letters of recommendation. Ms. Able wants to determine how to use this information in estimating how well each candidate would perform if given the job. Furthermore, she is interested in determining what sort of additional information she should obtain. How many personal interviews of each candidate should she try to get? Who should do the interviewing? How important are personal interviews compared with written credentials? Should aptitude or achievement tests be required?

Ms. Able is in the position of trying to make a decision on the basis of limited data. No matter whom she chooses for the job, she runs the risk of making a poor choice. But even though it is not possible always to make the best decision (as determined by hindsight), it is possible to make intelligent decisions. Making intelligent decisions involves making use of the kinds of information that do the most to predict job performance rather than spending time and money on obtaining information that helps very little or not at all in predicting performance. This may seem very obvious until you think about how often it is not done. If you have taken qualifying examinations that seek massive amounts of information about you that do not say much about how good you are for a job; if you have undergone interviews when people's minds have already been made up; if you have had to wait so long to hear the results of your interview that you were forced to make another commitment while waiting; then you may have some idea of how widespread poor decision-making procedures are in this area alone.

The art and science of statistics is the art and science of getting statistical data and using them efficiently for your purposes. Ms. Able's purpose is to decide on whom to hire; her problem is one of selecting the statistics most useful for her purpose and making good use of those statistics.

An understanding of probability and statistics can be of immeasurable help in enabling you to obtain information and to sift out the valuable information from the less valuable from the useless. It can also help you to know at what point to stop getting more information and make a decision. By making the best use of your information, you will make good decisions more frequently, and you may also be able to lessen the effects of bad decisions. In the long run, administrators who can do this will be far more effective than those who cannot.

The art of using statistical information applies not only to managerial decisions within an organization, but also to much broader areas of public policy. For example, some public officials are going to be concerned with meeting the energy needs of the United States (or southern Ohio, or of South Hadley) over the next 20 or 30 years. It will be overwhelmingly difficult, if not impossible, to determine with a high degree of accuracy what these needs will be. One would need to consider the future direction of national and regional economies, technological changes, changes in tastes, and changes in the political climate, just to mention a few variables. Even though each one of these variables is subject to much uncertainty, policy makers at the federal, state, and local levels, as well as those in private companies, will want predictions of these needs. It is important that these officials be able to determine how much confidence they can have in these predictions, and the probabilities that such predictions will be either too high or too low, and by how much. If they were lazy they could lean so heavily on the analysis and predictions of "experts" that they would in effect be letting the experts make the decisions for them. But public policy is too important to be left to the specialists. It is important that those involved in policy making be able to understand the reports of the experts, to communicate with them, and to ask them the relevant questions. This requires that they have a basic understanding of the statistics and the assumptions underlying the techniques that the experts use.

John Kenneth Galbraith has warned us about the dangers of a society ruled by technocrats; it is important that decisions in public policy be made not solely by specialists on the basis of their esoteric knowledge of technique. This is not to say that technocrats, specialists, or statisticians are of evil intent, nor that we do not need them and their skills very much. But it is essential that public policy makers understand what the specialists are saying. As a student of public policy you must be prepared to ask the experts specific and pertinent questions, and to interpret their answers intelligently.

There is hardly an area of government or public-sector enterprise that is not either actively engaged in research or making use (or trying to make use) of the research done by others. And whenever one is engaged in empirical research the problems of statistical inference are very much present. I will mention a few of the types of research carried on in both government and business that draw heavily on statistical knowledge.

Sample surveys. The field of sample-survey methods involves both statistical and nonstatistical techniques. The statistical techniques include the measurement of reliability and the reduction of errors through sampling design. The nonstatistical techniques include methods of obtaining information from nonrespondents, etc. In practice, it is very difficult to separate the statistical aspects from the nonstatistical aspects of a survey design. For example, you may be getting biased results because those who do not respond to your questionnaire differ substantially from those who do respond. The problem of measuring the size of this bias is largely a statistical problem, but the methods of getting information from the nonrespondents in order to reduce this bias require skills that go far beyond purely statistical skills. The problem of controlling this bias will require such a combination of statistical techniques that the survey researcher and the statistician had better be either the same person or two persons who understand each other, and each other's specialty, very well.

Experimental design. Many organizations conduct studies or make use of studies that are known as experimental designs. Experimental designs are methods of assessing the effect of one variable on another while controlling for the effect of other variables. Examples include studies on the effect of air pollution on the human body, and the effect of a new teaching method on the ability of students to learn. In many cases it is not possible to control for all the other relevant variables. In such cases it is necessary to resort to designs that fall short of the true experimental design. In assessing such studies, you should be knowledgeable enough to understand the extent to which the study succeeds or fails to live up to the experimental ideal.

Program evaluation. In determining whether or not to provide money for a public program, people may want to know how successful the program has been in the past in achieving its goals. If it is a new program, they might want to evaluate the performance of similar programs. Ideally, a program evaluation will consist of comparing the results of the program with results that would have occurred had there been no program. This may be an unobtainable ideal; however, something approaching such a comparison may be possible if the evaluators are endowed with ingenuity and skills of both a statistical and a nonstatistical nature.

Predictions. Even if you are not going to be participating in studies that form the bases for forecasts, the chances are nevertheless high that you will be working in an organization that makes use of such studies or forecasts. You will want to decide whether the forecasts made the best use of the information that went into them, and how much confidence you should place in them. All of this will require some amount of statistical analysis.

Operations Research techniques. Many managerial problems require more skills in mathematical analysis than those usually possessed by managers with traditional training in Business or Public Administration. The quantitative skills go under the name of "Operations Research" or "Management Science" techniques. They draw heavily on the statistical methods you will study in this book. Among the problems dealt with by Operations Research methods are those of scheduling work, controlling levels of inventory, providing and staffing facilities for clients or customers who are waiting in line, and many others.

The material in this book is designed to provide some of the basic skills and methods of thinking that a manager or policy maker will need to have. There is an additional purpose of the book that underlies that of providing skills. This is to develop your ability to think in terms of probability. When you make a decision you may be assuming that a certain situation is going to prevail in the future; but your assumption is in a sense only a bet, and you should be aware that your bet might be wrong. How likely, or probable, is your bet going to prove to be incorrect? How costly will it be if your bet does turn out to be wrong? And how might you change or adjust your decision to allow for these contingencies?

Thinking in terms of probability can be of immense value to you when making decisions ranging from major commitments such as the investment of millions of dollars all the way down to the scheduling of daily appointments and committee meetings.

Overview of the book. Chapter 1 will introduce you to some fundamental concepts of statistics and research. Chapters 2 and 3 discuss Descriptive Statistics, which are the data that statisticians work with. When you analyze the results of a sample, you will use descriptive statistics to describe the most relevant features of the sample. The following four chapters are intended to develop your understanding of and feeling for probability. Probability is not only a concept virtually essential in making decisions under conditions of uncertainty; it is also a necessary groundwork for understanding the problems of making inferences from the descriptive statistics that you obtain from sampling. The remaining chapters bring together the concepts of descriptive statistics and probability so that you will be able to make valid inferences from the information that you have obtained. These inferences may take the form of tests of hypotheses, estimations, or decisions to choose among several courses of action.

Statistical textbooks are typically not easy reading for most students who have not been exposed to the subject before. If you are typical, I would suggest reading the difficult passages several times. Even if you find a section heavy going, I would suggest that you read ahead to the end of the section and then go back to review the difficult passages. If you are doing the reading for a class, you may find it most helpful to do your first reading before the class period and the second reading after the class period. Above all, do as many exercises as you reasonably can. The exercises at the end of each section and at the end of each chapter cover the highlights of the sections and chapters that are of particular importance in understanding the current and subsequent material. The answers to selected exercises at the end of the book are provided to help you evaluate your own progress.

Some sections and problems that are a little more difficult are marked by an asterisk (*). Although your instructor may assign some of these sections and problems, be assured that you may not be alone in finding them difficult, and that the continuity of later sections with earlier sections will not be lost if you skip such sections or problems.

May I offer my hope that as you progress you not only shed any possible fears that you may have had about statistics, but that you may also begin to share some of my enthusiasm for a subject that is both intellectually fascinating and highly relevant to the problems of management and our society.

ONE

Statistics for the Public Manager

Outline of Chapter

1.1 The Uses and Misuses of Statistics

Statistics can be manipulated and misused to serve many purposes. The fact that this is so is a compelling reason for increasing your understanding of the subject.

1.2 Statistical Inference and Decision Making

We use statistical analysis to make inferences about the cases we do not observe.

1.3 Some Basic Concepts

Somc fundamental terms are defined to give you a foundation for understanding the nature of statistical inference.

Objectives for the Student

1. Become aware of some of the dangers and pitfalls in misusing statistical measures (1.1).
2. Appreciate the advantages of getting statistics from unbiased sources (1.1).
3. Become aware of the importance of making good use of information in making decisions (1.2).
4. Become aware of the way that implicit values and assumptions can become incorporated in policy recommendations (1.2).
5. Understand the concepts of population, sample, sampling error, and bias (1.3).

1.1 The Uses and Misuses of Statistics

A recent statement from a gas company pointed out that "November and December combined were about 42 percent colder than those months last year." There was no indication, however, of how one measures percentage increases in coldness. Would 5 degrees Fahrenheit be considered twice as cold (or half as warm) as 10 degrees Fahrenheit? Of course the percentage changes would be quite different if one were to use a Celsius scale. Perhaps the gas company measures "coldness" in terms of the amount of gas consumed for heating in these months, but if this was the case it was not indicated in the statement. When no measure of percentage changes in coldness is indicated, it is possible for a writer to come up with a number of apparently contradictory measures, and to select only the measure that best gets across to the reader the point that he wants to get across.

There is ample reason why many people harbor a deep distrust of statistics. When individuals are selecting and citing statistics in order to promote a particular point of view, the statistics can be very misleading. Percentage changes are presented without specifying the base on which the percentages are computed; individual interviews giving a favorable impression of a product are presented on TV by an advertiser who does not inform the viewer that only the most favorable interviews were selected for TV presentation; people cite cases to support their point of view and ignore or suppress cases that tend to refute their point of view.

The fact that statistics are so often misused, however, does not justify having a totally cynical attitude toward the subject. There are such things as good statistics, collected and presented by researchers who are careful to be clear and fair. Public managers need statistics on which to base their decisions, and they will need to know the difference between good statistics and bad statistics. Far from justifying a notion that we should not study statistics, the very fact that statistics can be so easily misused is an important reason for managers to acquire some sophistication about the subject.

EXERCISES

1–1. Identify some magazine advertisements or TV commercials that cite statistics that either are unclear about what they are measuring or give misleading impressions. It should not take you long to find a dozen or more.

1–2. A TV advertisement claims: "An independent unbiased survey shows that 3 out of 5 doctors recommend the use of Mufflerin pills for common headache complaints." Assuming that the claim made no false statements, point out any inferences that an unwary viewer is likely to make but are not warranted by the statement.

1.2 Statistical Inference and Decision Making

Many people think of statistics as the compilation of numerical data, but the subject involves far more than this. Statistical data are seldom gathered for their own sake; they are collected in order for people to make generalizations that go beyond the actual data, and to make decisions based on these generalizations. People are not ultimately interested in energy consumption in the past; they want to know what it will be in the future, and what to do about it. People are not ultimately interested in candidates' test scores, performance ratings, etc.; they want to know how well they will perform on the job. By the same token, statisticians are not ultimately interested in numerical data; they are interested in what the data mean for the future, and what decisions make the most sense in light of the data.

> The process of making generalizations or predictions on the basis of limited data is known as STATISTICAL INFERENCE.

All too often we encounter glaring examples of faulty reasoning in making inferences. Many of the most common examples of prejudice are cases of the misuse of information. For example, a statement such as "women are lousy drivers" is all too likely to be a conclusion based not on adequate information but on one or two casual and not necessarily representative observations. The same thing can be said of sweeping generalizations about blacks, Jews, fat people, other minority groups, and majority groups.

An administrator is called upon to make frequent decisions, and it is a characteristic of administration that the available information is seldom complete or adequate. To the extent that the administrator's effectiveness depends on the quality of his or her decisions, to that extent will the administrator's effectiveness be enhanced by an understanding of the statistical techniques of using incomplete information in making decisions.

For some types of decisions, administrators will have to go get the relevant information themselves. No matter how small and informal such information-gathering efforts may be, the principles of good statistical sampling procedure still apply.

For larger, long-range decisions, administrators are more likely to employ others to do the information gathering, or to read existing studies. But even here the manager must be familiar with the statistical problems in doing research. If you hire a consultant or have a technical staff already working for you, it is important that you communicate effectively with the consultant or staff. If you don't understand the way your experts think, and they do not understand your problem, then one of two results is likely to happen:

1. You may make an attempt to read their recommendation, you file it away, thank them for their efforts, and then do what you would have done anyway without their recommendation.
2. You may try to follow their recommendations to the letter without fully understanding the reasons for them.

The first of these courses of action is a total waste of the expertise for which you are paying. The second can get you into worse trouble than if you ignored their recommendation.

For example, as Director of Transportation, suppose you have hired a panel of experts to evaluate several plans for the construction of a highway between two cities. The panel, after considering proposals calling for expenditures of $15 million, $20 million, and $25 million, rejected the $15 million proposal because of inadequate provisions for safety, and also rejected the $25 million proposal on the grounds that it offered only a questionable increase in safety over the $20 million proposal, "an increase that fails to justify an additional $5 million in expense over the $20 million proposal." If you accept the panel's recommendation, you should be able to know how to weigh a "questionable increase" in safety against an additional cost of $5 million. A "questionable increase" involves an estimate of probabilities, and these probabilities need to be applied to the reduction of accidents, injuries, and deaths that may result from the additional expense. Analysis may reveal that, by accepting the panel's recommendation, you are implicitly assessing the value of a human life saved at below $25,000. Perhaps you would agree with this assessment; but if you do not, you should be able to determine which proposal does agree with your values.

In reading a research report, an executive would benefit from having an understanding of the conceptual problems involved in the study in order to better understand the conclusions of that study. If you are involved in making decisions or implementing decisions concerning a school desegregation program, for example, you will want to be familiar with research in the field such as the Coleman Report of 1966, which analyzes the educational gains and losses that accrue to minority and majority-group pupils when they attend integrated schools. The Coleman study embodies an elaborate statistical design, an understanding of which is essential to an adequate evaluation of the conclusions.

It may be no exaggeration to say that all the technical expertise that goes into studies such as the Coleman study will be largely wasted unless the public officials who read and make policy decisions on the basis of such reports have a reasonable grasp of the problems of research design involved in the reported study. The complexity of our social and economic problems is going to require an increasing number of researchers who have been trained to develop experimental designs for analyzing the relationships among many variables. An equally high priority, however, is the training of public officials and public managers who can understand and interpret the studies of these specialists.

EXERCISES

1–3. In 1928 the Democratic candidate for President, Alfred E. Smith, was overwhelmingly defeated by his Republican opponent. Smith was the first Roman Catholic to be nominated by a major political party. On the basis of this election, it was inferred by many that "no Catholic can be elected President."

a) Discuss the adequacy of the data on which this conclusion was based.
b) What voting patterns would you particularly look for to determine the effect of the candidate's religion on the outcome?
c) A Roman Catholic was elected President in 1960. Does this fact necessarily refute the inference drawn from the 1928 election? Why or why not?

1–4. An organization has passed a new rule requiring all employees to retire at the age of 65, so that more young blood can be instilled into the organization.

a) What explicit or implicit assumptions are being made by this rule?
b) What kind of data or information would you think would be appropriate on which to base these assumptions?

1–5. The governor has asked for a $25 million increase in support of highway safety, on the grounds that such an increase could result in the saving of 50 or more lives per year and three times as many serious injuries. The minority leader has opposed the increase on the grounds that it is extravagant, and that he doubts that half that many lives will be saved, or serious injuries prevented.

a) Do the positions taken by the governor and the minority leader reflect any implicit value judgments on their parts?
b) In your opinion, does determining a policy on spending public money to save lives require one to set a quantitative value on a human life to be saved?
c) How might you assess the value of a human life for this purpose?

1.3 Some Basic Concepts

This book is intended for the general administrator and the student who is studying to become a manager in an organization that has a public sector or public-service orientation. Such a person is going to be involved in making decisions and setting policy on the basis of studies involving statistics, whether these are published books, in-house surveys, or information gathered from very limited observation. On the basis of this information the decision maker will draw a general conclusion, or inference.

> A POPULATION OR UNIVERSE is the set of all items of interest to the research.

> A SAMPLE is a subset of the population.

One of the major tasks involved in a research effort is that of making the sample as representative of the relevant population as possible. There are many reasons why a sample may not be representative of the population, but these can be classified into two main reasons. One of these is chance. Purely by chance the individuals whom you select to include in your sample may tend to be older than average, richer than average, more alienated than average, etc. This is especially true for very small samples. In fact, when we take a small sample we would be very surprised if the sample was perfectly representative of the universe; we would expect measures from the sample (such as averages, proportions, etc.) to be somewhat different from their corresponding values in the population. For any given sample size, there is a certain "expected error," which measures how much we expect our sample results to be inaccurate.

> We define SAMPLING ERROR as error due to chance or probability. It is a measure of how far off we expect the sample value to be from the universe value it is trying to estimate. Sampling error tends to get smaller as the sample size increases, so that with very large samples the sampling error becomes close to zero.

Another way of describing sampling error is to say it is a measure of unreliability, or of our lack of faith in the sample results.

If sampling error were the only problem involved in statistical inference, we would be able to solve most sampling problems by simply using bigger samples. There is a popular misconception that bigger samples are necessarily more "scientific" than smaller ones. But, unfortunately, there is another kind of error that will not go away with an increase in the size of the sample. This error is known as systematic error or bias.

> SYSTEMATIC ERROR OR BIAS is a persistent error in the process of selecting individuals for the sample or measuring individuals in the sample. It will not disappear as the sample size increases.

For example, if you wanted to estimate the average income of graduates of the class of 1966 from Miami University, the returns from a questionnaire sent out to all the graduates whose mailing addresses are known could give you a very biased picture. The error in the process would be the selection of individuals whose addresses were known and who returned the questionnaire. If you suspect that those whose addresses are easily found may have different incomes from those who have disappeared from sight; if you suspect that those who return the questionnaires (or have their secretaries do it) differ in income from those who would rather not think about such things; if you suspect that the incomes people report tend to be inflated (or deflated), then you have several reasons for suspecting a bias in your sampling procedure. A bigger sample will not cure it.

The two kinds of error can be illustrated by two extreme examples.

EXAMPLE 1.1 A random sample of voters showed that 60% favored Miss Adams in her city-wide campaign to be prothonotary. A prediction was made on the basis of this sample that Miss Adams would carry the city with 60% of the vote.

Unfortunately for Miss Adams's peace of mind, she discovered that the sample consisted of only 5 voters, 3 of whom said they would support Miss Adams. Her instincts told her that no matter how randomly or carefully the sample was drawn, that there was a huge amount of error that could be attributed to chance. Although you cannot always trust your instincts in statistical work, in this case Miss Adams's instincts were quite right. When the sample size is very small, the sampling error is very large. ■

If the prediction made in this example strikes you as absurd, it is hardly any more absurd than the one in the following example. The chief difference is that the next example is an actual case of statistical malfeasance which contributed to the demise of a major literary periodical.

EXAMPLE 1.2 In 1936 the *Literary Digest* predicted the results of the presidential election on the basis of a sample of over 2 million responses. The poll, perhaps the largest in the history of presidential elections, predicted that Alfred Landon would defeat Franklin Roosevelt with over 54% of the vote. The election results, however, showed Mr. Roosevelt getting over 60% of the vote in one of the biggest landslides of the twentieth century.

What was wrong with this survey? A sample of that size would have reduced the sampling error to a negligible amount. The error had to be a bias, or systematic error. In fact, the sample had been selected from lists of telephone numbers, automobile registrations, and magazine subscriptions. A generalization from such a sample would have been valid only if the generalization had been made to telephone and magazine subscribers and automobile owners; it was not valid when made to the voters in general. ■

In many cases it is practically impossible to draw a sample from the population that you want. Sometimes this is simply because the population you want does not exist. This is the case when you are making predictions about the future. For example, if you want to estimate the demand for natural gas by the citizens of Tuscaloosa in 5 years, the population of interest is unavailable for sampling—it does not exist yet. The population from which you draw your sample will necessarily be one other than the one you want. It will require all of the skills (both statistical and nonstatistical) of a research team to obtain a sample from a population that closely resembles that future population about which you want information.

Suppose, for example, that a municipality is planning to build a recreational facility for young adults. To simplify the problem, suppose that it has been decided to build either a skating rink or a bowling alley, and that the choice between the two will be decided on the basis of which one is estimated to have the greater potential usage.

How might we get information on the amount of potential demand for each kind of facility? In answering this question, it is necessary to determine what is the population of interest. This is not an easy question to answer, because we have to decide on the range of ages of potential users, their residence or place of employment, the years that we expect each proposed facility to be in use, etc. In any case, we know that the population is a future one, and the means of estimating its characteristics are not going to be easy to find. If we take a sample, from what population should we draw the sample? If we sample the young adults of today, we will not be getting the future users themselves. If we sample the young children of today who will be the young adults when the proposed facility is built, we will be getting people who are going to change greatly in their recreational habits and tastes between now and then. Even though these may largely be the same people we are interested in, it is not the same population for our purposes. If we look at figures on the usage of skating rinks and bowling alleys that have already been built in other areas, we will be sampling a population that differs both in age and place from that we are interested in. Perhaps we might look at all three of these populations and combine the results in some way to make our estimate.

There is no simple answer to this question. It is sufficient here to pose the problem, which must be dealt with not only by the experts who may be hired to conduct the research, but also by the officials who will have to act on the basis of the recommendations coming out of the research. Both the experts and the administrators are going to have to work together and understand something of the other's problems and points of view.

EXERCISES

1–6. An investigator wants to determine the effect of exercise on the speed of recovery of patients after an operation. In a municipal hospital, he makes careful observations on 20 patients who have undergone surgery and been subjected to a heavy exercise routine thereafter, and 25 other patients who have undergone the same operation but who have been kept in bed with little exercise.

What is the population, and what is the sample? What biases would you look for in evaluating the study?

1–7. Interviews of patients who have suffered from a rare disease the previous year indicated that all those interviewed recovered from the disease.

a) Do these results warrant the inference that the recovery rate from this disease is 100%?

b) What bias would you suggest exists in the selection process?

1–8. A supervisor interviewed 10 of his direct subordinates and found that 9 of them expressed general satisfaction with their jobs. If the purpose of the interviews was to determine the proportion of this supervisor's subordinates who were satisfied with their jobs, criticize the approach taken.

Summary of Chapter

When reading any set of statistics it is important to ask yourself who obtained these statistics; who selected the statistics that were to be presented; and, did the people who obtained or presented the statistics have a special interest in persuading people to adopt a particular point of view? Unless the statistics you observe were both obtained and selected for presentation by individuals whose primary responsibility was to do a fair and unbiased study, you should look very closely and critically at the methods of collection and selection.

One of the most important aspects of a manager's job is to use information intelligently in making decisions. When the information comes from statistical studies, it is important that the manager have a grasp of the statistical concepts that are involved in the study in order to make the best possible use of the information. An understanding of the methodology underlying a policy recommendation may be necessary for giving an intelligent critique of the recommendation and for determining any implicit value judgments that go into the recommendation.

A good statistical study will make an explicit effort to control the size of both biases and sampling errors.

REVIEW EXERCISES FOR CHAPTER 1

1–9. An advertiser claims that tests show that an antiseptic spray kills millions of germs upon contact. What inference do you believe the advertiser is trying to get the viewer to make? If the statement is true, does the inference follow?

1–10. A public opinion poll samples 1762 voters across a state in order to predict the outcome of the election for governor.

a) What is the population?
b) What is the sample?
c) What possible biases might the study involve?
d) For what reason might there be sampling error?

1–11. Tom Wilson made the following comment: "Don't ever work for a woman boss if you can help it. When I worked for Miss Nitpick, she was so fussy about all of the petty details of my job that I was never able to devote my energies to more important things. I would quit before I'd work for another woman!"

Criticize the way Tom made use of sample information to draw an inference. Consider especially the problems of bias and sampling error.

SUGGESTED READINGS

COLEMAN, JAMES S., and ERNEST Q. CAMPBELL, et al. *Equality of Educational Opportunity.* Washington, D.C.: Government Printing Office, 1966.

HUFF, DARRELL. *How to Lie With Statistics.* New York: W. W. Norton and Company, Inc., 1954.

MORONEY, M. J. *Facts from Figures,* 3rd ed. Hammondsworth, Middlesex: Penguin Books, Ltd., 1956. Chapters 1, 20.

RIVLIN, ALICE. *Systematic Thinking for Social Action.* Washington, D.C.: The Brookings Institution, 1971.

TANUR, JUDITH, ed. *Statistics: A Guide to the Unknown.* San Francisco, Cal.: Holden-Day, Inc., 1972.

TWO

Descriptive Statistics for One Variable

Outline of Chapter

2.1 Variables and Measurement

The manager's information will be largely based on measurements of variables. She or he should be aware of the problems involved in measurement.

2.2 Validity of Measurement

It is often very difficult to measure the thing you want to measure.

2.3 Grouping of Data

In order to obtain greater clarity from a large amount of data, the manager will generally want the data to be presented in a reasonably condensed form.

2.4 Graphical Methods of Presenting Data

A graph can give a vivid picture of important variables.

2.5 Measures of Location, or Central Tendency

A single number is often called upon to be representative of a whole body of data.

2.6 Measures of Variability, or Dispersion

A number can be used to measure how much individuals tend to vary from each other.

Objectives for the Student

1. Understand the concepts of variable and value of a variable.
2. Appreciate the problems of obtaining a valid measurement of a variable.
3. Appreciate the problems involved in grouping data and the ways that are used to deal with these problems.
4. Distinguish between an ordinary bar chart and a histogram.
5. Know how to find the mode, the median, the arithmetic mean, and the weighted mean of a set of data, and to estimate the arithmetic mean of grouped data.
6. Recognize situations for which one or another measure of location is appropriate.
7. Know how to compute the variance and the standard deviation of a set of data.

Before proceeding to the matter of drawing inferences from a sample to a population, you should become familiar with some basic principles in presenting the information in the sample itself. The sample information is the raw material with which a researcher works. There are both tabular and graphical ways of presenting this information. We can describe samples in an even more concise way by computing averages and measures of variability. These descriptive measures, particularly the arithmetic mean, variance, and standard deviation, are used extensively as a basis for drawing inferences about populations.

2.1 Variables and Measurement

In statistical analysis we make observations. Our observations may be individual persons, organizations, nations, pairs of persons, freight car loadings, pollution readings, number of automobiles, etc.

> A VARIABLE is a characteristic of the observation that can be classified into at least two categories.

> A VALUE of a variable is one of the categories into which the variable can be classified.

Examples of variables and possible sets of their values are presented in Table 2.1.

Notice that it is possible to define a variable in ways that permit division into 2, 3, or more different possible categories. We could have classified "religion" into 2 categories, 3 categories, or many categories. H_1 and H_2 illustrate two ways we could define the variable "height." The notation $X = 0, 1, 2, 3, \ldots$ says that the variable X can take on a value equal to any nonnegative whole number, or integer. The dots (. . .) can simply be read "and so forth." Variable Y is one's age measured by the

Table 2.1. *EXAMPLES OF VARIABLES AND POSSIBLE VALUES*

NAME OF VARIABLE	VALUE
R (Religion)	Protestant, Roman Catholic, Jewish, Other
S (Sex)	Male, Female
C (Conservatism)	Reactionary, Conservative, Middle-of-the-Road, Liberal, Radical
H_1 (Height)	Short, Medium, Tall
H_2 (Height in inches)	$0, 1, 2, 3, \ldots$
X (Age as of last birthday)	$X = 0, 1, 2, 3, \ldots$
Y (True age)	$0 < Y < \infty$
Z (Number of auto accidents)	$Z = 0, 1, 2, 3, \ldots$
W (Family income)	W = Below \$10,000; at least \$10,000

actual time elapsed since birth (in years, days, minutes, etc.), so that a person's Y value increases continuously as time passes. The notation $0 < Y < \infty$ indicates that Y can be *any* nonnegative number.

EXAMPLE 2.1 In what different ways might the variable "Quality of Air" be defined that would call for different sets of values?

Discussion: For the specialist in air quality control, the variable should be defined in a manner to make fine distinctions, such as "quantity of particulate matter per cubic foot of air." Such a definition would permit any positive number as a value, such as 42.8, 116.9, etc. For the person involved in making decisions on whether to call a pollution alert, the values of the variable might be Satisfactory, Partial Alert, Total Alert. ■

> A MEASUREMENT is the procedure by which one determines what value to assign to a variable.

We can measure one's religion by asking the individual to identify his or her religion; we can measure conservatism either by a similar self-identification question or by a test that is designed to measure "conservatism."

> The recorded value of a single measurement is called an OBSERVED VALUE. The set of observed values is denoted by the plural noun DATA.

EXAMPLE 2.2 How might we measure the variable "race"?

Discussion: One method commonly used is the observation of an interviewer. She or he simply categorizes the interviewee into one of a list of several categories, such as Caucasian, Negroid, Mongolian, Other, Not Sure on the basis of appearance to the interviewer. Another method is to ask a person, "What is your race?" and record whatever response is given. This method might reveal self-perception as much as race. For example, the difference between the response "black" and the response "Negro" might not be so much a racial difference as a difference of perception about oneself. ■

EXERCISES

2–1. A Life Insurance Company needs to determine whether applicants are "insurable" risks. It needs to determine the actuarial risk of a person's dying within a specified period of time. In order to do this, the company wants to find out certain information about the person.

a) List some of the variables about which an insurance company would reasonably want information.
b) List possible sets of values of these variables.

2–2. In screening applicants for police training, what are some of the variables one would reasonably look for in the applicants? Indicate the set of values that each of these variables could take on.

2–3. The word "discrimination" means the drawing of a distinction. When an employer practices discrimination in hiring, promoting, etc., he is taking action on the basis of variables whose values we feel are irrelevant to one's job performance or potential.

Make up a list of variables frequently used as a basis for actions such as hiring, which you believe are irrelevant to job considerations.

2–4. What kinds of measurements might we employ to determine an individual's

a) height?
b) income?
c) job satisfaction?
d) quality of performance on the job?
e) knowledge attained in an academic course?

2–5. Five employees are asked to state the two months of the year they would prefer most for their vacation. Give a possible set of observations that could result from this study.

2.2 Validity of Measurement

One of the most fundamental problems in research is the fact that the variable one *wants* to measure is often not the variable one *actually* measures. Assuming that you had a clear idea of what you meant by "conservatism," you still might not be able to develop a measurement that truly measures "conservatism." You might not be satisfied that the variable C, for which the researcher has developed a measure, accurately represents what we or other readers have in mind when we say "conservatism."

The variable we want to measure must be operationally defined.

> A variable is OPERATIONALLY DEFINED when the procedures for measuring the variable are clearly specified, so that two or more individuals would follow a uniform procedure in measuring the variable.

> When an operationally defined variable truly measures the concept we want to measure, we say that the measured variable is a VALID indicator of the conceptual variable.

The problem of validity can be illustrated by attempts to measure the concept "intelligence." The IQ (Intelligence Quotient) is a measure obtained by dividing an individual's score on a standard test by that individual's chronological age. Is IQ truly a measure of intelligence? Or does it also measure cultural orientation, motivation, skill at taking written tests, and the state of the test taker's health? To the extent that IQ is a measurement of these other things it is not a valid measurement of intelligence.

EXAMPLE 2.3 A Recreation Center for the town of Fish Haven was completed 5 years ago. The nearby town of Tuna may build a similar facility, but not until a study is made of the success of Fish Haven's facility. Suggest one or two operational definitions of the "success" of the center, and discuss the validity of each.

Discussion: One method of measuring "success" would be to send questionnaires to all or a sample of Fish Haven residents, which ask a series of questions about feelings and attitudes toward the center.

The questionnaires could be scored on the basis of the number of favorable answers, and "success" could be operationally defined as the average score of the responses to questionnaires. One of the problems of validity would be that the respondents would not necessarily be the chief users of the facility, and you might wish your measure of success to reflect more the attitudes of the users rather than of those returning the questionnaires.

An objection to the above approach would be that it is more valid to observe actual behavior than people's statements about how they feel. This might lead one to define "success" as the amount of usage the facility received. This would be valid for one who was satisfied to equate success with amount of usage. However, if one's notion of success included the total satisfaction it afforded the users, then this definition too, might not be wholly valid. ■

> If a variable can take on only a finite number of possible values within a limited range of values, the variable is said to be a DISCRETE variable. If a variable can take on an infinite number of possible values (all real numbers) within such a range, the variable is said to be a CONTINUOUS variable.

More informally, we can say that there are gaps or spaces between possible values of a discrete variable, but the values of a continuous variable between any two points can be represented by a solid line between these two points. There are some variables that can readily be defined either as discrete or continuous, depending on how one defines and handles them. Age, for example, is a discrete variable if defined as "age at most recent birthday"; it is continuous if defined as total time elapsed since birth.

EXAMPLE 2.4 Would the variable "Time Spent Interviewing a Client" be discrete or continuous?

Solution: It could be either. In concept, the variable "time" is continuous; there is an infinite number of moments between any two points of time. But if we operationally define the variable as "time measured to the nearest minute," then time becomes a discrete variable. ■

Another useful way of classifying variables is by level of measurement.

> A NOMINAL-LEVEL VARIABLE is one whose values are categories with no magnitude specified or implied. Religion, nationality, race, and sex are examples of variables generally treated as nominal-level variables.

> An ORDINAL-LEVEL VARIABLE is one whose values are ordered (from lowest to highest, or from highest to lowest). Variables implying rank, such as military or civil-service grade, and measures of preference can be treated as ordinal-level variables.

> An INTERVAL-LEVEL VARIABLE is one whose values can be ordered *and* have measurable differences between them. The Consumer Price Index can be treated as an interval-level variable, since we can determine the difference between any two values.

> A RATIO-LEVEL VARIABLE is one having all the characteristics of an interval-level variable *plus* a natural origin, or zero point. Variables involving numbers of people or dollars, as well as distances and weights, can generally be treated as ratio-level variables.

EXERCISES

2–6. The following conceptual variables are frequently discussed in the literature on administration or social science. They are very difficult to measure. Choose 2 or 3 with which you are most familiar and give an operational definition that comes reasonably close to being a valid measurement.

Alienation	Social Class	Self-actualization
Productivity	Achievement (of a student)	Merit (of an employee)
Job Satisfaction		

2–7. A student is doing a study in which he wants to compare the standard of living in the United States with that of several other industrialized nations.

a) What operationally defined variable might be a good indicator of standard of living?
b) Where might the student obtain the information he wants?

2–8. Suppose the student in Exercise 2–7 decided that he would like to measure an even more important variable than standard of living, that is, quality of life. How might you operationally define "quality of life" in these industrialized countries?

Is the problem of finding an operational definition of "quality of life" easier or more difficult than that of finding an operational definition of "standard of living"? From your answer to this question, which of these two variables would you expect to see more widely used and quoted?

2–9. Among the devices commonly used to evaluate public programs is the "cost–effectiveness" ratio. Of the two elements in this ratio, generally the more difficult one to evaluate is the effectiveness of the program. Suggest operational definitions to measure the effectiveness of

a) a county health agency
b) the Department of Defense
c) a city fire department
d) a city police department
e) a school of public administration

2–10. An example of an operational definition with which many of us are intimately familiar is that of "overweight." The number of pounds overweight is determined by subtracting the "ideal" weight for one's height, as shown in a Table of Ideal Weights for Men and Women, from one's actual weight.

a) Is such a measurement a valid measurement of overweight?
b) What other things might such a measurement measure besides the things implied in the concept "overweight"?

2–11. Would the following variables be generally treated as discrete or continuous?

a) number of employees absent
b) time for a fire station to respond to an alarm
c) gas consumption in a city in a year
d) time to complete a speed test
e) $X = 0$ if time to complete a speed test is under 5 minutes
$X = 1$ if time to complete a speed test is 5 minutes or more

2–12. Would the following variables normally be classified as Nominal, Ordinal, Interval, or Ratio?

a) Hog production in Iowa
b) Attitude toward a proposed law
c) Occupation
d) Consumer Price Index
e) Social Security Number

***2–13.** What assumptions are necessary if you consider that score on an aptitude test is an interval-level variable?

2.3 Grouping of Data

In presenting data it is important to be as concise as possible, as long as you do not leave out essential information. The reader seldom wants to see information on every individual observation, and in most cases will get a clearer picture if your data are summarized in a concise manner. One of the most convenient ways of summarizing data is grouping; grouping can be done for discrete or continuous variables of any level of measurement.

Suppose you are considering offering the employees in your organization a Family Medical Plan. In order to determine the cost of such a program, you would like to have information on the number of dependents of your employees. From the records, you obtain figures for each of the 372 people in the organization. The data might look like this:

Name	*Number of Dependents*
Alexander, R.	2
Allen, H. D.	1
Anderson, V. R.	6
Auker, H. N.	1
and so forth for 372 employees	

Everything you need to know about number of dependents is there, but there is more data there than you need. Furthermore it is not arranged in a manner that would be helpful to your decision. Too much data arranged in a useless order is something like a disorderly stack of papers on an executive's desk; all the information he needs is there somewhere, but he cannot get his fingers on the precise things he wants.

A better form of presentation would involve omission of the names, which are extraneous information, and grouping of the data into a frequency table, or frequency distribution, as presented in Table 2.2.

Table 2.2. *FREQUENCY TABLE Number of Dependents of Employees in ABC Agency*

NUMBER OF DEPENDENTS	NUMBER OF CASES (FREQUENCY)
0	88
1	102
2	61
3	76
4	18
5	11
6	9
7	3
8	0
9	1
10	2
11	1
	372

The reader can then readily determine how many employees have 0, 1, 2, etc. dependents.

It may be desirable to condense the information still further by lumping the number of dependents into broader classes, as shown in Table 2.3.

Table 2.3. *FREQUENCY TABLE: GROUPED DATA Number of Dependents of Employees in ABC Agency*

NUMBER OF DEPENDENTS	NUMBER OF CASES
0 or 1	190
2 or 3	137
4 or 5	29
6 or 7	12
8 or 9	1
10 or 11	3
12 or more	0
	372

Table 2.4. *FREQUENCY TABLE: GROUPED DATA*
Grouped Data for a Continuous Variable

WEIGHTS	FREQUENCY
At least 100 pounds but less than 120 pounds	12
At least 120 pounds but less than 140 pounds	28
At least 140 pounds but less than 160 pounds	16
At least 160 pounds but less than 180 pounds	13
At least 180 pounds but less than 200 pounds	6
Total	75

In grouping data a troublesome problem is that of defining the classes so that each of the cases is clearly included in one and only one class. For example, if we were grouping people by weight, the classes

100 to 120 pounds
120 to 140 pounds
etc.

would be overlapping, since a person weighing exactly 120 pounds could be included in either the first or the second class. The problem can be solved by adding the words "at least" (which is inclusive) and "less than" (which is exclusive) as presented in Table 2.4.

The presentation in Table 2.4 is proper for a continuous variable. As I have indicated, a continuous variable has no "spaces" between its possible values. There is therefore no space between the lowest possible value in the second class, which is clearly 120 pounds, and the highest value in the first class.

> We define the LOWER CLASS LIMIT as the lowest value that a variable can take on within a class.
> We define the UPPER CLASS LIMIT as the highest value that a variable can take on within a class.

In the case of the second class of Table 2.4, the lower class limit is 120 pounds. In the case of a continuous variable, the upper limit of one class is the same as (or "touches") the lower limit of the next class above it. The upper limit of the first class in Table 2.4 is therefore also 120 pounds.

With a discrete variable the problem is somewhat different. In this case the upper class limit of one class is not the same as the lower class limit of the next class above it. Look at Table 2.5, which presents grouped data for a discrete variable.

> The CLASS INTERVAL OR CLASS WIDTH of a class is defined as the difference between the lower limit of a class and the lower limit of the class above it.

The class width for the data in Table 2.4 is 20 pounds; the class width for the data in Table 2.5 is $10.00.

Another concept that will be particularly useful in making computations from grouped data is that of the midpoint of the class.

> The CLASS MIDPOINT OR CLASS MARK is the value midway between the lower and upper limits of the class.

In Table 2.4, the midpoint of the first class is 110 pounds; in Table 2.5., the midpoint of the first class is $144.495.

Following are three general rules that can serve as a useful guide in presenting grouped data:

1. *Classes should generally be of the same width.* There are situations where it is impractical to adhere strictly to this rule, such as a range of values so great as to call for an open class, for example, incomes of "$25,000 and above."

2. *The class interval should be small enough not to lose too much detailed information, yet large enough to achieve a considerable degree of compactness.* The size of the interval will depend to some extent on the readers for whom the table is prepared; if they are generally going to read the table carefully in detail, small intervals are better; if the readers are typically interested only in summaries, large intervals are better. Almost always, however, the intervals should be such as to make the number of classes somewhere between four and twelve.

Table 2.5. *FREQUENCY TABLE: GROUPED DATA Amount of Paycheck for 72 Wage Earners in Sanitation Department*

AMOUNT OF PAYCHECK	FREQUENCY
$140.00 to 149.99	4
150.00 to 159.99	17
160.00 to 169.99	28
170.00 to 179.99	14
180.00 to 189.99	7
190.00 to 199.99	2
	72

3. *The class limits should be determined in such a manner that there will be no tendency for the values to cluster at the lower end, or at the upper end, of a class.* In a distribution of ages, for example, it may be unwise to select classes such as

20 to 24 years of age
25 to 29 years of age
30 to 34 years of age
etc.

if ages are recorded from interviews or questionnaires. The reason for this is that there is a tendency to state one's age in round numbers, so that a disproportionate number of those recorded in a class such as the 30 to 34 year old class will have given their ages as 30. Because many statistical procedures take the midpoint of a class as representative of the whole class, it is better to set up your classes so that the greatest concentrations, if they exist, will be near the midpoint, not one of the ends, of the class. In the case of ages, therefore, it might be better to set up the classes as:

Class	*Midpoint*
18 to 22 years	20
23 to 27 years	25
28 to 32 years	30
etc.	

EXAMPLE 2.5 Set up a frequency table for the following 50 monthly salaries of Teachers in a School District.

$1220	$1100	$ 800	$1120	$ 970
1000	1400	1100	1400	935
870	1135	1137	1333	870
1280	1250	1440	1170	1140
1100	1300	1350	1035	900
1200	1250	1300	1390	1000
850	890	1175	1150	1075
990	900	1200	1080	1280
800	900	1190	900	790
1000	1175	1315	1200	1300

Discussion: The lowest value is $790. The highest value is $1440. A class interval of $100 would be a reasonable size, since about seven classes would cover the range of values. However, you should avoid setting up the classes as

$700 to $799
800 to 899
etc.

because of the large number of salaries that are multiples of 100. These should be at the midpoint, not the lower limit, of the classes. Therefore, it would be better to set up your table as follows.

Class	*Frequency*
\$ 750 to \$ 849	3
850 to 949	9
950 to 1049	6
1050 to 1149	9
1150 to 1249	9
1250 to 1349	9
1350 to 1449	5

EXERCISES

2–14. Which of the following classification schemes would you suggest be changed? How would you change them?

a) Distance from Place of Work

0 to 1 mile
1 mile to 2 miles
2 miles to 3 miles
etc.

b) City Agency by Number of Employees

0 to 9
10 to 19
20 to 29
etc.

c) Water Consumption by Millions of Gallons

At least 0 but less than 0.9
At least 1.0 but less than 1.9
At least 2.0 but less than 2.9
etc.

2–15. The Per Capita Taxes in the 50 states in fiscal 1975 were as follows:

Alabama	\$307.50	Indiana	\$ 349.08
Alaska	576.27	Iowa	370.02
Arizona	421.94	Kansas	339.23
Arkansas	308.43	Kentucky	378.01
California	451.48	Louisiana	403.24
Colorado	341.92	Maine	348.46
Connecticut	342.12	Maryland	422.33
Delaware	580.99	Massachusetts	380.67
Florida	334.00	Michigan	380.69
Georgia	314.21	Minnesota	515.09
Hawaii	363.50	Mississippi	339.89
Idaho	363.60	Missouri	273.56
Illinois	395.65	Montana	311.11

Nebraska	$274.78	Rhode Island	$377.73
Nevada	450.71	South Carolina	339.45
New Hampshire	210.77	South Dakota	250.55
New Jersey	287.17	Tennessee	275.08
New Mexico	452.97	Texas	297.23
New York	493.33	Utah	330.68
North Carolina	348.64	Vermont	348.58
North Dakota	415.18	Virginia	215.29
Ohio	282.48	Washington	245.61
Oklahoma	325.86	West Virginia	337.29
Oregon	346.60	Wisconsin	206.34
Pennsylvania	400.22	Wyoming	360.65

Source: Census Bureau

a) Group these data into a frequency table.
b) What information does your table bring out that the original figures fail to show?
c) What information is lost when you group the data?

2.4 Graphical Methods of Presenting Data

Description of a set of data can be vividly accomplished through graphical or pictorial means. Although there are many kinds of graphs that portray data, the most useful for the presentation of grouped data are the bar chart and a special type of bar chart known as a histogram.

The bar chart can be used to display discrete or continuous data of any level of measurement. It is not limited to grouped data. In the bar chart shown in Figure 2.1, the countries are arranged alphabetically, and

Figure 2.1. *LIFE EXPECTANCIES*
Selected Countries

Source: UN Yearbook, 1974.

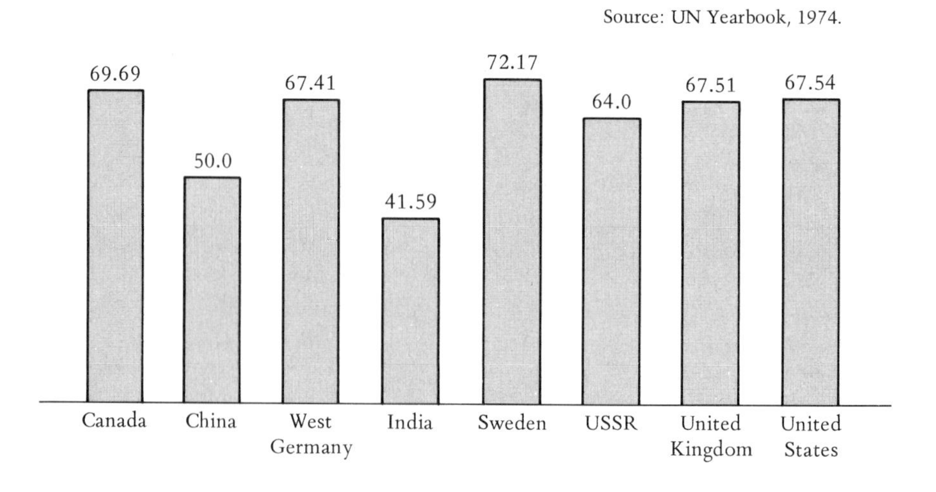

Table 2.6. *FREQUENCY TABLE GROUPING DATA FROM FIGURE 2.1*

LIFE EXPECTANCIES	FREQUENCIES
At least 35 but less than 45	1
At least 45 but less than 55	1
At least 55 but less than 65	1
At least 65 but less than 75	5

position on the scale has no quantitative significance. The heights of the bars are proportional to the life expectancies.

A bar chart such as that shown in Figure 2.1 is particularly useful when one of the scales indicates a nominal variable. From such a chart we could develop a frequency table of life expectancies, determining the number of countries that had life expectancies within various classes. For the very limited data from Figure 2.1, this could be done in Table 2.6.

A frequency table, such as Table 2.6, can be graphed most readily as a histogram. The histogram is appropriate for displaying grouped data when the data are concerned with interval-level or ratio-level variables. The frequency of each class is shown by bars whose widths represent the class intervals on the horizontal scale and whose heights are proportional to the class frequencies. Figure 2.2 is a histogram depicting the states in the United States grouped by population.

EXERCISES

2–16. Construct a histogram from the data in Table 2.6.

2–17. Why is a histogram not an appropriate graphical technique for displaying grouped data that are nominal or ordinal?

Figure 2.2. *HISTOGRAM REPRESENTING U.S. STATES BY POPULATION*

Source: 1970 Census.

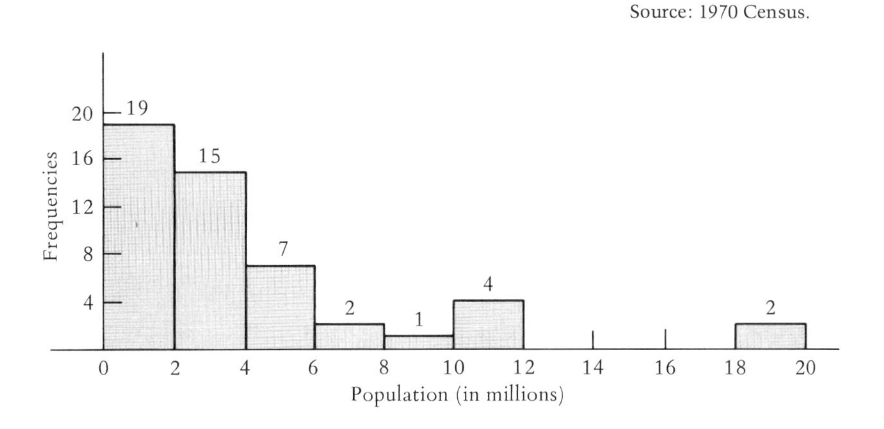

2.5 Measures of Location, or Central Tendency (Averages)

Grouping the data has the advantage of reducing what may be a long and cumbersome list of observations into a more compact display. Very often, however, even a frequency distribution is too cumbersome; the reader may want to see the data reduced to a single *number* that represents the general order of magnitude of the data. Such a number is known as a measure of location, since it "locates" the data on a scale.

> A MEASURE OF LOCATION, or AVERAGE, is a single number that describes the general order of magnitude of the data. Such a number is sometimes referred to as a MEASURE OF CENTRAL TENDENCY.

Unfortunately, most sets of data cannot be adequately described by a single number. There are therefore a number of different averages that could be used to describe the data; the choice of which average to use will vary with the nature of the data and the needs of the person using the average.

Do experienced managers achieve better results in motivating employees than inexperienced managers? Do women have fewer driving accidents than men? Do large school districts lose fewer days in teacher-strikes than small school districts?

These questions would be meaningless unless we had some single numerical value, or average, to measure groups in which individuals vary. We know that some experienced managers are better at motivating employees than others, that women differ from each other in frequency of accidents, as do men, and that large school districts, as well as small ones vary in number of days lost in strikes. The answers to the question about driving accidents, for example, cannot be adequately answered in a single word for all men and women; it can be answered only in terms of averages.

EXAMPLE 2.6 A social worker is required to handle on the average at least 8 cases per day. On 15 successive days the numbers of cases were:

$$6, 7, 7, 12, 6, 8, 5, 11, 6, 5, 6, 16, 13, 7, 8$$

Did the social worker achieve an average of at least 8? It all depends on which average we use. We could use the mode, the median, or the arithmetic mean. The following discussions of these various averages will refer to this example of the social worker. ■

THE MODE

> The MODE is the most commonly occurring observed value.

The mode is therefore most "typical," since it accurately portrays more observations than any other value. There is one serious problem with using a mode, however, in that sometimes it does not exist. If no values occur more than once there will be no mode. If there are ties for values with the most occurrences, there will be two or more modes. The social worker's mode, or modal number of days, is 6. If his average were computed as a mode, he would not have achieved his requirement of 8 visits per day.

> If the data are grouped, the MODE can be defined as the midpoint of the class with the greatest frequency.

THE MEDIAN

> When the observations are arranged in order of magnitude from lowest to highest, the middle value is known as the MEDIAN. If there is an even number of items, then the median is generally considered to be the value midway between the two middle observed values.

Arranging the social worker's days in ascending order of number of cases yields

$$5, 5, 6, 6, 6, 6, 7, 7, 7, 8, 8, 11, 12, 13, 16$$

The median is the 8th observation, which is 7. Using the median as an average, the social worker would find that his average was below the required 8 visits.

EXAMPLE 2.7 In a recent measles epidemic, one county in the state reported 22 cases, one county reported 9 cases, one county reported 5 cases, two counties reported 1 case, and three counties reported no cases.

Find the mode and median number of cases for the eight counties in the state.

Solution: Arrange the number of cases per county in increasing order, as

0, 0, 0, 1, 1, 5, 9, 22

The mode is zero, the most frequently occurring value.

The median is halfway between the 4th and 5th values arranged in order, which is 1.

When the mode is way off at one end, as in this case, any statement about its being typical can be very deceptive. To say that "the typical county had no cases of measles" would obscure the fact that there was an epidemic. ■

The Arithmetic Mean

> The ARITHMETIC MEAN, or MEAN, of a set of data is the total of all observed values divided by the number of observations.

It is an "equal share" type of average in the sense that if all the observed values were to share equally in the total, then each share would be the mean. The mean can be represented as follows.

Formula for arithmetic mean

$$\overline{X} = \sum_{i=1}^{n} X_i = \frac{X_1 + X_2 + X_3 + \cdots + X_n}{n} \qquad \textbf{2.1}$$

Formula 2.1 introduces some notation that may be unfamiliar to you. The letter X refers to the numbers, or observations, that are to be averaged. $\overline{X}$ (read "X bar") is the arithmetic mean. If we had used a different letter, such as Y, for the observations, then the mean would be represented by $\overline{Y}$. The small i (called a subscript of X) is a "counter," which identifies the 1st X when $i = 1$, the 2nd X when $i = 2$, on to the last, or nth, X, when $i = n$. The number is the total number of observations. The Greek capital sigma (Σ) says "the summation of" or "add them up." Formula 2.1 therefore says to add up all the X_i, starting with $i = 1$, the number at the bottom of the sigma, and go up to $i = n$, the number at the top of the sigma. After adding them up, Formula 2.1 says that you should divide by n, and the resulting quotient is $\overline{X}$, the arithmetic mean. All of this you probably did in high school, but not following this notation. It is recommended, however, that you become familiar with this notation and others presented in this book, not only for the comprehension of material presented here, but also because an increasing number of reports and articles use such notation and assume the reader can understand it.

Very often the formula for the mean is given in the simpler form

$$\overline{X} = \frac{\Sigma X}{n}$$

with the subscript i left out. This is perfectly all right if there is only one variable to be summed. The notation ΣX can simply be read as "the sum of all the X's." But when we deal with several variables at a time, it is often unclear which variables are to be added up unless the subscripts are included.

The mean number of cases for the social worker is the sum of all the 15 observed values divided by 15, or

$$\overline{X} = \frac{6 + 7 + 7 + 12 + 6 + 8 + 5 + 11 + 6 + 5 + 6 + 16 + 13 + 7 + 8}{15}$$

$$= \frac{123}{15} = 8.2$$

One disadvantage of the mean as a measure of location is that, unlike the mode, it may not be typical, or descriptive, of any single observation. In many cases the mean would be a nonsensical value in describing a particular observation. On no day did the social worker handle 8.2 cases. When we read that the "average" family consists of 4.1 persons, we have difficulty picturing a family that is average. The mean may also be atypical of the data in that it can be strongly influenced by extreme values. To illustrate this possibility, let us suppose that five people's incomes are given as:

$6,000, $7,000, $8,000, $9,000, and $100,000

The mean of these values is $26,000, a figure that is nowhere near any of the observed values, because that $100,000 figure pulled it so far from the other values. The median income ($8,000) is far more descriptive.

On the other hand, the mean has properties that make it a very useful measure for many purposes. By the very fact that it takes into consideration all values, extreme as well as typical, it contains more information than the mode (which looks only at where the values bunch most closely), and the median (which looks only at the rank ordering of the values).

A very important property of the mean is that it can be given a long-run interpretation that you cannot give to the other averages. If the social worker continues to visit a mean of 8.2 families a day, in the long run his mean for 10 days will be 82; for 100 days it will be 820, etc. The daily arithmetic mean will tell you how many families he will visit in a month or a year. If one is interested in how many visits the social worker makes in the long run, the best daily average to use is the arithmetic mean.

EXAMPLE 2.6 (continued) Suppose the average number of visits made by the social worker were defined as the mode. How might he meet the 8-visit average requirement in the next 50 days and still do very little work?

Solution: He could visit 8 families fairly often, and confine the number of visits on other days to 0 up to 7, taking care that none of these other numbers has as high a frequency as 8 visits. One specific solution would be:

No. of Visits	*Frequency*
0	9
1	8
2	7
3	6
4	5
6	1
7	0
8	10

In 50 days he would then be visiting only 166 families yet still maintain a mode of 8 days. ■

EXAMPLE 2.6 (continued) Suppose the average number of visits made by the social worker were defined as the median. How might he meet the 8-visit requirement with very little work in the next 50 days?

Solution: He could make zero visits on 24 days and 8 visits on 26 days. The median would be 8, and the total number of visits would be 208. ■

Another property of the arithmetic mean is that the sum of the deviations of all the observations from the mean is zero, or

$$\sum_{i=1}^{n} (X_i - \overline{X}) = 0 \qquad \textbf{2.2}$$

Let us illustrate this property with the data for the social worker (where $\overline{X} = 8.2$).

X_i	$X_i - \overline{X}$
$X_1 = 6$	-2.2
$X_2 = 7$	-1.2
$X_3 = 7$	-1.2
$X_4 = 12$	$+3.8$
$X_5 = 6$	-2.2
$X_6 = 8$	-0.2
$X_7 = 5$	-3.2
$X_8 = 11$	$+2.8$
$X_9 = 6$	-2.2
$X_{10} = 5$	-3.2
$X_{11} = 6$	-2.2
$X_{12} = 16$	$+7.8$
$X_{13} = 13$	$+4.8$
$X_{14} = 7$	-1.2
$X_{15} = 8$	-0.2
	0

$$\Sigma(X_i - \overline{X}) = -19.2 + 19.2 = 0$$

The Mean of Grouped Data Since grouping of data involves the loss of precise information on individual observations, the calculation of a mean from a frequency table will be only an approximation. In calculating such a mean it is necessary to make the assumption that the mean of data in the class falls at the class midpoint. This assumption will not be valid unless care is taken that the class limits are selected in a way that avoids a systematic tendency for the data to cluster near the lower or upper limit of the class. The midpoint assumption enables us to estimate the mean of grouped data by the following:

Formula for mean of grouped data

$$\overline{X}_f = \frac{\sum_{i=1}^{k} f_i m_i}{\sum_{i=1}^{k} f_i} \qquad 2.3$$

where $\overline{X}$ is the mean of grouped data (or frequency distribution)
f_i is the number of cases (or frequency) in the ith class
m_i is the midpoint of the ith class
k is the number of classes

and

$$\sum_{i=1}^{k} f_i = n$$

Let us calculate the approximate mean of the data in Table 2.4, the weights of 75 individuals. Table 2.7 shows these calculations.

THE WEIGHTED MEAN The concept of the weighted mean is an extension of the arithmetic mean, which is sometimes referred to as the unweighted mean.

A WEIGHTED MEAN is a mean that places a different emphasis (or weight) on the different observations. It is denoted by

$$\overline{X}_W = \frac{\sum_{i=1}^{k} X_i W_i}{\sum_{i=1}^{k} W_i} \quad \textbf{2.4}$$

where $\overline{X}_W$ is the weighted mean
X_i is the observed values
W_i are the weights of the observed values
k is the number of observed values

Suppose, for example, that in a Public Management course you scored 60 on a quiz, 80 on the midterm examination, and 100 on the final examination. If the professor were to compute your average as the arithmetic mean (unweighted) of these scores, your average would be

$$\frac{60 + 80 + 100}{3} = 80$$

You might feel cause to protest the calculation of your average in this way, and for good reason. If a final examination is more important than a midterm examination, and a midterm examination is more important than a quiz, then the scores on these examinations should be given different weights. Let us assign a weight of 10 to the quiz, 30 to the mid-

Table 2.7. *CALCULATION OF MEAN OF GROUPED DATA FROM TABLE 2.4*

CLASS (Pounds)	MIDPOINT OF CLASS (m_i)	FREQUENCY (f_i)	$m_i f_i$
100 up to 120	110	12	1320
120 up to 140	130	28	3640
140 up to 160	150	16	2400
160 up to 180	170	13	2210
180 up to 200	190	6	1140
$n = 75$		75	10710

$\Sigma M_i f_i = 10710$

$$X = \frac{\Sigma M_i f_i}{n} = \frac{10710}{75} = 142.8$$

Table 2.8. *CALCULATION OF WEIGHTED MEAN*
Scores on Three Examinations

X_i	W_i	$X_i W_i$	
68	10	600	$\Sigma W_i = 100$
80	30	2400	$\Sigma X_i W_i = 9000$
100	60	6000	$\overline{X}_W = \frac{9000}{100} = 90$
	100	9000	

term, and 60 to the final examination. Table 2.8 shows the calculation of the weighted mean to be 90.

EXAMPLE 2.8 The mean income in Aberdeen County is \$12,000; the mean income of Bradley County is \$10,000; and the mean income of Colfax County is \$8,000. Assume the populations of these three counties are respectively, 100,000, 200,000, and 300,000 people. What is the mean income of the population in all three counties?

Solution: You should calculate a weighted mean. The unweighted mean, \$10,000, would give each county equal weight, but a more populous county should count more heavily than a less populous county. The calculation would be as follows:

X_i	W_i	$X_i W_i$
\$12,000	100,000	1,200,000,000
10,000	200,000	2,000,000,000
8,000	300,000	2,400,000,000
	$\Sigma W_i = 600,000$	$\Sigma X_i W_i = \$5,600,000,000$

\$9,333.33

THE "BEST" AVERAGE This section has discussed a variety of measures of location. It is not always easy to determine which of these to use on any occasion. Some averages are best for some purposes; some for others.

When the data are of an interval or ratio level, the arithmetic mean is most commonly used, partly because of the wide variety of statistical procedures that have been developed making use of the mean. An exception to this rule is the case of grouped data with an open class, such as "$25,000 and over." Since the midpoint of such a class cannot be readily defined, the median is generally calculated rather than the mean from data so grouped.

If one is estimating or predicting a value, the mode is the average that has the best chance of being exactly correct, of "hitting it on the nose." On the other hand, if you are interested in minimizing your error (how much you are off in either direction) the median is the best average, since the sum of the deviations, ignoring whether they are positive or negative, is less from the median than from any other value. Also, the median has the appealing feature that you have an equal chance of overestimating or underestimating. On the other hand, the total *size* of your overestimations and underestimations will tend to cancel each other out in the long run if you use the arithmetic mean as your estimator. It is this feature that makes the mean particularly attractive if you are interested in a "long-run average."

In the course of this book the average employed more often than any other is the arithmetic mean; occasionally you will encounter the median. The mode was introduced to point out the variety of averages that you can use, and that each has limitations and advantages. The "best average" for all purposes does not exist. The choice depends on the purposes of the person using the average, and the kinds of questions being asked.

EXERCISES

2–18. The minimum legal ages for the purchase of alcoholic beverages in the 50 states and the District of Columbia at the end of 1976 are as follows:

	Minimum Age		*Minimum Age*
Alabama	19	California	21
Alaska	19	Colorado	21
Arizona	19	Connecticut	18
Arkansas	21	Delaware	20

D.C.	21	New Hampshire	18
Florida	18	New Jersey	18
Georgia	18	New Mexico	21
Hawaii	18	New York	18
Idaho	19	North Carolina	21
Illinois	21	North Dakota	21
Indiana	21	Ohio	21
Iowa	18	Oklahoma	21
Kansas	21	Oregon	21
Kentucky	21	Pennsylvania	21
Louisiana	18	Rhode Island	18
Maine	18	South Carolina	21
Maryland	21	South Dakota	21
Massachusetts	18	Tennessee	18
Michigan	18	Texas	18
Minnesota	19	Utah	21
Mississippi	21	Vermont	18
Missouri	21	Virginia	21
Montana	18	Washington	21
Nebraska	19	West Virginia	18
Nevada	21	Wisconsin	18
		Wyoming	19

a) Find the mode, median, and the arithmetic mean of the minimum age.
b) Give a brief verbal interpretation of each of the averages you found in (a).
c) If 4 states with minimum ages of 21 changed their minimum age to 18, what would be the effect on the mode, the median, and the mean?

2–19. Which average would be most appropriate if one wished to determine
a) the average number of births per 1000 families
b) the average future life span of a 35-year-old woman (1) from the point of view of a 35-year-old woman, and (2) from the point of view of a life insurance company
c) the amount of money spent on food by a "typical" urban American family

2–20. The amounts of time (in minutes) that it took a volunteer Fire Company to reach 10 fires, were:

1.6, 5.4, 2.8, 7.7, 3.5, 3.0, 7.7, 2.5, 3.9, and 7.0

Find the mode, the median, and the mean time to reach the 10 fires.

2–21. Two students took 5 quizzes. These were their scores:

Murphy	*Schwartz*
0	60
20	70
100	80
100	80
100	90

Compare the two students with respect to their modes, medians, and arithmetic means.

2–22. The following information was reported about 20 employees who worked overtime during a snowstorm last January.

"Half the employees worked less than 8 hours overtime; half worked more than 8 hours. More worked 6 hours than any other number of hours. The shortest amount of overtime worked was 2 hours; the longest 24 hours. The total of hours overtime was 204 hours." From this information, what were the mode, median, and arithmetic mean number of hours of overtime?

2–23. The following data were collected on 10 employees at the Internal Revenue Service.

(1) *Name*	*(2)* *Sex*	*(3)* *Age*	*(4)* *Views on Unionization*
Chapman	Female	42	For
Maxwell	Male	46	Neutral
Moore	Female	31	For
Trask	Female	37	Neutral
Welte	Female	56	Against

a) What averages, if any, can be calculated on the variable "Name" in Column 1?

b) What averages, if any, can be calculated on the variable "Sex" in Column 2?

c) What averages, if any, can be calculated on the variable "Age" in Column 3?

d) What averages, if any, can be calculated on the variable "Views" in Column 4?

2–24. If you want to calculate the average area of a state in the U.S., what averages could you calculate on the basis of the information that the total area of the fifty states is 3,615,122 square miles?

2–25. The counties in New Hampshire and their 1970 populations are:

Belknap	32,367	Hillsborough	223,941
Carroll	18,548	Merrimack	80,925
Cheshire	52,364	Rochingham	138,951
Coos	34,291	Strafford	70,431
Grafton	54,914	Sullivan	30,949

a) Calculate the median and mean.

b) Comment on the advantages and disadvantages of each of these measures as a description of New Hampshire's county populations.

2–26. Five experts were asked to estimate how long a research project would take to complete. Their estimates are:

Expert	*Estimated Number of Weeks*
A	4.4
B	6.4
C	3.7
D	4.6
E	5.5

a) Calculate the arithmetic mean of these 5 estimates.

b) Experts A and B are considered more knowledgeable than the other 3. Calculate a weighted mean of the estimates of the 5 experts, giving each of A's and B's estimates twice as much weight as each of the other 3 experts.

2–27. Calculate the mean of the following grouped data:

Number of Teachers	*Number of Schools*
0 to 19	2
20 to 39	10
40 to 59	20
60 to 79	8

2–28. Calculate the median and mean age of the members of your family and comment on the usefulness of your answers.

2–29. Given the following frequency table:

County Budget	*Number of Counties*
$0 to 499,999.99	3
500,000.00 to 999,999.99	7
1,000,000.00 to 1,499,999.99	6
1,500,000.00 to 1,999,999.99	2
2,000,000.00 to 2,499,999.99	3
2,500,000.00 and over	4

a) What is the mode of the distribution?
b) In what class does the median fall?
c) The existence of the open class "$2,500,000.00 and over" makes it impossible to estimate a mean unless we assume a mean value for the class. Assume that the mean of this class is $6,000,000.00 and calculate the mean of the distribution.

2–30. A professor, in writing a statistics book, sets himself a goal of writing at least 800 words a day. He knows that his output will vary from day to day; on some days he knows he will fall short of 800, and on some days he hopes to exceed 800. He will be satisfied if he averages 800. Which average would be most appropriate in measuring his output? Why?

2.6 Measures of Variability, or Dispersion

The reason we have a problem choosing an average from such a number of possible averages stems from the fact that individual observations vary from each other. If everyone were alike, social scientists would have a much easier time measuring groups of people. The mode, the median, and the mean of any measurements would all be the same. The more variety we encounter in our individual observations, the less adequate any single average becomes as a descriptive measure. It is important, therefore, after having selected and computed an average that we also look at the degree to which individuals vary from that average. This is done using measures of variability, or measures of dispersion.

Statements such as "blacks have lower incomes than whites," "men are taller than women," "children are attracted by the parent of the opposite sex," etc., are statements about averages, not about individuals. Most people are familiar with averages in some form; comparatively few, however, have had much exposure to the concept of variability or how it can be measured.

Just as with averages, there are many measures of dispersion, each having its advantages and weaknesses. Of the many measures of dispersion by far the most widely used are the variance and the standard deviation.

If 5 customers entered a post office, and the times to serve them were 2, 5, 3, 12, and 8 minutes, you can describe the general order of magnitude of the times using one of the averages discussed in Section 2.5. For example, the mean is

$$\overline{X} = \frac{2 + 5 + 3 + 12 + 8}{5} = 6 \text{ minutes}$$

But although the mean tells us something about the central tendency, it gives no idea of how much these times vary. Without a measure of variability, one might get the erroneous impression that each of the people took about 6 minutes.

We can measure variability in a number of ways. A simple way is to find the difference between the highest and lowest values. Another is to calculate the mean of the deviations of the individuals from their mean. The most useful measures for statistical work, however, involve the summing of the *squares* of the deviations of individual values from the mean.

> The VARIANCE of a set of n observations, denoted by s^2, is defined to be the sum of the squares of the deviations of the observations from their mean, divided by $n - 1$.
>
> $$s^2 = \frac{\sum_{i=1}^{n} (X_i - \overline{X})^2}{n - 1} \qquad \textbf{2.5a}$$
>
> or, more simply
>
> $$s^2 = \frac{\Sigma(X - \overline{X})^2}{n - 1} \qquad \textbf{2.5b}$$

The variance is an expression involving squares, or second powers; in order to reduce this measure to the original units (the first power) we take the square root of the variance and obtain the standard deviation.

> The STANDARD DEVIATION of a set of n observations, denoted by s, is defined to be the square root of the variance. **2.6**
>
> $$s = \sqrt{s^2} = \sqrt{\frac{\Sigma(X - \overline{X})^2}{n - 1}}$$

Students are typically much more confused by discussions of dispersion than by discussions of averages. This is partly because most of us have had much more exposure to the concept of averages than to that of variability. Even after looking at the definition of the standard deviation, a student is likely to say, "I can see how you compute it, but what *is* it?"

The variance and standard deviation are attempts to measure how "spread out" the data are. If you look at Formulas 2.5 and 2.6 you can see that increasing the deviations of values from their means will tend to increase the value of $\Sigma(X - \overline{X})^2$, and hence to increase the variance and the standard deviation.

The measures may become more meaningful to you if you use them to compare the dispersions of two sets of data. Suppose you want to compare the salaries of employees in Agency A with those in Agency B. If Agency A has a mean salary of $185 a week and B a mean of $195 a week, you will get a picture of a generally higher salary scale in B. However, if you also learn that the standard deviation in A is $15 and the standard deviation in B is $25, you can get a new slant on the two agencies. B has more variability, or inequality. Therefore, although B's salaries are generally located higher than A's, you might also expect more extreme salaries in B, both at the low end and at the high end of the scale.

There is a special rule that can also help to give you a practical feeling for the concept of standard deviation. You may have heard of the Normal Probability Distribution. In Chapter 6 the discussion of the normal distribution will cover rules for estimating probabilities using standard deviations. If a distribution of data is normal (which is very unusual) or approximately normal (which is much more common) then we can say that about ⅔ of the values will fall within one standard deviation of the mean, about 95% of the observations will fall within two standard deviations of the mean, and nearly all observations will fall within three standard deviations of the mean.

Try this rule for a variable you are familiar with, the heights of adult males, which are approximately normally distributed. If the mean height is 69 inches and the standard deviation is 3 inches, then about ⅔ of adult males will be within 3 inches (one standard deviation) and about 95% will be within 6 inches of the mean 69 inches. On the other hand, if the standard deviation of male heights were, say, 5 inches, then ⅔ would be within 5 inches of the mean and 95% within 10 inches of the mean. As you vary the assumption of the standard deviation, you can see how the ⅔ or 95% range will vary also. Although the ⅔ and 95% figures are accurate only for the normal distribution, it is generally true for any distribution of data that higher standard deviations will mean that any given percentage of the data (50%, 95%) will be spread over a greater range.

EXAMPLE 2.9 Compute the variance of the numbers 2, 4, and 6.
Solution: The calculations are shown in Table 2.9.

Table 2.9. *CALCULATION OF MEAN, VARIANCE, AND STANDARD DEVIATION*
Numbers 2, 4, and 6

(1) X	(2) $X - \overline{X}$	(3) $(X - \overline{X})^2$	(4) X^2
2	−2	4	4
4	0	0	16
6	2	4	36
12	0	8	56

$$X = \frac{12}{3} = 4.0$$

$$s^2 = \frac{\Sigma(X - \overline{X})^2}{n - 1} = \frac{8}{2} = 4.0 \qquad s = \sqrt{s^2} = \sqrt{4.0} = 2.0$$

EXAMPLE 2.10 On 3 successive tests Ben received scores of 40, 90, and 80. On the same tests Beatrice received scores of 60, 80, and 70. Although both students achieved the same mean score of 70, they differed in their consistency.

Find the variance and standard deviation of each student's scores.

Solution: Let Ben's score be X and Beatrice's be Y. The calculations are shown in Table 2.10.

In Formula 2.5, had we divided the sum of the squared deviations by n instead of $n - 1$ we would have computed the mean of the squared deviations. When n is a large number, the variance differs very little from the mean of the squared deviations. The reason that we define s^2 with a

Table 2.10. *CALCULATION OF VARIANCE AND STANDARD DEVIATION*
Scores of Two Students

X	$X - \overline{X}$	$(X - \overline{X})^2$	Y	$Y - \overline{Y}$	$(Y - \overline{Y})^2$
40	−30	900	60	−10	100
90	20	400	80	10	100
80	10	100	70	0	0
		1400			200

$$s^2 = \frac{\Sigma(X - \overline{X})^2}{n - 1} = \frac{1400}{2} = 700$$

$$s_X = \sqrt{700} = 26.5$$

$$s_Y^2 = \Sigma(Y - \overline{Y})^2 = \frac{200}{2} = 100$$

$$s_Y = \sqrt{100} = 10.0$$

denominator of $n - 1$ rather than n is that the variance so defined is especially useful in estimating the variability in the universe from which a sample was drawn. This feature of the variance will be discussed more fully in the treatment of sampling in Chapter 7.

Formula 2.5 is a definitional formula for the variance but does not lend itself easily to computation. It can be shown that the formula for the variance is equivalent to Formula 2.7, which is better suited for hand calculation.

Computational formula for the variance

$$s^2 = \frac{\Sigma X^2}{n - 1} - \frac{(\Sigma X)^2}{n(n - 1)} \qquad \textbf{2.7}$$

EXAMPLE 2.9 (continued) Compute the variance of the numbers 2, 4, and 6 using Formula 2.7.

Solution: Column 4 of Table 2.9 gives us $\Sigma X^2 = 56$. Thus, by Formula 2.7, the calculation of s^2 is

$$s^2 = \frac{\Sigma X^2}{n - 1} - \frac{(\Sigma X)^2}{n(n - 1)} = \frac{56}{2} - \frac{(12)^2}{3(2)} = 4.0$$

which agrees with the result using Formula 2.5.

The standard deviation can generally be more efficiently computed by the following which is derived from Formula 2.6.

Computational formula for standard deviation

$$s = \sqrt{\frac{\Sigma X^2}{n - 1} - \frac{(\Sigma X)^2}{n(n - 1)}} \qquad \textbf{2.8}$$

The standard deviation of grouped data is defined by

$$s = \sqrt{\frac{\Sigma f(X - \overline{X})^2}{\Sigma f - 1}} \qquad \textbf{2.9}$$

This is often more easily calculated by

$$s = \sqrt{\frac{\Sigma fX^2}{\Sigma f - 1} - \frac{(\Sigma fX)^2}{\Sigma f(\Sigma f - 1)}} \qquad \textbf{2.10}$$

EXAMPLE 2.6 (continued) Calculate the variance and standard deviation of the data for the number of cases of the social worker on 15 successive days.

Solution: The computations are shown in Table 2.11.

EXERCISES

2–31. Calculate the variance and standard deviation of the data in Exercise 2.18 on the minimum ages for purchase of alcoholic beverages in 50 states. You may use either Formula 2.5 or 2.7, but this example will illustrate the advantages of the computational Formula 2.7.

2–32. a) Calculate the variance, and the standard deviation of the numbers 1, 2, 3, 4, 5, 6, 7, and 8.

b) Find the variance and standard deviation of the numbers 0 and 1.

2–33. The number of fatal accidents in City A on 5 successive months was 3, 4, 8, 3, and 6. The number of fatal accidents in City B during the same months was 4, 3, 7, 2, and 4.

Table 2.11. *CALCULATION OF MEASURES OF DISPERSION*

X	$X - \overline{X}$	$(X - \overline{X})^2$	X^2
6	−2.2	4.84	36
7	−1.2	1.44	49
7	−1.2	1.44	49
12	3.8	14.44	144
6	−2.2	4.84	36
8	−0.2	0.04	64
5	−3.2	10.24	25
11	2.8	7.84	121
6	−2.2	4.84	36
5	−3.2	10.24	25
6	−2.2	4.84	36
16	7.8	60.84	256
13	4.8	23.04	169
7	−1.2	1.44	49
88	−0.2	0.04	64
123		150.40	1159

$\Sigma(X - \overline{X})^2 = 150.40 \qquad \Sigma X^2 = 1161$

(2.5) $$s^2 = \frac{\Sigma(X - \overline{X})^2}{n - 1} = \frac{150.40}{14} = 10.74$$

(2.7) $$s^2 = \frac{\Sigma X^2}{n - 1} - \frac{(\Sigma X)^2}{n(n - 1)} = \frac{1159}{14} - \frac{(123)^2}{15(14)} = 10.74$$

(2.6) $$s = \sqrt{10.74} = 3.28$$

a) Compare the variability in fatal accidents between the two cities, by comparing their variances and standard deviations.
b) Is it possible for one set of data to have a higher variance than another set but still have a lower standard deviation? Why?

Summary of Chapter

We can describe a set of individuals by the use of variables, such as age, height, income, etc. It is often difficult to find a variable, or to measure the value of the variable, that truly reflects what you want to measure.

In presenting data to a reader or an audience, one might want to group the data in order to give a clearer and more concise picture of what the data say. In grouping data, however, one should take care to avoid practices that can give a misleading picture. This is also true in giving graphical presentations.

An even more concise way of summarizing data is to use a single number that represents the averages or central tendency, and a single number that measures the degree of variability. Measures of central tendency include the mode, the median, and the arithmetic mean. Measures of variability include the variance and the standard deviation. The mean, variance, and standard deviation will be used extensively throughout the rest of this book.

REVIEW EXERCISES FOR CHAPTER 2

2–34. What are the principal advantages and disadvantages of grouping data?

2–35. A legislator has proposed a surtax on all incomes above $30,000. Do you think such a tax would increase or decrease the standard deviation of income? Why?

2–36. A writer recently developed the concept of the "madness level of a society." His operational definition of a society's level of madness in any year was the marriage rate for that year.

Discuss the validity of this measure.

2–37. In a 40-mile-an-hour speed zone, 50 randomly selected cars were checked for speed. Their speeds (in miles per hour) were as follows:

42	33	47	39	40
38	38	38	38	25
57	39	37	39	52
40	38	38	34	40
35	35	38	35	45
33	39	39	39	38
45	41	36	38	38
38	39	38	39	39
39	39	39	36	41
70	40	30	40	38

a) Compute the mode, median, and mean.
b) Group the data into a frequency table.
c) Compute the mean of the grouped data and compare it with the mean of the original data you calculated in (a).
d) If the data were gathered to determine how well motorists were abiding by speed laws, would any of the averages you calculated in (a) be valid indicators of the law-abiding behavior of motorists? Discuss.

2–38. Find the standard deviation of the motorists' speeds in Exercise 2–37.

2–39. If data are grouped, it is impossible to determine exactly where the median is located unless one assumes a particular distribution of the observations within a class. It is possible, however, to determine which class contains the median. The sum of the frequencies, and the sum of the frequencies *up to and including* this class, will be at least half of the total frequencies.

Given the following frequency table:

Number of Traffic Fatalities	*Frequency*
0 to 9	4
10 to 19	14
20 to 29	22
30 to 39	27
40 to 49	7
50 and above	4

a) Which class contains the median?
b) Does the open class "50 and above" present a problem in finding the class containing the median?

2–40. The number of patients admitted to a municipal hospital in the past 130 days has the following frequency distribution:

Number of Patients Admitted	*Number of Days*
0 to 9	4
10 to 19	23
20 to 29	40
30 to 39	32
40 to 49	20
50 to 59	10
60 to 69	1

a) Compute the mean number of patients admitted.
b) Compute the standard deviation of number of patients admitted.

2–41. Sociologists have tried to develop operational measures of family status (or to what extent families stick together) in an urban area. One variable that has been used to measure family status is "percentage of the adult female population that is employed." Comment on the use of this measure. Is it easier to obtain than other possible measures? Is it a valid measure? Can you think of other measures of this concept that might be used?

SUGGESTED READINGS

FREUND, JOHN E. *Modern Elementary Statistics,* 4th ed. Englewood Cliffs, N.J.: Prentice-Hall, 1972.

JOHNSON, ROBERT. *Elementary Statistics,* 2nd ed. North Scituate, Mass.: Duxbury Press, 1976. Chapters 2 and 3.

MENDENHALL, WILLIAM; Lyman Ott; and R. Larson. *Statistics: A Tool for the Social Sciences.* North Scituate, Mass.: Duxbury Press, 1974. Chapters 3 and 4.

RUNYAN, RICHARD P. *Descriptive Statistics: A Contemporary Approach.* Reading, Mass.: Addison-Wesley Publishing Co., 1977.

THREE

Descriptive Measures Relating Two or More Variables

Outline of Chapter

3.1 The Nature of Multivariate Statistical Analysis

Most statistical studies are concerned with two or more variables acting together.

3.2 Two Nominal Variables

There are several ways to measure the strength of a relationship between two variables that classify individuals into qualitative categories.

3.3 Two Ordinal Variables

It is possible to measure both the strength and the direction of the relationship between two variables that rank-order their values.

3.4 Two Interval or Ratio Variables

Scatter diagrams, correlation techniques, and regression techniques are used to display and measure relationships between variables that assign specific magnitudes to individuals.

*3.5 Three or More Variables

A multiple regression equation indicates the relationship between three or more variables, and combinations of any two variables while holding others constant.

Objectives for the Student

1. Understand why we need measures that relate two or more variables.
2. Calculate and interpret the chi-square statistic and the contingency coefficient from a contingency table.
3. Calculate and interpret the Spearman rank correlation coefficient.
4. Plot and interpret a scatter diagram.
5. Calculate and interpret a linear correlation coefficient.
6. Calculate and interpret a linear regression equation.
7. Interpret the coefficients in a multiple regression equation.

Chapter 3 extends the discussion of Chapter 2 to descriptive measures that relate two or more variables to each other. Since nearly all statistical problems relevant to management involve more than one variable, it is essential that a person working with statistics have an understanding of these bivariate and multivariate measures. This chapter's main emphasis is on measures relating two variables; the last section is a brief discussion on the problem of controlling for additional variables.

An extensive analysis of measures relating three or more variables belongs in a more advanced treatise on statistics. See, for example, Kleinbaum and Kupper, *Applied Regression Analysis and Other Multivariate Methods.*

3.1 The Nature of Multivariate Statistical Analysis

Most of the questions to which administrators and social scientists seek answers involve the relationship between two or more variables. Questions involving such relationships include:

How does age affect absentee rates?
In what types of organizations does an authoritarian structure increase effectiveness?
Will enacting stricter gun-control laws reduce the homicide rate?
Will the passage and enforcement of stricter laws against discrimination in hiring reduce the difference in employment rates of whites and blacks?
If I give the people in my department different percentage raises according to my perception of their merit, will I increase the effectiveness of the department more than if I give everyone the same percentage increase?
Does cigarette smoking increase the probability of one's getting heart disease?
What effect would an antismoking campaign by the HEW Department have on the smoking habits of the American people?

All of these questions involve at least two variables, and most of them involve more variables than one might think. For example, research on the effect of gun-control laws on homicide would probably involve looking at homicide rates in areas with differing degrees of strictness of laws. But these areas will also differ in other characteristics, such as unemployment rate, racial and ethnic mix, median income, etc. Since these other variables could very well have as much effect on the homicide rate as do gun-control laws, we had better take these into account or "control for them."

In order to control for these variables, we must either bring them specifically into the analysis or else deal with them by randomization of assignment, a method that can practically eliminate the unwanted effects of these other variables if the sample is sufficiently large.

In making a study involving several variables, unfortunately we do not generally enjoy the "ideal" conditions of a scientific laboratory. A laboratory is a place well-suited to keeping variables under control so that one can isolate the effect of one variable on another without the disturbing effect of unwanted changes that would be taking place in a less-controlled environment. For example, a scientist may have facilities to hold temperature and atmospheric pressure constant while observing the effect of another variable, such as humidity, on the outcome of an experiment. One variable can be changed at a time; everything else is held constant.

It is sometimes possible to achieve a situation comparable to that of a controlled physical experiment when working with human subjects. We may be able to hold age constant, sex constant, or social class constant by an experimental design that matches experimental groups on these characteristics. But much research in public policy or administration cannot be so controlled. The variables will not "stand still" for us. We are forced to look at gun-control laws in areas where they already exist, and these areas will generally not be identical with respect to racial mix, unemployment rate, etc. When everything is varying at once, we can isolate the effects of two variables on each other only through statistical methods that come under the general heading of "multivariate analysis" techniques.

EXERCISES

3–1. The Chief of Police in a city has requested that the size of the police force be increased by 10% in order better to control crime. A question posed by several members of the city council is, "How much, if any, would increasing the size of the force reduce crime?"

a) Would a study of different localities differing in size of police force and crime rate be useful in answering such a question?

b) In such a study, what variables might you include in an analysis?

3.2 Two Nominal Variables

The techniques for analyzing two, three, or more variables simultaneously are generally not simple; for this reason Chapter 2 began the discussion of descriptive statistics by looking at a single variable at a time; this chapter will extend the discussion to analysis of two variables ("bivariate analysis"). The data in this bivariate analysis will be *individual* observations, just as in Chapter 2. However, each individual observation will consist of a *pair* of values. The first element of each pair will be the value of the first variable for this observation, the second element of each pair will be the value of the second variable.

This section discusses measures of association of two nominal variables.

> The classification of individuals in two ways, or by two variables, the categories of each variable being exhaustive and mutually exclusive, is known as a CROSS-PARTITIONING of the two variables. A two-dimensional table displaying the frequencies of cases in each cell of the cross-partitioning is known as a CONTINGENCY TABLE.

EXAMPLE 3.1 The 80 employees of the RST Bureau have answered a question about whether they believe that an employees' union should be formed. Table 3.1 shows how the employees can simultaneously be classified by two variables, their job classification and their responses to the question on the union.

The numbers in the "cells" are the frequencies of cases. For example, there are 4 employees who are both administrative and favorable to the union. To find the number of employees who are administrative, add the frequencies in the cells across the first row, obtaining 4 + 16 = 20, and place the total at the end of the first row in the right margin. By adding the cell frequencies across the rows you obtain the total number of observations of each value of the variable "job classification," and by adding the cell frequencies down the columns you get the total number of observations of each value of the variable "views on the proposed union." ■

Table 3.1 is a presentation of grouped data, as much as the frequency tables presented in Section 2.3 were. The primary difference is that the contingency table is two dimensional and gives a picture of the relationship between two variables. It is possible to develop descriptive measures of this relationship, much as we developed measures on a single variable in the previous chapter.

In order to describe the relationship, we could start by a comparison of proportions or percentages. From Table 3.1 you can see that only 20% (4 out of 20) of the administrative employees had favorable views about a union, 40% (12 out of 30) of the secretarial employees had favorable views, and 67% (20 out of 30) of the custodial employees held favorable views. This considerable difference in percent suggests a strong relationship between job classification and views on the union. Had there

Table 3.1. *CONTINGENCY TABLE*
Cross-Partitioning of 80 Employees by Job Classification and Views on Proposed Union

Job Class	Views on Proposed Union		
	Favorable	Unfavorable	
Administrative	4	16	20
Secretarial	12	18	30
Custodial	20	10	30
	36	44	

been no relationship between the two variables, we would have expected these percentages to be more nearly equal.

We can also compare the percentages in the other direction (down the columns). Of those favorable to unionization, only 11% (4 out of 36) were administrative; 33% (12 out of 36) were secretarial and 56% (20 out of 36) were custodial. On the other hand, of those whose views were unfavorable, 36% (16 out of 44) were administrative, 41% (18 out of 44) were secretarial, and only 23% (10 out of 44) were custodial.

At this point you might feel like an executive who has been beseiged with too much data. "These differences in percentages suggest that there is a relationship between the two variables," you might be saying, "but there are too many numbers. Can't you boil this down to a single number that tells me how strong or important the relationship is?" If you say this, you may have come to appreciate the value of a single measure that describes the relationship between two variables.

As with measures of location and dispersion, there is a variety of measures that have been developed, all of which have advantages and disadvantages. One of these is the coefficient of contingency. This involves the use of the χ^2 (chi-square) statistic.

The Chi-Square Statistic

The chi-square statistic, represented by the Greek letter χ raised to the 2nd power, is a measure of how much, proportionally, the frequencies in the cells differ from the frequencies you would expect if there were absolutely no relationship between the variables.

If job classification and views on a union were unrelated to each other, how many people would you "expect" to be in the upper left cell? To answer this question, note that the percentage of *all* employees who favor a union is 45% (36 out of 80). If job classification had nothing to do with one's views on the union, then we would "expect," on the average, 45% of the 20 administrative employees, or 9 of them, to favor a union. The expected frequency in the upper left cell would be 11. Of the 30 secretarial employees we would also "expect" 45% (or 13.5) to favor a union, so that the expected frequency in the second-row/first-column cell is 13.5, etc.

The word "expect" has been put in quotation marks to indicate that the word is used in the specialized sense of a long-run expectation, that is, the mean number of frequencies in each of the cells if you continued to take many samples of 80 employees and cross-partitioned them by these two variables. In a single sampling, we would not be surprised if the frequency in a cell departed considerably from the "expected" frequency.

The EXPECTED FREQUENCY OF OCCURRENCE (f_e) in a cell of a contingency table, assuming the variables are unrelated to each other, is computed by dividing the product of the total of the row in which the cell appears and the total of the column in which the cell appears by the total frequency:

$$f_e = \frac{(\text{row total})(\text{column total})}{\text{total of frequencies}} \qquad \textbf{3.1}$$

By use of Formula 3.1 you can compute the expected frequencies, f_e, for all 6 cells of Table 3.1. For example, the f_e in the upper left cell is calculated by

$$\frac{20 \times 36}{80} = 9$$

Table 3.2 presents the row totals; the column totals; the grand total; the observed frequency in each cell; and the expected frequency in each cell, calculated using Formula 3.1 and shown in a circle to the right of the observed frequency. Notice that the expected frequencies add up to the same row and column totals as the observed frequencies.

Inspection of Table 3.2 reveals considerable differences between some of the expected frequencies (f_e) and the observed frequencies, f. The χ^2 statistic is a single measure representing the magnitude of these differences. It is defined by:

The formula for chi-square

$$\chi^2 = \Sigma\frac{(f - f_e)^2}{f_e} \qquad \textbf{3.2}$$

where f = observed frequency in each cell
f_e = expected frequency in each cell

The calculation of χ^2 from the data in Table 3.2, using Formula 3.2, is shown in Table 3.3.

The χ^2 value, (11.04 in Table 3.3) has little meaning as it sits by itself. When combined with other measures, however, it can help to describe the degree of association.

Table 3.2. *CONTINGENCY TABLE OF DATA FROM TABLE 3.1*
Observed Frequencies, Expected Frequencies, Row and Column Totals

	Views on Union		
	Favorable	Unfavorable	
Administrative	4 (9)	16 (11)	20
Secretarial	12 (13.5)	18 (16.5)	30
Custodial	20 (13.5)	10 (16.5)	30
	36	44	80

Table 3.3. *CALCULATION OF* χ^2 *FROM DATA IN TABLE 3.2*

Cell	f	f_e	$f - f_e$	$(f - f_e)^2$	$\frac{(f - f_e)^2}{f_e}$
1, 1	4	9	−5	25	2.78
1, 2	16	11	5	25	2.27
2, 1	12	13.5	−1.5	−2.25	.17
2, 2	18	16.5	1.5	2.25	.13
3, 1	20	13.5	6.5	42.25	3.13
3, 2	10	16.5	−6.5	42.25	2.56
					11.04

$$\chi^2 = \Sigma \frac{(f - f_e)^2}{f_e} = 11.04$$

The COEFFICIENT OF CONTINGENCY is defined by

$$C = \sqrt{\frac{\chi^2}{\chi^2 + n}} \quad \textbf{3.3}$$

where χ^2 is computed from a contingency table
n is the total number of observations

The coefficient of contingency cannot be less than 0 nor as much as 1. A value of C very close to 0 would generally be considered a weak relationship, and a value of C very close to the maximum of 1 would generally be considered a strong relationship; but such interpretations must be made with much caution, since a relationship that is considered strong in one context might be considered weak in another. The coefficient of contingency in the present example is

$$C = \sqrt{\frac{11.04}{11.04 + 80}} = .34$$

Whether one considers this relationship to be strong or weak depends on the context of the problem. If you were to do a study on the relationship between hair color and views on a union, a coefficient of contingency of .34 might be surprisingly high; if the study had been on the relationship between sex and sexual preference, a coefficient of contingency of .34 would probably be considered low.

EXAMPLE 3.2 In a study of 187 participants in a Health Maintenance Organization (HMO) the participants were classified by employment status, by sex, and by satisfaction with the HMO. The coefficient of contingency between employment status and satisfaction with the HMO was .22; the coefficient of contingency between sex and satisfaction with the HMO was .16. Can we conclude that, among these 187 participants, the relationship between employment status and satisfaction is a stronger one than that between sex and satisfaction with the HMO?

Discussion: Assuming the data are valid indicators of the variables we are trying to measure, we do have evidence that, at least among those in this sample, there is a stronger relationship between employment status and satisfaction than between sex and satisfaction. But such a conclusion can be drawn only if both coefficients of contingency were based on the same sample number and a contingency table of the same dimensions. ■

When we want to compare relationships involving different sample sizes and/or different numbers of categories for each variable, we run into problems if we try to use the coefficient of contingency. The maximum value that C can take on is $\sqrt{(k - 1)/k}$, where k is either the number of rows or the number of columns, whichever is smaller. From the data in Example 3.1, the number of rows is 3, the number of columns is 2, k is therefore 2, and the maximum value that C could possibly take on is $\sqrt{1/2} = .707$. On the other hand, data from a 4×4 table would have a maximum C value of $\sqrt{3/4} = .866$. A comparison of coefficients of contingency from tables of these two sizes would not be valid, nor would it be valid to compare coefficients of contingency based on different sample sizes.

Interpretation of Measures of Association

When a measure of association is equal to zero, the data show no degree of association with each other. One would expect to see measures of association of close to zero for pairs of unrelated variables such as eye color and sex, or social security number and political preference.

When a measure of association is at its maximum, the data show perfect association. In this case, one variable totally "explains" the other; that is, information on one variable gives perfect information on the other. When this occurs, a more careful examination of the way the variables were defined may reveal that the two "variables" might actually be two names for the same variable.

EXERCISES

3–2. Given the following contingency table:

	Employed	Unemployed
Male	52	8
Female	28	12

a) Find the number of males, of females, of employed, and of unemployed.
b) Find the difference between the employment rates of males and of females.
c) Find the expected frequencies in each of the 4 cells, assuming that sex and employment status are unrelated.
d) Calculate the χ^2 statistic.
e) Calculate *C*, the coefficient of contingency.

3–3. In the case of a perfect association, one variable totally explains the other. Consider the following contingency table, showing hypothetical data on two variables for 60 married people.

		Sex of Spouse	
		Male	Female
Sex	Male	0	36
	Female	24	0

a) Calculate the expected frequencies, assuming independence.
b) Calculate χ^2.
c) Calculate the coefficient of contingency *C*.
d) Would it be safe to say that the relationship between these two variables is a strong one? Why or why not?

3–4. Given the following data:

	Income below $5000	Income from $5000 to $10,000	Income above $10,000
Disabled	20	15	5
Not Disabled	10	25	45

a) Compute the expected frequencies for all the cells, assuming the variables are unrelated.
b) Compute the χ^2 statistic.
c) Compute the coefficient of contingency.

3–5. If you compare the coefficient of contingency calculated in Exercise 3–2 with the coefficient of contingency calculated in Exercise 3–4, would this comparison be a valid basis for saying that one of the relationships was stronger than the other? Why or why not?

3.3 Two Ordinal Variables

The measures of association of nominal variables lack the property of conveying a sense of *direction* to the relationship. The coefficient of contingency, for example, may tell you something about the magnitude of the relationship between job classification and views on a union, but it does not tell you which way the relationship goes; it does not say whether administrative employees are more favorable toward a union than custodial employees, or the other way about.

When the values of two variables can be put in a rank order, however, we can describe their relationship not only with measures suggesting how *strong* the relationship is, but also what is the *direction* of the relationship. A positive relationship means that high values of one tend to be associated with high values of the other, and low values of one will tend to be associated with low values of the other. If the relationship is negative, on the other hand, then there will be more high–low and low–high associations.

For example, a study might reveal a strong relationship between number of children and a measure of mental stability. A contingency coefficient might be used to measure the *strength* of this relationship. But it would not tell us the *direction* of the relationship; whether it is positive or negative. It is important to know whether people with more children tend to have *greater* mental stability or *less.*

A number of statistics have been developed to measure the relationship between two ordinal variables. This section will present the Spearman rank correlation coefficient.

Spearman Rank Correlation Coefficient

If a set of observations is ranked on two different variables, we can measure the degree of relationship between the two sets of ranks by the Spearman rank correlation coefficient.

> The SPEARMAN RANK CORRELATION COEFFICIENT is defined by
>
> $$r_s = 1 - \frac{6\sum_{i=1}^{n} d_i^2}{n(n^2 - 1)} \qquad \textbf{3.4}$$
>
> where d_i is the difference in the two ranks of the ith observation and n is the number of paired observations.

EXAMPLE 3.3 Five middle-level managers are given effectiveness ratings by their bosses and by their subordinates. The five are ranked by both their bosses' rating *(X)* and by their subordinates' rating *(Y)*. The ratings, 5 being the highest and 1 being the lowest, are as follows:

	X	*Y*
McVey	5	4
Osgood	4	2
Taylor	3	5
Evans	2	3
Kelson	1	1

Calculate a measure of association between the two sets of ranks. Do the bosses and the subordinates tend to agree or to disagree?

Solution: The computation is displayed in Table 3.4. Notice that the Spearman coefficient indicates a positive relationship. It shows that the bosses and the subordinates tend to be more similar than dissimilar in their ratings. ■

One of the major problems in describing paired rankings is the problem of ties. Many rankings are full of ties, and in the case of grouped data all of the observations in a cell are tied with each other. In the computation of the Spearman rank correlation coefficient, tied observations are given the mean of the ranks of all the observations in the tie had they

Table 3.4. *CALCULATION OF SPEARMAN RANK CORRELATION COEFFICIENT*

MANAGER	RANK BY BOSS X	MEAN RANK BY SUBORDINATES Y	$X - Y = d$	d^2
McVey	5	4	1	1
Osgood	4	2	2	4
Taylor	3	5	−2	4
Evans	2	3	−1	1
Kelson	1	1	0	0
			0	10

$$r_s = 1 - \frac{6\Sigma d^2}{n(n^2 - 1)} = 1 - \frac{6(10)}{5(25 - 1)} = +.500$$

been ranked individually in an arbitrary order. For example, if the first two ranks are tied, each one is assigned the rank of 1.5 (the mean of 1 and 2); if the next three places are tied, they are each given a rank of 4 (the mean of 3, 4, and 5) etc. This method of dealing with ties insures that the sum of the ranks will be the same as if they had not been tied.

EXAMPLE 3.4 Twelve employees were evaluated by a supervisor on an Effectiveness Rating Form. Question #14 on the form was "How long have you known this person?" Question #19 asked the supervisor to give an overall evaluation of the employee by checking one of 7 categories ranked in descending order. The answers the supervisor gave for the 12 employees are shown in Table 3.5.

Find the Spearman rank correlation coefficient to describe the relationship between the supervisor's response to these two questions.

Solution: The ranks of the employees, from lowest to highest, are:

Rank	*Question 14*	*Rank*	*Question 19*
1.5	F,K (tied)	1	J
3	D	2	C
5	B,I,L (tied)	3	A
7.5	E,G (tied)	4.5	E,H (tied)
9	H	6	G
11	A,C,J (tied)	8.5	D,F,I,K (tied)
		11.5	B,L (tied)

The calculation of r_s is shown in Table 3.6.

Table 3.5. *RESPONSES OF SUPERVISOR TO TWO QUESTIONS ON TWELVE EMPLOYEES*

EMPLOYEE	RESPONSE TO QUESTION #14	RESPONSE TO QUESTION #19
A	More than 20 years	Average
B	Between 5 and 10 years	Outstanding
C	More than 20 years	Below average
D	Between 2 and 5 years	Very good
E	Between 10 and 15 years	Above average
F	Less than 2 years	Very good
G	Between 10 and 15 years	Good
H	Between 15 and 20 years	Above average
I	Between 5 and 10 years	Very good
J	More than 20 years	Poor
K	Less than 2 years	Very good
L	Between 5 and 10 years	Outstanding

Table 3.6. *CALCULATION OF SPEARMAN RANK CORRELATION COEFFICIENT*
Data from Table 3.5

EMPLOYEE	RANK X	RANK Y	d	d^2
A	11.0	3.0	8.0	64.00
B	5.0	11.5	−6.5	42.25
C	11.0	2.0	9.0	81.00
D	3.0	8.5	−5.5	30.25
E	7.5	4.5	3.0	9.00
F	1.5	8.5	−7.0	49.00
G	7.5	6.0	1.5	2.25
H	9.0	4.5	4.5	20.25
I	5.0	8.5	−3.5	12.25
J	11.0	1.0	10.0	100.00
K	1.5	8.5	−7.0	49.00
L	5.0	11.0	−6.5	42.25
			0.0	501.50

$$r_s = 1 - \frac{6(501.50)}{12(144 - 1)} = -.75$$

The negative value, $r_s = -.75$, indicates that the supervisor has a tendency to give lower evaluations to employees he has known for a long time. ■

EXERCISES

3–6. The 50 states are ranked (from highest to lowest) in area and 1970 population as follows:

State	Area Rank	Pop. Rank	State	Area Rank	Pop. Rank
Alabama	29	21	Montana	4	43
Alaska	1	50	Nebraska	15	35
Arizona	6	33	Nevada	7	47
Arkansas	27	32	New Hampshire	44	41
California	3	1	New Jersey	46	8
Colorado	8	30	New Mexico	5	37
Connecticut	48	24	New York	30	2
Delaware	49	46	N. Carolina	28	12
Florida	22	9	N. Dakota	17	45
Georgia	21	15	Ohio	35	6
Hawaii	47	40	Oklahoma	19	27
Idaho	11	42	Oregon	10	31
Illinois	25	5	Pennsylvania	32	3
Indiana	38	11	Rhode Island	50	39
Iowa	24	25	S. Carolina	40	26
Kansas	13	28	S. Dakota	16	44
Kentucky	37	23	Tennessee	34	17
Louisiana	33	20	Texas	2	4
Maine	39	38	Utah	12	36
Maryland	42	18	Vermont	43	48
Massachusetts	45	10	Virginia	36	14
Michigan	23	7	Washington	20	22
Minnesota	14	19	W. Virginia	41	34
Mississippi	31	29	Wisconsin	26	16
Missouri	18	13	Wyoming	9	49

Calculate the Spearman rank correlation coefficient.

3–7. Calculate the Spearman rank correlation coefficient for the following three sets of paired data:

a)

Rank X	*Rank* Y
1	1
2	2
3	3
4	4
5	5

b)

Rank X	*Rank* Y
1	5
2	4
3	3
4	2
5	1

c)

Rank X	*Rank* Y
1	5
2	1
3	2
4	3
5	4

3–8. Twelve heavily populated cities were ranked, from lowest to highest, according to per capita income (X) and mean annual pollution index (Y), as follows:

City	X	Y	*City*	X	Y
A	11	8	G	9	11
B	1	6	H	8	10
C	12	5	I	4	9
D	3	1	J	10	12
E	7	4	K	5	2
F	6	7	L	2	3

a) Calculate the Spearman rank correlation coefficient.
b) Does your answer to (a) suggest any conclusions about pollution and per capita income?
c) Suppose you discovered that the twelve cities were selected for the study by an individual who is known to believe that economic prosperity is bad for the environment. Why is it important to investigate the manner in which this individual selected the twelve cities before you draw general conclusions?

3.4 Two Interval or Ratio Variables

Section 3.3 was concerned with measuring the relationship between two variables whose values can be rank ordered. But if the values of the variables can be plotted on a pair of rectangular coordinates, we have a useful device for displaying the data that is denied us when we study data that can only be categorized or ranked.

> When each paired observation can be plotted as a point with X and Y coordinates, the resulting plot is a SCATTER DIAGRAM.

Visual inspection of a scatter diagram can provide considerable insight into the nature of the relationship.

EXAMPLE 3.5 Judge Doe in the Juvenile Court claims that the amount of time he spends hearing a case depends strictly on factors that relate to the case itself rather than the pressure on him caused by a backlog of cases. One young student who was studying judicial processes, however, suspected that the size of the backlog of cases had a considerable influence on the amount of time Judge Doe spent on a case. She collected information on 16 cases, which is shown in Table 3.7. What conclusions can you draw from these data?

Discussion: The backlogs and times can be plotted in a scatter diagram, as shown in Figure 3.1. The scatter diagram gives a fairly good visual idea of how the backlog and time spent on cases vary together. In general, it appears that there is a negative relationship, a tendency for high X values to be associated with low Y values, and vice versa. If this is the relationship, then there is a suggestion of some kind of "Parkinson's Law" operating here. When the backlog is low, the results suggest that Judge Doe feels less pressure on his time, and he consumes more time on each case he hears. ■

Linear Correlation Analysis

The scatter diagram of Figure 3.1 suggests that there is a tendency for time spent on cases to decrease as the backlog of cases increases. But this suggestion is based only on subjective inspection of the diagram. If someone else says, "I don't see it that way," you would have little basis

Table 3.7. *DATA ON HEARING OF SIXTEEN JUVENILE CASES*
Number of Cases Backlog (X) and Minutes Spent (Y)

Observation	X	Y	Observation	X	Y
1	12	25	9	13	45
2	8	75	10	13	10
3	13	33	11	12	15
4	0	160	12	16	10
5	5	25	13	18	15
6	2	120	14	10	45
7	11	137	15	8	20
8	10	80	16	8	60

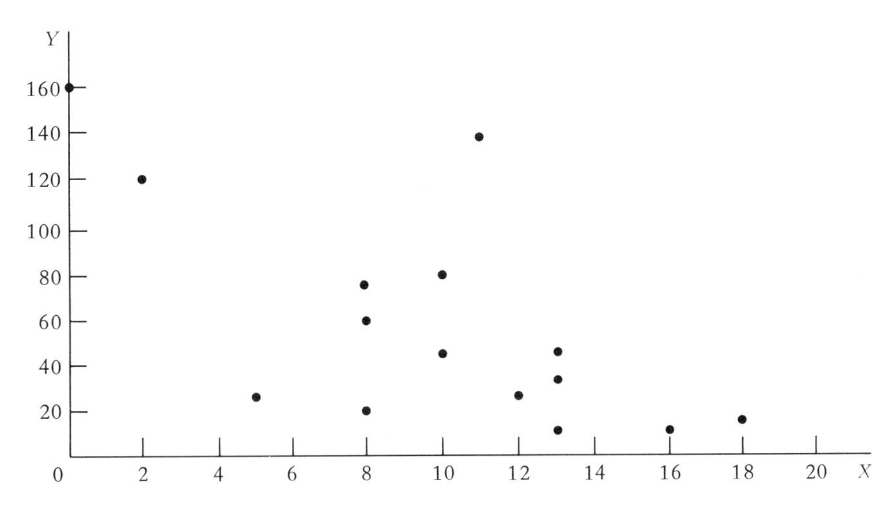

Figure 3.1. *SCATTER DIAGRAM*
Data from Table 3.7

to persuade him otherwise. In order to agree on the degree of relationship that exists, we need a measure of association.

The measure of association that is introduced in this section is based on the assumption that a *linear relationship* exists between the two variables.

> If there is a tendency for one variable to change by a constant amount as another variable changes, we say there is a LINEAR RELATIONSHIP between the two variables. A linear relationship between two variables can be expressed in the form
>
> $$Y = a + bX \qquad \text{or} \qquad X = -\frac{a}{b} + \frac{1}{b}Y$$

A well-known example of a linear relationship is that between Fahrenheit and Centigrade temperature readings. The relationship can be expressed as

$$F = 32 + {}^{9}/_{5}C \qquad \text{or} \qquad C = -17{}^{7}/_{9} + {}^{5}/_{9}F$$

For every degree of change in Centigrade, the Fahrenheit changes by a constant amount, 9/5 of a degree. For every degree of change in Fahrenheit, the Centigrade changes by 5/9 of a degree.

In the case of scatter diagrams, the pattern of the points will indicate the type of relationship, as illustrated in Figure 3.2.

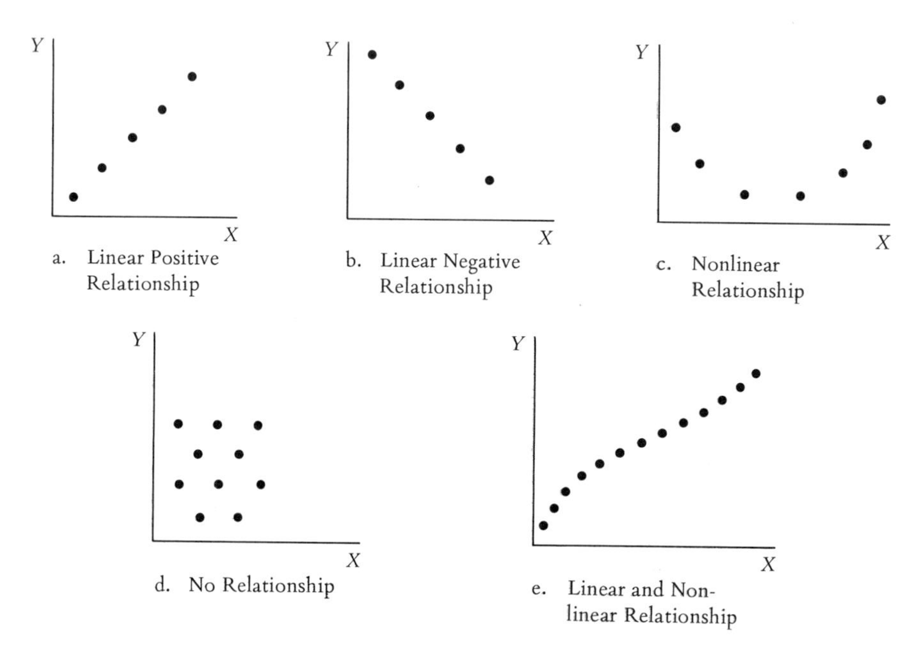

Figure 3.2. *ILLUSTRATIONS OF TYPES OF RELATIONSHIPS WITH SCATTER DIAGRAMS*

Most pairs of variables that we analyze are not as simple to describe as temperature or those illustrated in Figure 3.2. In most cases the relationship will be only partial. In the case of a partial relationship, we may be able to discern linear or nonlinear *components,* but there will be some scatter of points that cannot be explained by the mathematical relationship. For example, nobody claims that the size of case-backlog totally *determines* the amount of time spent on a case, but only that there is a *tendency* for the Y values to decrease as X increases. If the scatter diagram suggests that this tendency can be expressed as a linear relationship, we can describe this relationship by the Pearson coefficient of linear correlation.

> The PEARSON COEFFICIENT OF LINEAR CORRELATION, measures the linear relationship between two variables X and Y and is defined by
>
> $$r = \frac{\Sigma(X - \overline{X})(Y - \overline{Y})}{\sqrt{\Sigma(X - \overline{X})^2 \Sigma(Y - \overline{Y})^2}} \qquad 3.5$$

An interesting feature of this coefficient, often referred to as the Pearson r, can be seen from observing the numerator of the fraction. If the relationship is a positive one, we expect high X values to be associated with high Y values. When both the X values and Y values are above their respective means, both $(X - \overline{X})$ and $(Y - \overline{Y})$ will be positive, and the cross-products $(X - \overline{X})(Y - \overline{Y})$ will be positive. Also we would expect low X values to be associated with low Y values. When both X and Y values are below their respective means, both $(X - \overline{X})$ and $(Y - \overline{Y})$ will be negative, but their cross-products $(X - \overline{X})(Y - \overline{Y})$ will again be positive. Therefore, if the relationship is positive, we expect the numerator $\Sigma(X - \overline{X})(Y - \overline{Y})$ to be a summation of cross-products in which the positive elements predominate, so that the sum will be positive and r will be positive. On the other hand, if the relationship is negative, we expect the high X values to be associated with low Y values, and vice versa, so that the $(X - \overline{X})$ and $(Y - \overline{Y})$ differences will generally be of opposite sign. Since the product of numbers with unlike signs is negative, the negative $(X - \overline{X})(Y - \overline{Y})$ cross-products will tend to predominate in the numerator, so that the Pearson r will be negative.

Formula 3.5 is not generally an efficient formula for the calculation of r, particularly when $\overline{X}$ and $\overline{Y}$ are not integers and there are many observations. For computational purposes, therefore, we often use an equivalent formula.

Computational formula for Pearson *r*

$$r = \frac{n\Sigma XY - \Sigma X \Sigma Y}{\sqrt{n\Sigma X^2 - (\Sigma X)^2}\sqrt{n\Sigma Y^2 - (\Sigma Y)^2}} \qquad \textbf{3.6}$$

EXAMPLE 3.6 Find the Pearson coefficient of linear correlation for the following set of paired data, using both Formula 3.5 and Formula 3.6.

X	Y
2	1
4	5
6	3

Solution: Tables 3.8 and 3.9 display the calculations.

A careful comparison of these two tables should be very helpful if you need to acquire a little more familiarity with the notation in Formula 3.5 and Formula 3.6.

EXAMPLE 3.6 (continued) Calculate the Pearson coefficient of linear correlation for the data on case times and backlogs given in Table 3.7.

Table 3.8. *COMPUTATION OF r USING FORMULA 3.5*

X	Y	$(X - \overline{X})$	$(X - \overline{X})^2$	$(Y - \overline{Y})$	$(Y - \overline{Y})^2$	$(X - \overline{X})(Y - \overline{Y})$
2	1	−2	4	−2	4	4
4	5	0	0	2	0	0
6	3	2	4	0	4	0
12	9	0	8	0	8	4

$$\overline{X} = \frac{12}{3} = 4$$

$$\overline{Y} = \frac{9}{3} = 3$$

$$r = \frac{\Sigma(X - \overline{X})(Y - \overline{Y})}{\sqrt{\Sigma(X - \overline{X})^2 \Sigma(Y - \overline{Y})^2}} = \frac{4}{\sqrt{(8)(8)}} = \frac{1}{2}$$

Table 3.9. *COMPUTATION OF r USING FORMULA 3.6*

X	Y	XY	X^2	Y^2
2	1	2	4	1
4	5	20	16	25
6	3	18	36	9
12	9	40	56	35

$$r = \frac{n\Sigma XY - \Sigma X \Sigma Y}{\sqrt{n\Sigma X^2 - (\Sigma X)^2}\sqrt{n\Sigma Y^2 - (\Sigma Y)^2}}$$

$$= \frac{3(40) - (12)(9)}{\sqrt{3(56) - (12)^2}\sqrt{3(35) - (9)^2}}$$

$$= \frac{1}{2}$$

Solution: The calculations using Formula 3.5 are displayed in Table 3.10.

INTERPRETATION OF THE PEARSON *r*

Like the measures of association of ordinal variables discussed in Section 3.3, the Pearson coefficient of linear correlation is a measure both of the strength and the direction of the relationship. A value of +1 indicates a perfect positive relationship, a value of −1 a perfect negative relationship and a value of 0 a lack of a linear relationship. Remember, however, that the Pearson *r* measures only the *linear* relationship; it is not an appropriate measure if the relationship between the two variables is decidedly nonlinear. For this reason, it is desirable to draw a scatter diagram before doing any calculations; inspection of the scatter diagram may reveal considerable nonlinearity of the data.

Table 3.10. *CALCULATION OF PEARSON r*
Data from Table 3.7

X	Y	XY	X^2	Y^2
12	25	300	144	625
8	75	600	64	5,625
13	33	429	169	1,089
0	160	0	0	25,600
5	25	125	25	625
2	120	240	4	14,400
11	137	1,507	121	18,769
10	80	800	100	6,400
13	45	585	169	2,025
13	10	140	196	100
12	15	300	144	225
16	10	160	256	100
18	15	270	324	225
10	45	450	100	2,025
8	20	160	64	400
8	60	520	64	4,225
158	875	6586	1944	82,458

$$r = \frac{n\Sigma XY - \Sigma X \Sigma Y}{\sqrt{n\Sigma X^2 - (\Sigma X)^2}\sqrt{n\Sigma Y^2 - (\Sigma Y)^2}}$$

$$= \frac{16(6586) - (160)(875)}{\sqrt{16(1944) - 158^2}\sqrt{16(82458) - (875)^2}}$$

$$= -.564$$

EXERCISES

3–9. From your own experience or that of your organization or school, record a set of 20 or more paired observations on two interval-level or ratio-level variables. Plot a scatter diagram of your observations. Does this diagram give you a picture of the relationship between the two variables?

3–10. A proposal to require all managerial employees in the County Treasurer's Office to obtain M.P.A. or equivalent degrees is being considered. A survey of 16 employees was taken to determine whether amount of schooling had much effect on job performance. The following X, Y pairs were recorded, X being years of schooling and Y being a recent rating on the job.

X	Y	X	Y	X	Y	X	Y
12	8	18	8	10	8	11	2
16	5	17	9	16	10	14	5
13	4	12	5	14	5	16	8
12	9	12	2	18	10	9	2

Plot a scatter diagram of these data. What generalizations can you draw from the scatter diagram?

3–11. Data were recorded on 6 city employees on number of dollars (X) contributed to the party in power, and number of days (Y) of extra vacation per year received. The results are as follows:

X	Y	X	Y
50	6	100	12
0	0	250	30
25	3	125	15

Plot a scatter diagram of the data and draw your own conclusions.

3–12. A perfect linear relationship will occur if two variables are in fact measures of the same thing. Suppose, for example, that one was studying the relationship between the weekly earnings of hourly workers and the number of hours they worked. If the study were made on a set of people who all had the same hourly rate, the data might look something like this:

Hours (X)	*Earnings* (Y)
40	$100
32	80
30	75
20	50

Verify that the Pearson coefficient of linear correlation between these two variables is 1.

3–13. Given the ages and salaries of 5 employees:

Age	*Salary*
32	$10,000
36	16,000
30	13,000
34	12,000
38	19,000

Compute the Pearson r.

3–14. Subtract 30 from each of the ages in Exercise 3–13 and subtract $10,000 from each of the salaries. Then compute r on the "transformed" variables. How did subtracting a constant from the values of each variable by a constant affect the Pearson r?

3–15. From the data in Exercise 3–13 let $X' = \text{age}/2$ and let $Y' = \text{salary}/1000$.

Calculate the Pearson r on the transformed variables X' and Y'. How did dividing the values of each variable by a constant affect the Pearson r?

3–16. From the data in Exercise 3–13 let x = age − 34; let y = salary − \$13,000. Calculate the Pearson r. How did subtracting the mean from the values of each variable affect the Pearson r?

3–17. Data on 5 school districts show their spending per pupil (X) and their mean class size (Y):

District	X	Y
1	450	28
2	560	24
3	380	25
4	640	20
5	550	23

Compute the Pearson coefficient of linear correlation. Using the results suggested by Exercises 3–14 and 3–15, you can simplify your calculation by dividing each X by 10 and subtracting 20 from each Y.

3–18. Suppose X is age and Y is earnings. If you take a random sample of the people of a city who are between 15 and 75 years of age, do you think it would be appropriate to compute a Pearson coefficient of linear correlation? Why or why not?

Linear Regression

The study of how two or more variables vary together is called correlation analysis. The relationships described by coefficients of correlation are symmetrical, or two-way, relationships. The correlation between X and Y is identically the same thing as the correlation between Y and X.

Regression analysis, on the other hand, is a method of deriving an asymmetrical, or one-way, relationship. Given information on one variable, which we call the *independent variable,* we use regression techniques to predict the value of another variable, which we call the *dependent variable.* The purpose is not necessarily to make a perfect prediction; the purpose is to use information on the independent variable to decrease the error we are likely to make in predicting a value of the dependent variable. The different nature of regression analysis from correlation analysis is illustrated by two English prepositions. Correlation analysis of two variables is often referred to as the correlation of X *with* Y. Regression analysis of an independent variable X and a dependent variable Y is often referred to as the regression of Y *on* X.

EXAMPLE 3.7 To what extent does increasing the number of Nurses' Aides decrease the number of patients' complaints? In a study in a County General Hospital the number of nurses' aides assigned to a section of the hospital was varied from day to day. For simplicity's sake, only data for 8 days will be presented, although for such a study to have much meaning, it would have to be conducted over a much longer time period. The X values (number of Nurses' Aides) and Y values (patients' complaints) are as follows:

Day	X	Y	*Day*	X	Y
1	10	7	5	18	4
2	12	4	6	20	3
3	14	4	7	22	0
4	16	1	8	24	1

We can plot a scatter diagram of this information as we did in the previous section. ■

Inspection of Figure 3.3 suggests that there is a negative relationship between X (number of aides) and Y (number of complaints), and that the relationship might very well be linear. If this is the case, a straight line with a negative slope (moving downward to the right) could represent the general relationship.

Figure 3.3. *SCATTER DIAGRAM Number of Nurses Aides (X) and Complaints (Y)*

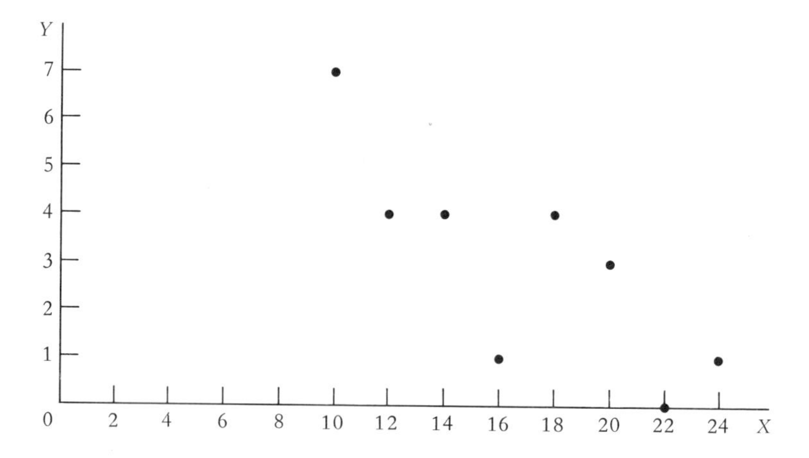

A straight line predicting values of Y from given values of X is known as a REGRESSION LINE and will have the form

$$\hat{Y} = a + bX$$

where $\hat{Y}$ is the predicted value of Y **3.7**

a is the Y intercept (where the line crosses the Y axis)

b is the slope of the line

The equation for this line is known as a LINEAR REGRESSION EQUATION. The coefficient of X is known as the LINEAR REGRESSION COEFFICIENT.

It is obvious that no straight line can pass through all the points in Figure 3.3. Since no line can "fit" all the points perfectly, we will have to settle for a line that comes as close as possible to being a perfect fit. But what does "as close as possible" mean? You could draw a straight line that seemed to you to be a best fit, but someone else could come up with another line that he thought was a better fit. There is no way, therefore, that we can agree on which line fits best until we agree on a criterion of what constitutes a best fit.

A reasonable criterion for a best fitting prediction would be that of minimizing the sum of the absolute values of the "errors," or differences between the actual Y values and the $\hat{Y}$ values predicted from X; that is, to minimize

$$\Sigma|Y - \hat{Y}|$$

Although this criterion has some intuitive appeal, the criterion that is generally used to determine "best fit" is that of minimizing the *squares* of the deviations; that is, to minimize

$$\Sigma(Y - \hat{Y})^2$$

The reason the "least squares" criterion is almost universally preferred to the "least absolute value" criterion is that a wide variety of statistical methods have been developed to analyze data in terms of total squared deviations from predicted values based on the least-squares criterion.

To find the equation for the least-squares line $\hat{Y}$, we need to solve for the two constants a and b. We can first solve for the regression coefficient b by,

$$b = \frac{\Sigma(X - \overline{X})(Y - \overline{Y})}{\Sigma(X - \overline{X})^2} \qquad \textbf{3.8}$$

When $\overline{X}$ and $\overline{Y}$ are not integers, the computation of this equation becomes very tedious. In this case it is generally more efficient to use Formula 3.9.

Computational formula for the linear regression coefficient

$$b = \frac{n\Sigma XY - \Sigma X \Sigma Y}{n\Sigma X^2 - (\Sigma X)^2} \tag{3.9}$$

Having found b, we can solve for a by

$$a = \frac{\Sigma Y - b\Sigma X}{n} \tag{3.10}$$

Table 3.11 shows the calculations for a and b from the data in Example 3.7.

These calculations give us an a of 9.273 and a b of $-.369$, so that

$$\hat{Y} = 9.273 - .369X$$

INTERPRETATION OF THE LINEAR REGRESSION EQUATION

The a value is the "Y intercept," that is, the value of $\hat{Y}$ when X is zero. If interpreted in terms of nurses' aides and complaints, the above results would mean that we predict 9.273 complaints if the number of

Table 3.11. *CALCULATION OF a AND b COEFFICIENTS Data from Example 3.7*

X	Y	XY	X^2
10	7	70	100
12	4	48	144
14	4	56	196
16	1	16	256
18	4	72	324
20	3	60	400
22	0	0	484
24	1	24	576
136	24	346	2480

$$b = \frac{n\Sigma XY - \Sigma X \Sigma Y}{n\Sigma X^2 - (\Sigma X)^2} = \frac{8(346) - (136)(24)}{8(2480) - (136)^2} = -.369$$

$$a = \frac{\Sigma Y - b\Sigma X}{n} = \frac{24 + .369(136)}{8} = 9.273$$

nurses' aides is zero. We should not make that interpretation in this problem, however, since the data on X extend only from 10 to 24; and, in general, interpretations of linear regression equations should be confined to points within the range of values being used. The a value in Example 3.7 therefore gets no interpretation except a mathematical one. It is the value of $\hat{Y}$ when X is zero.

The regression coefficient of $-.369$ means that for every unit increase in X, $\hat{Y}$ *decreases* by .369. Within the range of our data ($10 \leq X \leq 24$), we can say that, for every nurses' aide that is added, we predict, or expect, .369 fewer complaints.

Let us use the equation $\hat{Y} = 9.273 - .369X$ to predict Y for the values of the data, $X = 10, 12, \ldots, 24$. We can then compare these predicted values with the actual Y values. The difference between the two is known as "errors."

Notice that in the fourth column of Table 3.12, the algebraic sum of all the deviations, or errors, is zero. This will always be true of a least squares prediction equation. The least squares prediction equation therefore has a property similar to that of the arithmetic mean. The least squares equation is a kind of "neutral" predictor in the sense that its total positive errors are exactly balanced by its total negative errors.

Another thing to notice in the 4th column is the size of the errors. The largest one (in absolute values) is 2.369; all other predictions are off by less than 2. To get an idea of how large are the errors in the aggregate, we can add the sums of the squares of the errors, as shown in the last column of Table 3.12. The sum of the squared errors is seen to be 13.121. And because the line was fitted according to the criterion of minimizing

Table 3.12. *CALCULATION OF $\hat{Y}$ AND SUM OF SQUARED ERRORS*
Data from Table 3.11 and Prediction Equation $\hat{Y} = 9.273 - .369X$

X	$\hat{Y} = 9.273 - .369X$	Y	$Y - \hat{Y}$	$(Y - \hat{Y})^2$
10	5.583	7	1.417	2.008
12	4.845	4	−.845	0.714
14	4.107	4	−.107	0.014
16	3.369	1	−2.369	5.612
18	2.631	4	1.369	1.874
20	1.893	3	−1.107	1.225
22	1.155	0	−1.155	1.334
24	0.417	1	.583	0.340
			0	13.121

the sum of these squares, we know that no other linear equation would have given a sum of squared errors smaller than 13.121.

But how good is the fit? Is getting a sum of squared deviations of 13.121 a remarkable achievement, or not? We can give an answer to this question only if we compare this sum of squared errors with the sum we would have achieved had we not used information about X to predict Y. If we simply used the mean of Y, which is 3, to predict the Y values, we could compute the sum of the squared deviations about $\overline{Y}$ as shown in Table 3.13.

The sum of the squared deviations about $\overline{Y}$ in Table 3.13 is 36. This figure is the best we can do in predicting Y without benefit of X. The sum of the squared deviations from the mean is less than the sum of the squared deviations from any other constant value. But by using a linear regression line based on X to predict Y, we reduce the sum of the squared deviations from 36 to 13.121 (Table 3.12). The linear regression equation has therefore reduced the sum of the squared deviations by 36 − 13.121 = 22.879, and the proportional reduction of the sum of the squared deviations is 22.879/36 = .636, or 63.6%.

> The sum of the squared deviations of the Y values around the predicted values $\hat{Y}$, denoted by
>
> $$\Sigma(Y - \hat{Y})^2$$
>
> is known as the RESIDUAL SUM OF SQUARES.

Table 3.13. *COMPUTATION OF SUM OF SQUARED DEVIATIONS OF Y ABOUT $\overline{Y}$*
Data from Example 3.7

Y	$\overline{Y}$	$Y - \overline{Y}$	$(Y - \overline{Y})^2$
7	3	4	16
4	3	1	1
4	3	1	1
1	3	−2	4
4	3	1	1
3	3	0	0
0	3	−3	9
1	3	−2	4
			36

In this example, the residual sum of squares is 13.121. It is an indication of how much the regression equation *fails* to explain.

Comparison of Regression Analysis and Correlation Analysis

If you had calculated a Pearson coefficient of linear correlation on the data in Table 3.11 you would have found that r was .797. The square of .797 is .635, which is almost precisely the proportion of the total squared error that the linear regression equation reduced. The difference between the two values is due simply to rounding error.

The COEFFICIENT OF DETERMINATION measures the proportion of the total sum of squared errors of one variable that is reduced, or "explained" by a linear regression equation. **3.11**

$$r^2 = \frac{\Sigma(Y - \overline{Y})^2 - \Sigma(Y - \hat{Y})^2}{\Sigma(Y - \overline{Y})^2} = 1 - \frac{\Sigma(Y - \hat{Y})^2}{\Sigma(Y - \overline{Y})^2}$$

Numerically, the coefficient of determination is equal to the square of the Pearson coefficient of linear correlation between X and Y.

The coefficient of determination is a measure of association having an interpretative significance that we are unable to give to the measures of association previously discussed. It tells us what proportion of the total squared deviations of the independent variable is "explained" by or associated with the linear regression equation on the independent variable. An r^2 of 0 indicates that the independent variable is useless as a predictor; the regression equation failed to reduce the error at all. An r^2 of 1 indicates that the independent variable was a perfect predictor; it eliminated all of the error. In the example in this section, r^2 was .636, indicating that about 64% of the squared deviations of the dependent variable around its mean was eliminated by the regression equation.

Both correlation analysis and regression analysis are used when analyzing bivariate interval or ratio-level variables. The difference between the two is the way one looks at the variables. Correlation coefficients are measures of how both variables vary together. Correlation, which means "co-relation" is therefore symmetrical in its approach. If you examine Formula 3.5 or 3.6 you will see that interchanging X and Y throughout the expression will not affect the value of the coefficient. This symmetry is also a characteristic of the measures of association between nominal variables and between ordinal variables, which were discussed in Sections 3.2 and 3.3.

But regression analysis takes a different approach. It considers one variable as known, and calls that the independent variable. It develops an equation for predicting values of the other variable, the dependent variable, on the basis of assumed values of the independent variable. Regression analysis is therefore an asymmetrical approach, inasmuch as it treats the two variables differently.

A regression equation that predicted Y values from X values would be totally different from a regression equation that predicted X values from Y values, even though equations were calculated using the same data.

It is important, however, to avoid the temptation of inferring a "cause and effect" relationship simply because one variable is treated as independent and the other is treated as dependent. Changes in the independent variable do not necessarily "cause" changes in the dependent variable. The direction of causation might go the other way, or there might be a common factor causing both. One of the original studies in regression analysis developed an equation to predict the heights of sons based on the heights of their fathers. (Evidently the heights of mothers were considered irrelevant.) But one could just as easily develop an equation to "predict" the heights of fathers from the heights of sons, thus reversing the direction of cause and effect. One could even "predict" the heights of men from the heights of their sisters. The relationship so established would be based on factors that are a common cause of both variables.

Regression techniques are used chiefly where one of the variables is known or can be readily controlled and the other is less directly accessible. Studies predicting sales on the basis of packaging, predicting changes in the price level on the basis of changes in the money supply, and predicting test scores on the basis of preparation time, lend themselves readily to such methods.

EXERCISES

3–19. A major problem for organizations using heavy equipment is that of setting a policy on replacement. Generally, maintenance costs increase as equipment gets older; but acquisition and depreciation costs are higher if you replace newer equipment. Finding the optimal age at which to replace is a problem in Operations Research that requires, among other things, a prediction of maintenance costs on the basis of age. Suppose your organization uses a fleet of medium-weight trucks, and you are asked to predict the cost of maintenance on a truck based on the following information collected on eight trucks.

X (*Age*)	*Y* (*Cost of Maintenance*)
2	\$ 100
4	200
7	900
6	400
1	100
3	200
2	200
7	1100

a) Compute a linear regression equation, $\hat{Y} = a + bX$.
b) Suppose a study shows that it is advantageous to replace trucks when their maintenance cost exceeds \$600 a year. On the basis of your answer to (a), at what age would you recommend that trucks be replaced?
c) What proportion of the variation in maintenance cost is explained by age? Use sum of squared errors as your criterion of variation.

3–20. Given the following data:

X (*% Change in Money Supply*)	*Y* (*Inflation Rate, One Year Later*)
14	5
6	7
12	10
8	10
10	8

a) Find the least squares regression equation $\hat{Y} = a + bX$.
b) Interpret the meaning of the a and the b you computed in terms of this problem.
c) There is a widespread belief that expanding the money supply has an inflationary effect. Would a negative value of the regression coefficient b refute this belief? Discuss.

3–21. Assume that the sum of the squared deviations of the values of a variable Y about their mean $\overline{Y}$ was 200. A regression equation $\hat{Y} = a + bX$ is computed on an independent variable X. How good a predictor of Y would you say that the regression equation was if

a) the sum of the squared errors about $\hat{Y}$ was 200?
b) the sum of the squared errors about $\hat{Y}$ was 0?
c) the sum of the squared errors about $\hat{Y}$ was 100?

Would it be possible for the sum of the squared errors about $\hat{Y}$ to be 220? Why or why not?

3–22. Let Y be the electric power generated by a power plant in a day. If you were to predict Y from an equation of the form

$$\hat{Y} = a + bX$$

a) give an example of a variable X, that would be a good predictor of Y. Would the regression coefficient be positive, negative, or near zero?
b) give an example of a variable X_2 that would be a poor predictor of Y. Would the regression coefficient be positive, negative, or near zero?

3–23. Suppose you wished to develop a prediction equation for the value of a variable Y. You collected information on Y values paired with a variable X, and computed the prediction equation

$$\hat{Y}_1 = a_1 + b_1X_1$$

You also paired the Y values with another variable, X_2, and computed the prediction equation

$$\hat{Y}_2 = a_2 + b_2X_2$$

A statistician analyzed these two results and decided that X_2 was a better predictor of Y than X_1. What criterion could the statistician use to make such a statement?

3–24. Given the following pairs of X and Y variables, decide in each case the general direction of causation (if any). Among the possibilities are: X causes Y; Y causes X; both cause each other; common cause to both.

X	Y
Age	Income
Smoking	Incidence of Heart Disease
Teachers' Salaries	National Income
Teachers' Salaries	Number of Auto Accidents
Size of Police Force	Crime Rate
Birth Rate in 1980	Death Rate in 1980
Birth Rate in 1900	Death Rate in 1980
Rainfall in Bombay per Day	Number of hits Pete Rose gets per day

*3.5 Three or More Variables

The Pearson coefficient of linear correlation is a measure of the extent to which two variables vary together, but it does not measure how much this relationship is associated with a third, or fourth, variable that

*This section may be omitted without loss of continuity.

is related to the first two. Variables X and Y may vary together, but as they do, Z may also vary with them. Sometimes we would like to make Z stand still, and observe the effect that X and Y have on each other that is not attributable to changes in Z.

> A MULTIPLE LINEAR REGRESSION EQUATION has the form (for one dependent variable and three independent variables)
>
> $$\hat{Y} = a + b_1X_1 + b_2X_2 + b_3X_3$$

3.12

A multiple linear regression equation predicts the value of a dependent variable Y, given information on two or more independent variables X_1, X_2, etc. The equation is "linear" in that it contains only constants and first powers of the independent variables X (no squares, roots, cubes, cross-products of variables, etc.).

In the case of a simple regression equation involving only one independent variable, this restriction on the powers of X insures that the equation would plot as a straight line. When predicted by two independent variables, $\hat{Y}$ could be plotted as a plane; when there are three or more independent variables, a physical representation of the equation becomes impractical, the $\hat{Y}$ being known as a linear hyperplane.

The computation of the constants a, b_1, etc. is a bit complicated and will not be introduced here. Most multiple regression analysis would be horrendously laborious without the aid of computer programs that do the drudgery for you. But no computer program or printout can interpret the results for you. This section will therefore discuss interpretations.

The $\hat{Y}$ on the left side of Equation 3.12 is the predicted value of Y. Just as in the case of the simple linear equation discussed in Section 3.4, the equation was fitted to the data according to the "least squares" criterion; the sum of the squared deviations of the actual Y values from the predicted $\hat{Y}$ values will be less than from any other linear equation we could have computed to predict Y from these independent variables.

The first term on the right side of Equation 3.12 is the a constant, which indicates the value of $\hat{Y}$ when all of the independent variables are zero. Whether the a value has any interpretative significance in the context of the problem depends on whether the actual data involve X's that could all realistically be zero. If they cannot, then we give no interpretation at all to a except to say that it is a part of the predicting equation.

The next term, b_1X_1, says that as X_1 increases by one unit, *while X_2 and X_3 remain constant,* the predicted value $\hat{Y}$ will increase by the amount b_1. The coefficient b_1, therefore, expresses the effect of X_1 on $\hat{Y}$ when it is isolated from the effects of X_2 and X_3.

In a similar vein, the term b_2X_2 says that as X_2 increases by one unit, *while* X_1 *and* X_3 *remain constant,* the predicted value $\hat{Y}$ will increase by the amount b_2. The last term, b_3X_3 says that as X_3 increases by one unit, *while* X_1 *and* X_2 *remain constant,* the predicted value $\hat{Y}$ will increase by the amount b_3.

> These b coefficients of the X_1, X_2, and X_3 variables are known as PARTIAL REGRESSION COEFFICIENTS.

EXAMPLE 3.8 A study was made of the use of newly constructed rapid transit line by the families in an urban area. A multiple linear regression equation was fitted to the data that predicted the usage of rapid transit (in number of times per month) by members of a family on the basis of income, proximity to the line, and number of adults in the family. The equation was computed to be

$$\hat{Y} = 12.45 - .0003X_1 - 2.4X_2 + 3.4X_3 \quad \textbf{3.13}$$

where $\hat{Y}$ is the predicted number of times per month the members of the family use the rapid transit line
X_1 is the annual income of the family in dollars
X_2 is the distance in kilometers of the family residence from the nearest rapid transit station
X_3 is the number of people over 18 years old in the family

This is Equation 3.12, for which the constants a, b_1, b_2, and b_3 have been computed. Literally, the 12.45 means that we would predict that a family with zero income (X_1), living zero kilometers (X_2) from the nearest station, and having zero members over 18 years of age (X_3), would use the line 12.45 times a month. But since some of the independent variables cannot realistically take on values of zero, we should refrain from such an interpretation. All we can say is that 12.45 is the value of $\hat{Y}$ when all the X's are zero.

The $-.0003$ means that for every additional dollar of income, we predict a decrease of .0003 times per month in usage if distance from nearest station and family size were held constant. For example, if two families differed in income by $1000, we would predict that the family with the higher income would use the line 3 times less per month, if the two families were equal in distance from the nearest station and number of members over 18.

The coefficient -2.4 means that for every increase of one kilometer's distance from the nearest station, we would predict a decrease in usage of 2.4 times a month, if income and number of family members over 18 were held constant.

The coefficient $+3.4$ means that for every additional family member over 18 years of age, we predict an increase in usage of 3.4 times a month, if income and distance from the nearest station are held constant.

Given Equation 3.13, how would we predict the number of times a month a family would use the line if it had 3 members over 18, it lived 0.4 kilometers from the nearest station, and its annual income was $10,000?

Solution: Substituting the values above along with the constants of Equation 3.13 into Equation 3.12 gives

$$\hat{Y} = 12.45 - .0003(10{,}000) - 2.4(0.4) + 3.4(3) = 18.69$$

times per month. ■

EXAMPLE 3.9 In a study of age discrimination, a sample of 5 individuals was selected and the following information concerning Salary *(Y)*, Age (X_1), and Years of Job Experience (X_2) was collected.

Individual	Y	X_1	X_2
A	12,000	20	2
B	10,000	25	6
C	20,000	47	10
D	6,000	22	0
E	12,000	36	12

The multiple linear regression equation was computed from these figures:

$$\hat{Y} = -43.36 + 429.12X_1 - 138.37X_2 \qquad \textbf{3.14}$$

(a) Find the $\hat{Y}$ value for each of the five individuals on the basis of their ages and their job experience.
(b) Calculate the sum of the squared deviations of the actual salaries from the predicted values $\hat{Y}$.
(c) Calculate the sum of the squared deviations of the actual salaries Y from $\overline{Y}$, the mean of the Y's.
(d) Compare your answers in (b) and (c). What proportion of the total sum of squares of Y is "explained" by the multiple linear regression equation?
(e) From Equation 3.14, would you say that age was a more important variable in determining salary than experience? Why?

Solution and Discussion: (a) We calculate $\hat{Y}$ for individual A by substituting into Equation 3.14, as follows.

$$\hat{Y}_A = -43.36 + 429.12(20) - 138.37(2) = \$8{,}262.30$$

By the same method we predict the salaries of individuals B, C, D, and E as

$$\hat{Y}_B = -43.36 + 429.12(25) - 138.37(6) = \$9,854.42$$
$$\hat{Y}_C = -43.36 + 429.12(47) - 138.37(10) = \$18,741.58$$
$$\hat{Y}_D = -43.36 + 429.12(22) - 138.37(0) = \$9,397.28$$
$$\hat{Y}_E = -43.36 + 429.12(36) - 138.37(12) = \$13,744.52$$

(b) Calculate $\Sigma(Y - \hat{Y})^2$ as

$$\begin{aligned}(12,000 - 8262.30)^2 &= 13,970,401\\(10,000 - 9854.42)^2 &= 21,194\\(20,000 - 18,741.58)^2 &= 1,583,621\\(6,000 - 9397.28)^2 &= 11,541,511\\(12,000 - 13,744.52)^2 &= \underline{3,043,350}\\&\ 30,160,077\end{aligned}$$

(c) The mean of the Y's is \$10,000. The sum of the squared deviations, $\Sigma(Y - \overline{Y})^2$, is

$$\begin{aligned}(12,000 - 10,000)^2 &= 4,000,000\\(10,000 - 10,000)^2 &= 0\\(20,000 - 10,000)^2 &= 100,000,000\\(6,000 - 10,000)^2 &= 16,000,000\\(12,000 - 10,000)^2 &= \underline{4,000,000}\\&\ 124,000,000\end{aligned}$$

(d) The total sum of the squares of the Y values around $\overline{Y}$ is 124,000,000. The sum of the squares of the Y values around the multiple linear regression line is 30,160,077. This last sum of squares is what is "unexplained" by the regression equation. The "explained" sum of squares is

$$124,000,000 - 30,160,077 = 93,839,923$$

so the regression equation has explained

$$\frac{93,839,923}{124,000,000} = .757$$

or 75.7% of the total sum of squared deviations about the mean.

(e) You can see that one year of age (holding experience constant) counts for more than one year of experience (holding age constant) because of the size of the two partial regression coefficients.

Notice that the partial regression coefficients do not reflect the simple effect of a variable X on Y. If you examine the data, you can see that as experience increases, so in general does salary. Then how can b_2 be negative? Because it is adjusted for age. It can be shown that the simple linear regression equation expressing the effect of experience on salary has a regression coefficient of +653. But the multiple regression coefficient shows that the increase in salary associated with experience can really be attributed to age. The negative value indicates that for a given age it is actually better to have less experience. ■

EXERCISES

***3–25.** A multiple linear regression equation expressing the effects of years of experience, years of schooling, and aptitude on job performance was computed as follows:

$$\hat{Y} = -2.8 + 2.04X_1 + 3.18X_2 + .93X_3$$

where $\hat{Y}$ is the predicted job performance rating, as measured by ratings of supervisors and peers
X_1 is number of years of experience working in the field
X_2 is number of years of schooling completed
X_3 is score on a managerial aptitude test

Interpret the meanings of each of the numbers in the multiple linear regression equation in terms of the example.

***3–26.** Using the multiple linear regression equation in Exercise 3–25 predict the job performance rating of a person who has had 10 years of experience in the field, has completed 16 years of schooling, and scored 82 on the managerial aptitude test.

3–27. A brochure asking for charitable donations pointed out that a study showed that there is a positive correlation between the amount of money people give to charities and the amount of money that they spend on liquor.

Would such evidence indicate that if you gave more to charity, you could expect, in general, to spend more on liquor?

***3–28.** The "demand curve" that is presented in most elementary economics courses is based on the assumption of *ceteris paribus*, or "everything else being equal." The slope of the curve represents the change in the amount demanded for a unit change in price, if all other variables are held constant.

Suppose an analyst formulated the demand for a particular product as

$$Y = 14.3 - 6.4X_1 + 2.1X_2$$

where Y = amount demanded of a product in a period of time
X_1 = price of the product
X_2 = price of the product of the leading competitor

Interpret the meaning of the coefficient −6.4. Is this an indicator of how much sales one expects to lose if he raises his price one unit? Why or why not?

Summary of Chapter

Most statistical analysis involves studying the relationship between two or more variables. This chapter has discussed some of the measures that are widely used to describe statistical data in terms of how two or more variables are related. When two variables are nominal, the data can be recorded in a contingency table. The chi-square statistic and the contingency coefficient are measures that describe the relationship between two such variables. When the variables are ordinal, the Spearman rank correlation coefficient can be used to indicate both the strength and the direction of the relationship. For two quantitative variables, the Pearson r measures the linear relationship between them, and a linear regression equation predicts the value of one variable from information on the other. A multiple regression equation is used to predict the value of a dependent variable from information on two or more independent variables. Interpretation of the partial regression coefficients was discussed, but computational procedures were not covered.

REVIEW EXERCISES FOR CHAPTER 3

3–29. Because of emergency conditions a state agency issued an order requiring all of its employees to work until 6:00 P.M. for a two-week period. Out of 120 employees in the agency, 40 signed a petition protesting the order on the grounds that it imposed a hardship on those who were needed at home to look after children and perform other family-support activities. Of the 40 employees who signed, 24 were women. Of the 80 who did not sign, 16 were women. Construct a contingency table and calculate χ^2 and the coefficient of contingency.

3–30. Two supervisors, A and B, ranked the secretaries in the office in order of their job performance. Their rankings, from "best" (1) to "worst" (6), were as follows:

A	*B*
1. Annette	1. George
2. Suzanne	2. Annette
3. Martha	3. Suzanne
4. Betty L.	4. Betty B.
5. George	5. Martha
6. Betty B.	6. Betty L.

a) Do the two supervisors tend to agree, or to disagree, in their rankings?
b) Calculate the Spearman rank correlation coefficient to support your answer.

3–31. A study of 6 working women showed earnings and ages of youngest children as follows:

Earnings	*Age of Youngest Child*
\$6000	5
8000	8
1500	3
4000	1
7500	9

a) Calculate the Pearson coefficient of linear correlation.

b) What interpretation can you give to the coefficient in terms of this example?

3–32. Would it be reasonable to use the data in Exercise 3–31 to calculate a linear regression equation? If not, why not? If so, which variable would you treat as the independent variable, and for what purpose might the equation be used?

3–33. Given, the following data on 4 heads of families:

Y	X_1	X_2
\$ 6,000	13	2
20,000	17	4
16,000	18	1
10,000	8	5

where Y = earnings of head of family
X_1 = years of schooling of head of family
X_2 = number of children

a) Find the linear regression equation of Y on X_1. Interpret the meaning of the regression coefficient you calculated.

b) Calculate the linear regression equation of Y on X_2. Interpret the meaning of the regression coefficient you calculated.

***3–34.** In Exercise 3–33(a), the regression coefficient of X_1 differs from the partial regression coefficient of X_1, holding X_2 constant. In terms of the example, distinguish between the meanings of the two coefficients.

3–35. On 15 successive July days the high Fahrenheit temperatures for the day in Horsechester were recorded along with the number of arrests. The results were:

Date	*High Temperature*	*No. of Arrests*
July 1	72	12
July 2	75	18
July 3	78	20

July 4	82	13
July 5	88	25
July 6	96	30
July 7	98	28
July 8	68	24
July 9	70	16
July 10	74	12
July 11	78	16
July 12	80	8
July 13	76	7
July 14	66	4
July 15	70	7

a) Plot a scatter diagram of the data points to see whether the relationship between high temperature and number of arrests is approximately linear.
b) Compute an appropriate measure of the relationship between high temperature and number of arrests.

SUGGESTED READINGS

FREUND, JOHN E. *Modern Elementary Statistics,* 4th ed. Englewood Cliffs, N.J.: Prentice-Hall, Inc., 1973.

HOEL, P. G. *Elementary Statistics,* 4th ed. New York: John Wiley and Sons, Inc., 1976.

JOHNSON, ROBERT. *Elementary Statistics,* 2nd ed. North Scituate, Mass.: Duxbury Press, 1976. Chapter 4.

KLEINBAUM, DAVID G., and LAWRENCE L. KUPPER. *Applied Regression Analysis and Other Multivariate Methods.* North Scituate, Mass.: Duxbury Press, 1978.

LOETHER, HERMAN J., and DONALD G. McTAVISH. *Descriptive and Inferential Statistics: An Introduction.* London: Allyn and Bacon, Inc., 1976. Chapter 6 to 9.

MOSTELLER, FREDERICK, et al. *Statistics by Example.* Reading, Mass.: Addison-Wesley Publishing Co., Inc., 1976.

FOUR

Fundamentals of Probability

Outline of Chapter

4.1 Experiments, Sets, and Events

The possible outcomes of an experiment or any uncertain situation can be formally specified.

4.2 The Nature of Probability

There are several approaches to assigning probabilities to events.

4.3 Probabilities of Compound Events

Once we have decided on the probabilities of simple events, some mathematical results will help us to determine probabilities of compound events.

*4.4 Bayes' Theorem

Bayes' Theorem is used to make inferences from a result or a symptom back to a prior event or cause.

Objectives for the Student

1. Become familiar with the concepts of sample space and event.
2. Understand the concepts of union, intersection, complement, empty set, and universal set.
3. Be able to define probability from a mathematical point of view.
4. Distinguish between *a priori*, empirical, and subjective probability.
5. Know the basic rules for determining the probabilities of compound events.
6. Understand the concept of independent events.

*7. Know Bayes' theorem and its purpose.

Chapters 1 and 2 presented methods of describing data. Discussed were such measures as the relationship between the size of backlog of cases and time spent by a judge hearing a case. But the measure of the relationship applied only to the cases that were observed. The coefficient of linear correlation described only the association between backlog and time spent on 16 recorded cases. How would the relationship be affected if one looked at a 17th, 18th, or a 500th case? We do not know, because we did not get information on those cases. Yet this is the kind of thing we need to know. One does not generally get information on 16 cases just because one is interested in describing 16 cases. One does this to make generalizations about the way a judge, or other judges, react to case backlog—now and in the future. The reason that a sample of 16 cases may be important data is that properly chosen samples behave with a certain degree of regularity or predictability, according to the mathematical rules of probability. An understanding of probability will enable you to know what generalizations you can and cannot make from a sample, how much confidence you can put in the generalizations you do make, and whether you need more information.

When I was in Junior High School, a colleague said to me only half in jest: "We study the past in history classes, and we study the present in current events classes, but we have no classes that study the future. But I'm going to be living in the future, and that's what I want to learn about." I thought I had the answer to that one; we can't study the future because we have no information about the future. But I was wrong. Every time you get a Weather Report, every time you read an economic projection, every time you look at a proposed budget you are getting information on the future. The only difference between information on the past and information on the future is that information on the future is less certain. It needs to be qualified in terms of probability. People are becoming increasingly aware of this need to express statements about the future in terms of probability, as suggested by the fact that weather reports no longer make flat predictions about rain, but qualify their predictions in terms of probability.

A portion of your day is likely to be devoted to reading descriptive statistics in the forms of reports, records, surveys, etc. But such reading is incidental to the more important part of your job as a manager in the public sector: to make decisions involving the future. Since managers live in a world of considerable uncertainty, you are sometimes going to be wrong in your decisions. This is inevitable. But there are many things that you can do to increase the probability of making good decisions and to decrease the ill-effects of the poor decisions you are still inevitably going to make from time to time. These things include: obtaining the best information that you can; knowing when to decide whether you have enough information to make a decision or to get more data; considering all the consequences of each possible course of action; and assessing the probabilities of these consequences. Essential to a successful approach to all of these tasks is an ability to think in terms of probability. The manager who does not consider probabilities may reveal this failing in several possible ways. Such a manager may be overly meticulous and thorough in obtaining information and thereby habitually avoid taking action until the time for effective action is past, or, on the other hand, may jump to conclusions and take hasty action on the basis of too little information. Both extremes reflect an inability to assess the value of information and an inability to think in terms of probability.

Probability is a way of looking into the future or into the unknown. In looking at the future we can ask two questions:

1. What possible things can happen?
2. How likely, or how probable, is it that each of these things will happen?

The first of these questions is concerned with experiments and the possible results; the second is concerned with the probabilities of these results.

4.1 Experiments, Sets, and Events

When you think of the word "experiment" you probably picture in your mind something like a scientist experimenting with chemical compounds in a laboratory, or a psychologist experimenting with hamsters in a maze. The word is used here in a more general sense to include any procedure that specifies a way of observing results.

> An EXPERIMENT is a specified procedure for obtaining and observing an outcome, or a result. The set of all possible outcomes of an experiment is known as a SAMPLE SPACE.

EXAMPLE 4.1 We will observe the sex of the next baby to be born in Magee Hospital. What is the sample space?

Solution: The sample space consists of the two possible outcomes, boy and girl. Symbolically we can express this sample space as

$$S = \{B, G\} \blacksquare$$

EXAMPLE 4.2 What is the sample space if we observe the sexes of the next two babies to be born in Magee Hospital?

Solution: Express each outcome as a set of ordered pairs, the first member of the pair being the sex of the first baby observed, the second member of the pair being the sex of the second baby. The sample space is

$$S = \{(B,B), (B,G), (G,B), (G,G)\}$$

There are four outcomes in the sample space. Notice that it does not matter in which order you list the outcomes, but the ordering of the letters within the parentheses is important. ■

EXAMPLE 4.3 Two persons are applying for jobs. What is the sample space for the possible decisions that could be made?

Solution: The four possible outcomes are:

1. first person rejected; second person rejected
2. first person rejected; second person hired
3. first person hired; second person rejected
4. first person hired; second person hired

The sample space can be represented as

$$S = \{(R,R), (R,H), (H,R), (H,H)\} \blacksquare$$

Tree Diagrams

In these examples, the sample space was represented by a listing. A very useful pictorial device for displaying a sample space is a tree diagram. Figure 4.1 is a tree diagram for the sample space of Example 4.3. The junctions on the tree are called "nodes," the lines emanating to the right of each node are called "branches," and a set of branches from the first node at the far left to any final branch on the right is called a "path."

> Any set of outcomes in a sample space is known as an EVENT.

For example, the sample space for the two job applicants contains many events, among which are:

1. Nobody is hired.
2. Exactly one person is hired.
3. At least one person is hired.
4. The same action is taken on both.
5. Not more than one person is hired.
6. Not more than two persons are hired.
7. More than two persons are hired.

> An event that consists of exactly one outcome is known as an ELEMENT or an ELEMENTARY EVENT of the sample space.

Figure 4.1. *TREE DIAGRAM The Sample Space of Decisions to Hire (H) or Reject (R) Two Job Applicants*

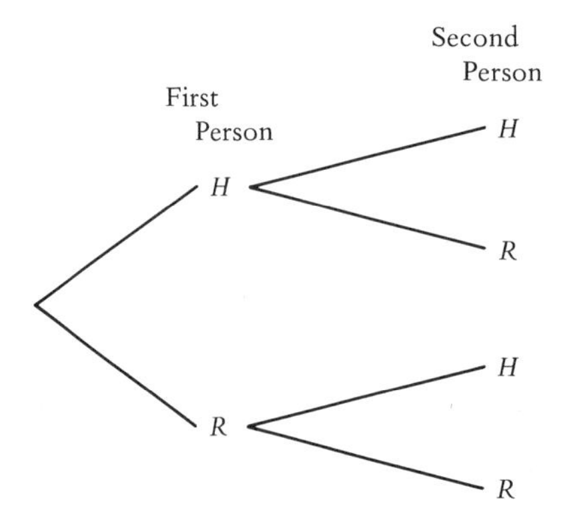

The concept of "element" is closely tied with our method of observation. The specified method of observation does not permit an elementary event to be subdivided into two or more outcomes.

EXAMPLE 4.4 Define the seven events listed above in terms of the elements they contain.

Solution: The events are as follows:

Event	*Elements in the Event*
(1) Nobody is hired.	$\{(R,R)\}$
(2) Exactly one person is hired.	$\{(R,H), (H,R)\}$
(3) At least one person is hired.	$\{(R,H), (H,R), (H,H)\}$
(4) The same action is taken on both.	$\{(R,R), (H,H)\}$
(5) Not more than one person is hired.	$\{(R,R), (R,H), (H,R)\}$
(6) Not more than two persons are hired.	$\{(R,R), (R,H), (H,R), (H,H)\}$
(7) More than two persons are hired.	No elements.

Notice in particular Events 6 and 7. Event 6 contains all the elements in the sample space. It is an event that will occur with certainty. Event 7 contains no elements. It is an impossible event, but nevertheless it is still considered to be an event. ■

SETS

We have already used the word "set" several times in this section. The sample space was defined as the set of all possible outcomes of an experiment. An event was defined as a set of possible outcomes. A set can be loosely defined as a conceptual collection of things or members. It is "conceptual" because the collection is in one's mind; the collection does not have to be physically in one place. Examples of sets are:

The set of people in the Commonwealth of Massachusetts
The set of postal clerks in the United States
The set of atoms in the universe
The set of human beings who have not yet been born
The set of bachelors who have wives
The set of positive integers

It is possible for a set to have no elements, one element, two elements, etc., on up to an infinite number of elements.

> When every element of a set A is also an element of a set B, then A is defined to be a SUBSET of B, or
>
> $$A \subseteq B$$ **4.1**

This relationship can be illustrated by a Venn diagram, which is shown in Figure 4.2.

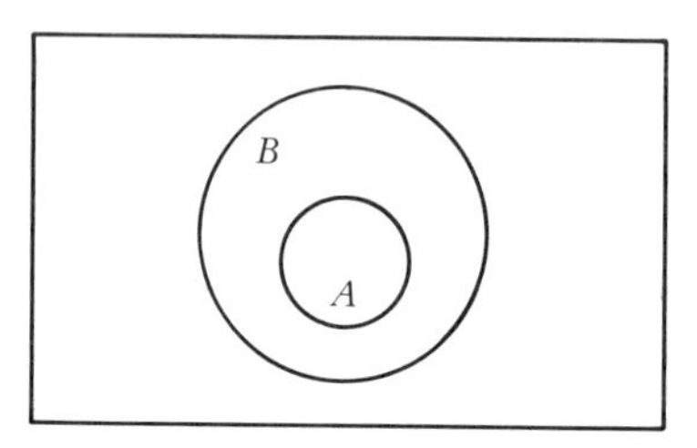

Figure 4.2. *VENN DIAGRAM*
Depicting $A \subseteq B$

A VENN DIAGRAM is a pictorial method of displaying sets and subsets. One set is represented by a large rectangle, or circle, and subsets of that set are represented by circles inside the rectangle or circle.

It is possible that not only is every element of A an element of B ($A \subseteq B$) but also that every element of B is an element of A ($B \subseteq A$). In this case each set is a subset of the other, and we say that $A = B$.

EXAMPLE 4.5 All the students taking Statistics 201B are enrolled in a Public Administration program. If all the students enrolled in the Public Administration program are also taking Statistics 201B, depict these relationships symbolically.

Solution: Let A be the set of Statistics 201B students and let B be the set of students enrolled in the Public Administration program. Then

$$A \subseteq B \quad \text{and} \quad B \subseteq A$$

hence $A = B$. ■

The UNION of two sets A and B is defined to be the set of all elements that are in A *or* B (including those in both). The union of three or more sets, A, B, C, etc., is the set of all elements that are in any one of A, B, C, etc. (including those in more than one).

To get into the union of sets A, B, C, etc. it is necessary to be a member of at least one of them.

The union of two sets A and B is written as

$A \cup B$ (read "A union B" or "the union of A and B")

The union of many sets A to Z can be written as

$$A \cup B \quad C \cup \ldots \cup Z$$

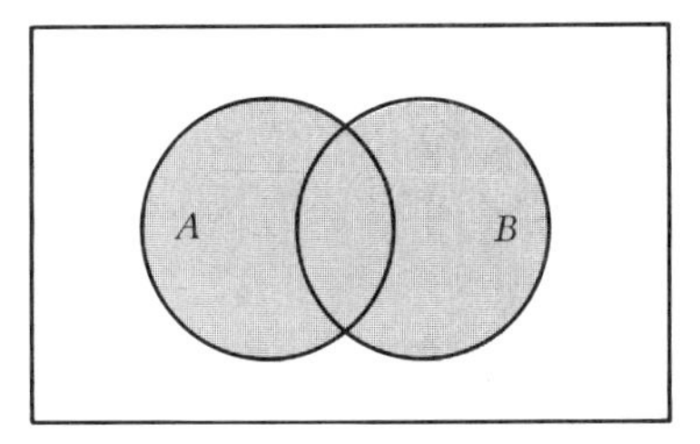

4.3. *VENN DIAGRAM*
Depicting $A \cup B$

By representing A and B as intersecting circles, a Venn diagram would represent $A \cup B$ as the entire shaded area in Figure 4.3.

For example, assume a special Fellowship Fund had been set aside for "minorities or women." If you are either a member of a minority group *or* a woman, you are eligible to apply. Letting A be the set of minority people and B be the set of women, we can represent the set of eligible people as $A \cap B$.

> The INTERSECTION of two sets A and B is defined to be the set of elements that are in both A and B. The intersection of three or more sets A, B, C, etc. is defined to be the set of elements that are in all of them.

To get into the intersection of sets A, B, C, etc., it is necessary to be a member of *all* of them.

The intersection of two sets A and B is written as

$A \cap B$ (read "A intersect B" or "the intersection of A and B")

The intersection of sets A to Z is written as

$$A \cap B \cap \ldots \cap Z$$

A Venn diagram would depict the intersection of A and B as the shaded area in Figure 4.4.

Figure 4.4. *VENN DIAGRAM*
Depicting

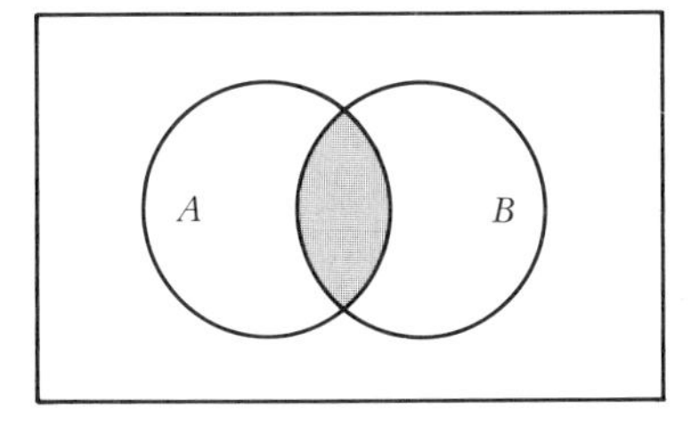

For example, if the Fellowship Fund had been designated for "minortiy women," then you would need to be both a member of a minority group *and* a woman to be eligible. The set of eligible people would be $A \cap B$.

Suppose a college is searching for a new Dean of Liberal Arts. In response to the wishes of the Board of Trustees, the Search Committee has defined the requirements for the position as "white male between the ages of 40 and 50 with a Ph.D."

Let A be the set of white people.
Let B be the set of males.
Let C be the set of people between 40 and 50.
Let D be the set of people with a Ph.D.

Then the successful candidate must be a member of the set $A \cap B \cap C \cap D$. Without doubt the requirements will have sharply limited the number of eligible candidates.

> In any discourse involving sets there is an overall frame of reference, a set of all elements that are relevant to the discourse. This set is known as the UNIVERSAL SET. All other sets are subsets of the universal set.

The universal set is depicted in a Venn diagram as a large rectangle which contains all the other sets. All other sets are shown as circles within the rectangle.

> The COMPLEMENT of a set A, written as A', is the set of all elements in the universal set S which are *not* in the set A.

The complement of a set A is shown as the shaded area in Figure 4.5.

For example, if S is the set of natural-born citizens of the United States and A is the set of natural-born citizens who are under 35 years of age, then A' is the set of all natural-born citizens who are not under 35 years old. The U.S. Constitution requires that the President be a member of A'.

The union of a set and its complement is the universal set:

$$A \cup A' = S \qquad \textbf{4.2}$$

where A is any set and S is the universal set.

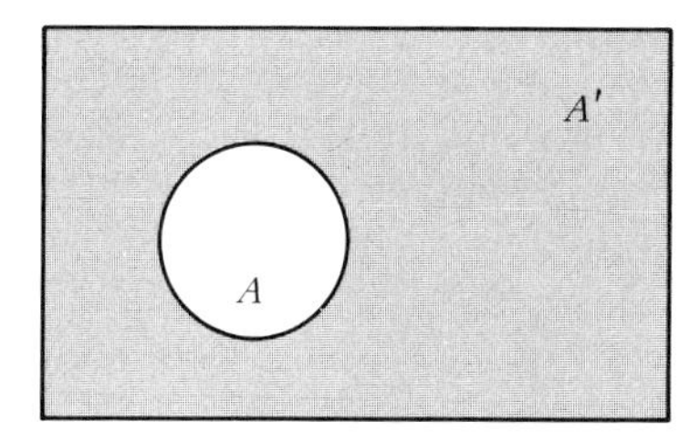

Figure 4.5. *VENN DIAGRAM The Complement of a Set A*

> If a set has no elements, it is known as the EMPTY SET. The empty set will be designated by the Greek letter ϕ (phi).

Let the set A be the set of people receiving Medicare benefits and the set B be the set of people under 15 years of age. The proper display of these sets in a Venn diagram should show the intersection of these two sets, even though we may believe that the intersection is the empty set. This display is shown in Figure 4.6.

SETS AND EVENTS

The concept of sets was introduced into this section because sample spaces and events *are* sets. *An event is a subset of the sample space.* We can use the set notation to describe events, as follows.

Symbol	*Represented Event*	*Description of Event as a Set*
S	sample space	The set of all possible outcomes, the universal set
A	A occurs	The set of all outcomes that are contained in A
B	B occurs	The set of all outcomes that are contained in B
$A \cup B$	A or B (or both) occur	The set of all outcomes that are contained in A or B (or both)
$A \cap B$	A and B occur	The set of all outcomes that are contained in both A and B
A'	A does not occur	The set of all outcomes that are *not* contained in A
B'	B does not occur	The set of all outcomes that are *not* contained in B
ϕ	an impossible event	The empty set

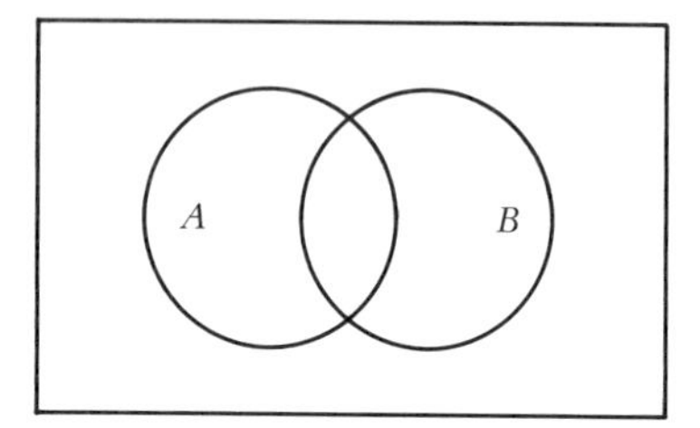

Figure 4.6. *VENN DIAGRAM Proper Display of Sets A and B, whether $A \cap B$ is the Empty Set or Not*

> If the intersection of two events, A and B, is the empty set (or an impossible event), that is if $A \cap B = \phi$, then we say that A and B are MUTUALLY EXCLUSIVE EVENTS.

Any pair of events A and A' are by definition mutually exclusive.

EXERCISES

4–1. Two proposals are before a state legislature: one is to enact a state income tax, and the other is to increase the allocation to state-supported colleges. If we observe which proposals pass and which do not, define the sample space. Depict this sample space by a tree diagram.

4–2. Letting R = Republican and D = Democrat, express the following events symbolically by listing all elements:

a) President is Republican, Senate is Democratic, House is Democratic.
b) President is Republican, Senate is Republican, House is Democratic.
c) President, Senate, and House are all of the same party.
d) President is of one party, Senate and House are of the other party.
e) One of the Houses is of the same party as the President.
f) At least one of the Houses is of the same party as the President.

4–3. Letting B = Black and W = White, express in symbolic form the set of possible racial mixes of a married couple.

4–4. Depict the set of racial mixes in Exercise 4–3 by a tree diagram.

4–5. Let S be the set of voters, A be the set of Republican voters, and B be the set of voters who voted for Benny McCoy. Describe in words, the following sets, and show them by shading areas in a Venn diagram.

a) $A \cup B'$
b) $A \cup B$
c) $A \cap B$
d) B'
e) $(A \cap B)'$

4–6. Let S be the set of 4 people: Mrs. Allen, a married female; Mr. Kern, a single male; Miss Hunt, a single female; and Ms. Lee, a single female. Let A be the set of females in S, and B be the set of married people in S. List the elements in the following sets.

a) A
b) $A \cup B$
c) $A \cup A'$
d) $A \cap A'$
e) $A \cap B$
f) $A' \cup B'$

4–7. Referring to Exercise 4–6, consider an experiment which involves the selection of 1 person from the set of 4. Let A be the event "a female is selected" and B be the event "a married person is selected." List the outcomes comprising the following events.

a) A
b) A'
c) B
d) $A \cap B$
e) $A \cap B'$
f) $(A \cup B)'$

4–8. Let A be the event "the fire alarm sounds" and B be the event "there is a fire." Describe the following events.

a) $A \cup B$
b) $A \cap B$
c) $A \cap A'$
d) $A \cap B'$
e) $A' \cap B'$

4–9. Give an example of a pair of mutually exclusive events

a) when they are mutually exclusive by definition;
b) when they are not mutually exclusive by definition but from your own observation.

4.2 The Nature of Probability

Up to this point, the discussion has been about the set of possible outcomes and events. In this section we introduce the problem of determining how *probable* (how likely) each outcome or event is.

When someone talks about "the probability of rain" or "the probability of a recession next year" we may disagree with the numbers being used, but we seldom question the use of the word "probability." It may surprise you, therefore, to learn that the word is very difficult (if not impossible) to define, and that there is considerable disagreement among highly knowledgeable people as to what the word does mean. This is not to say, however, that the concept is useless; like electricity, probability is easier to use than to define.

This section will present the mathematical concept of probability and three interpretations that relate the mathematical concept of probability to the real world.

Mathematical Probability

To a mathematician, probability is a number that is assigned to an event in a sample space according to certain rules:

The PROBABILITY of A, designated as $P(A)$, is a number assigned to any event A in the sample space S, such that

$$0 \leq P(A) \leq 1 \text{ for every event } A \text{ in } S \quad \textbf{4.3}$$

that is, that $P(A)$ is between 0 and 1 inclusive for every event A in S.

$$P(S) = 1 \quad \textbf{4.4}$$

If A_1 and A_2 are events in S, and $A_1 \cap A_2 = \phi$, then

$$P(A_1 \cup A_2) = P(A_1) + P(A_2)$$

is the MATHEMATICAL DEFINITION OF PROBABILITY **4.5**

The symbol $\leq$ means "less than or equal to." You will encounter similar symbols later on:

$<$ means "less than"
$>$ means "greater than"
$\geq$ means "greater than or equal to"

In less formal terminology, the above three conditions can be restated as:

1. The probability of every event is at least zero but not more than one.
2. The probability that some event in the sample space will occur is one.
3. If A_1 and A_2 are mutually exclusive events (i.e., it is impossible for both of them to occur), then the probability that either A_1 or A_2 will occur is the sum of probabilities of the two events.

As an exercise, check that the three conditions as formally stated are equivalent to the three conditions stated in less formal terms.

It can be shown from Equation 4.5 that for any set of k mutually exclusive events $A_1, A_2, \ldots, A_k$, the probability that any one of them will occur is **4.6**

$$P(A_1 \cup A_2 \cup \ldots \cup A_k) = P(A_1) + P(A_2) + \ldots + P(A_k)$$

EXAMPLE 4.6 Let A_1 be the event that you will become Vice President of the United States; let A_2 be the event that you will not become Vice President of the United States. What is the sample space?

Solution: The sample space consists of two events: you will become Vice President (A_1), and you will not become Vice President (A_2). Therefore

$$S = \{A_1, A_2\}$$

If someone assigned the probabilities of

$$P(A_1) = .7 \qquad \text{and} \qquad P(A_2) = .3$$

to these events, would this assignment be in accordance with the three conditions imposed by the mathematical definition of probability?

Solution: Yes, the assignment is in accordance with the rules.

1. All probabilities are between 0 and 1, satisfying Equation 4.3.
2. The probability that one of these events will occur is 1, satisfying Equation 4.4.
3. The events A_1 and A_2 are mutually exclusive, and the sum of their probabilities is $.7 + .3 = 1$, satisfying Equation 4.5. ■

The probabilities assigned in the above example may satisfy the conditions for mathematical probability, but they may not satisfy you. You would like the probabilities to have meaning in the real world. Three approaches toward a definition of probability that meets our realistic expectations are discussed in the rest of this section.

A Priori Probability

A traditional method of defining probability, one that appears in many classical works in statistics, is something like this:

> If an experiment can lead to N equally likely outcomes and the event A consists of $N(A)$ of these outcomes, then the *A PRIORI* PROBABILITY of A is given by
>
> $$P(A) = \frac{N(A)}{N}$$ **4.7**

The phrase "equally likely" is another way of saying "equally probable." If all the outcomes are equally likely, and there are N of these outcomes, then it can be shown from Equation 4.6 that each outcome has a probability of $1/N$, and if there are $N(A)$ mutually exclusive outcomes comprising event A, then it also follows from Equation 4.6 that

$$P(A) = \underbrace{\frac{1}{N} + \frac{1}{N} + \cdots + \frac{1}{N}}_{N(A) \text{ terms}} = \frac{N(A)}{N}$$

Equation 4.7 therefore follows from the assumption of equally likely outcomes.

Where did the assumption of equally likely outcomes come from? Experimentation might have suggested that one outcome was more likely than another, but the "equally likely" assumption is not based on experimentation. It is *prior* to any experience; it is arrived at *a priori*. It is an intuitive assumption, based on general human experience. For example, why do we assign a probability of 1/2 to "heads" when we flip a coin? Simply because intuition tells us there is no reason not to assume that heads and tails are equally likely.

EXAMPLE 4.7 A card is drawn from a bridge deck consisting of 52 cards. What is the probability that the card is an ace?

Solution: If the 52 cards are all of the same size, shape, and consistency, and are well shuffled, we may have reason to assume *a priori*, that each card has a probability of 1/52 of being selected. Four of these are aces. From Equation 4.7 it can be shown that

$$P(A) = \frac{4}{52} = \frac{1}{13} \quad ■$$

EXAMPLE 4.8 On the roll of a balanced die, what is the probability of rolling a 5 or a 6?

Solution: Assuming, *a priori*, that each of the 6 faces of the die has an equal probability of appearing, the event "5 or 6" consists of two outcomes each with a probability of 1/6. From Equation 4.5 we get

$$P(5 \text{ or } 6) = \frac{1}{6} + \frac{1}{6} = \frac{1}{3} \quad ■$$

EXAMPLE 4.9 A person is selected at random from a set of 3 people. Jones is one of the three. What is the probability that Jones is selected?

Solution: The method of random selection is one that carefully insures that each elementary event is equally probable. Knowledge that the selection process is random enables us to make the assumptions necessary for Equation 4.7. Jones is one of the three, and the probability of Jones is 1/3. ■

EMPIRICAL PROBABILITY BASED ON RELATIVE FREQUENCY

Perhaps the best place to find examples using *a priori* probability is in games of chance. In fact, much of the early development of the theory of probability got its examples (and its inspiration) from the gambling halls of Europe. Many of the examples that you will find in books on probability and statistics involve flipping coins, rolling dice, drawing cards, and other such productive pastimes. Such examples are not as far from the practical concerns of a manager as you might think, however, since many of the decisions you make will be "gambles," the success or failure of which may depend on events as chancy as the outcomes of dice or cards. A major difference between the probabilities faced by the gambler and those by the administrator is that the probabilities facing the latter can seldom be derived from *a priori* assumptions.

The probabilities that you will use to make most of your decisions will be based on empirical *facts,* on the basis of observations from the past. How do you derive probabilities from past observations? If an experiment has been performed n times under very similar conditions and $n(A)$ is the number of times that event A has occurred, then the fraction $n(A)/n$ can serve as a guide to determining the probability that A will occur on each observation that is made, as long as the conditions are essentially the same.

EXAMPLE 4.10 It has been determined that unless there are at least 2.5 inches of rainfall in the next 3 months, Arid City will be forced to issue mandatory curtailments on water consumption. How would you determine the probability that this event will occur?

Discussion: You could look at the record of rainfall over the past 5, 10, 20, or more years. In order to determine probabilities based on relative frequencies, you need to base your conclusions on observations made under conditions very similar to the present situation. You may want to confine your observations to the same three-month period in previous years, if you believe that rainfall follows seasonal patterns. You may not want to go back more than 20 years if you suspect that climatic or environmental changes may have produced rainfall conditions today that are unlike those of more than 20 years ago.

Suppose you collect data for 20 years that you believe represent conditions similar to the present one, and in 7 of those years this 3-month period failed to produce the 2.5 inches that you are looking for.

You now have a reasonable basis for estimating the probability of mandatory curtailments to be 7/20 or .35. ■

However, in a short number of trials, the relative frequency of occurrences of an event may be far from the relative frequency in the long run. For example, on the first two occasions I visited the city of Atlanta it was snowing. It would be a mistake, however, on the basis of two days of snow in two observations, to assume that it snows every day in Atlanta. The empirical approach to probability looks at relative frequencies in the *long run.*

> EMPIRICAL PROBABILITY BASED ON RELATIVE FREQUENCY is the proportion of times an event is observed to occur in a series of experiments in the long run under constant conditions.

In any finite number of observations the observed relative frequency may differ from the true probability, but as the number of observations increases, the relative frequency of occurrences gets very close to the probability.

EXAMPLE 4.11 How would you assess the probability that a randomly selected 65-year-old man will survive for 20 years or longer?

Discussion: The Public Health Service of HEW publishes an annual *Vital Statistics of the United States,* giving mortality rates from which you can derive the required probability. It is important to assume that the conditions underlying the mortality rate are going to remain essentially the same over the next 20 years as they were in the period on which the statistics were obtained. In looking up recorded mortality rates, you are obtaining relative frequencies based on a very large number of observations. ■

Subjective Probability

Empirical probability based on relative frequency requires a long series of observations under conditions that do not change. In most cases, as administrators know perhaps better than anyone else, we do not have the conditions necessary to determine probability on the basis of relative frequency. Every situation is different in some way from every other one, and the basis for computing long-run frequencies is often simply not there. But we still need to take action, and our actions will be based on our best guess as to how something will turn out. We still need to know or estimate the probabilities. On what information or assumptions will the probabilities be based?

When no formal method of assessing probability is practical, we may resort to subjective probability.

> SUBJECTIVE PROBABILITY is the degree of belief (ranging from 0 to 1) that an individual places in the occurrence of an event.

A subjective probability of 0 means the individual believes the event is impossible; a subjective probability of 1 means that the individual believes the event is certain to occur.

The concept of subjective probability may seem to some to be a cop-out. It specifies no criterion or basis on which the individual must determine the probability. But the reason for the cop-out is a good one. The kinds of information that can and should go into the assessment of a probability are far too numerous and varied for a statistician to spell out in advance. If an economist assigns a probability of .4 to the event that a recession will occur within two years, this probability may reflect a large number of factors, which would include data on many facets of the economy, an assessment of the political climate, and the economic philosophy of the economist. Some *a priori* and relative frequencies may be in there too. But we do not dissect his thought processes. What comes out of all the economist's studies is the probability of .4, which represents his judgment. No matter how much hard evidence went into the judgment, it is still subjective—an individual judgment—and other economists can and will come up with different probabilities.

Subjective probability is a very practical way of looking at probability for those who need to make decisions based on their assessment of the future. Administrators are seldom specialists; your job will more likely involve finding and consulting with experts than becoming an expert yourself. When you want to know the probability that an event will occur, your estimate of this will be based on a variety of factors, including the views of people you consider to be knowledgeable about the subject.

EXAMPLE 4.12 You have received a bid from the WJF firm to manufacture a component part on a missile system at a cost of $10 million, on a cost-plus basis. Having been sensitized to the problems of "cost overruns," you are very concerned about the probability that the contractor will end up billing you for more than 50% above the original estimate. How would you assess the probability, if awarded the contract, that WJF would go more than 50% above the $10 million?

Discussion: The most naive approach would be the *a priori* approach: in the absence of any other information, assign equal (.50) probabilities to the events A (overrun 50% or less) and A' (overrun more than 50%).

The relative-frequency approach would be to look at WJF's record in the past on similar contracts; what proportion of their contract resulted in overruns of more than 50%?

The problem with this approach is that you may have trouble finding a large number of contracts that WJF performed *under the same essential conditions.* The present situation is unlike any other in some ways, and it is possible that the uniqueness is such that you cannot consider past relative frequency as a valid indicator.

What you will undoubtedly end up doing is assessing the probability based on a whole array of factors: past performance of WJF, recent publicity about overruns, the fact that WJF made some changes in its top management last year, prospective problems at WJF in negotiating a contract with the union, and other considerations. After having weighed all the evidence, let us say you come up with a subjective probability of .8. But why do we call this probability "subjective" if you took great care to look at all the facts in as objective a manner as possible? Because there is no objective procedure (that is, no generally accepted procedure the accuracy of which is verifiable) spelled out as to *how* you should weigh these facts, and another person, being equally "objective" in his evaluation, may very well arrive at a different probability. The subjective probability, therefore, depends on which subject is assessing the probability. ■

EXERCISES

4–10. What method of assigning probability would you use to determine

a) the probability of holding the winning lottery ticket if you held 2 tickets and there were 1000 tickets outstanding?
b) the probability that at least 150 vehicles will pass through a toll gate between 3:00 and 4:00 P.M. on Tuesdays?
c) the probability that the next governor to be elected in your state will be a Democrat?
d) the probability that it will rain next April 12 in Nashville?
e) the probability that Congress will raise its pay next year?
f) the probability that Sam Leever will be late to work tomorrow?

***4–11.** The concept of probability as presented in this and most statistics books is applied to events resulting from experiments. Such events one generally thinks of as future events. Do you believe the concept can be applied to an event in the past about which you are uncertain, such as the probability that an individual committed a crime? Discuss, taking consolation from the fact that profound writers on the subject of probability disagree about this.

4–12. The personnel working in the Budgeting and Accounting office are classified into 4 categories, as follows:

Job Category	*Number of Persons*
Managerial	2
Professional	3
Secretarial	6
Manual Labor	4
Total	15

One persons wins the office pool each week.

a) Assume, *a priori,* that each person has an equal probability of winning the pool. What is the probability that a manual laborer will win the pool?

b) If you take note of the fact that 7 times in the past 10 weeks a professional person has won the pool, how would you assess the probability that a professional person will win the pool this week?

c) If, in addition to the information in (b), you learned from a manager that his horoscope predicted that he would win the office pool, how would you now assess the probability that a professional person will win the pool this week?

4–13. A die has 6 faces, numbered from 1 to 6. The die is rolled, and one of the faces will turn up.

a) What is the probability that the number 5 will turn up?

b) What method of assigning probability did you use to answer (a)?

4–14. a) If you had rolled the die in Exercise 4–13 1000 times and a 5 turned up 200 times, what probability would you assign to the event the number 5 will turn up on the next roll?

b) What method of assigning probability did you use to answer (a)?

4–15. If you pick a state at random from the 50 states of the U.S.:

a) What is the probability the state's name will start with W?

b) What is the probability the state's name will start with M?

c) What is the probability the state's name will consist of 2 words?

(For a list of states, refer to Exercise 3–6.)

4–16. In a bridge deck, of the 52 cards

13 are spades, a black suit
13 are clubs, a black suit
13 are hearts, a red suit
13 are diamonds, a red suit

If a card is drawn at random, what probability would you assign *a priori* to the cards being

a) spade?

b) diamond?

c) a red suit?
d) either a heart or club?

4.3 Probabilities of Compound Events

The previous section presented several approaches to establishing $P(A)$, $P(B)$, etc., the probability of the occurrence of an event A, B, etc. Once the probability of an event is known or assumed, we can use mathematics to determine the probabilities of events that are combinations of A, B, etc.

Given $P(A)$, we can find $P(A')$, the probability that the event A does not occur. Since A' means that A does not occur, then A and A' are mutually exclusive events. Then, from Equation 4.5,

$$P(A \cup A') = P(A) + P(A')$$

Since all possibilities are in either A or A', then $A \cup A' = S$. Therefore

$$P(S) = P(A) + P(A')$$

Substituting from Equation 4.4 gives us

$$1 = P(A) + P(A')$$

and we find that:

> The probability that an event A does not occur is 1 minus the probability that A does occur, that is
>
> $$P(A') = 1 - P(A)$$

4.8

EXAMPLE 4.13 There are 200 graduates of a School of Public Administration who were interviewed to find out their employment status. They were classified in a contingency table as shown in Table 4.1.

We choose an individual at random from the 200. The random-

Table 4.1. *CONTINGENCY TABLE Cross Partitioning of 200 Graduates of P.A. School by Sex and Job Status*

	WORKING FULL-TIME	WORKING PART-TIME	NOT WORKING	
Men	80	0	40	120
Women	20	40	20	80
	100	40	60	200

ness of the choice insures that each individual has an equal probability of being selected. This probability will be 1/200.

Determine the probabilities of selecting individuals in each of the six cells in Table 4.1.

Solution: Since each person has an equal probability of 1/200 of being selected, we can use Equation 4.7 to determine the probabilities for each of the cells. In every case we divide the number or frequency in each cell by 200, and obtain the probabilities shown in Table 4.2.

Table 4.2. *JOINT PROBABILITIES Derived from Table 4.1*

	WORKING FULL-TIME	WORKING PART-TIME	NOT WORKING	
Men	.4	0	.2	.6
Women	.1	.2	.1	.4
	.5	.2	.3	

> The probabilities in the cells of a contingency table are called joint probabilities. JOINT PROBABILITIES are the probabilities of two or more events occurring together. In set terminology, the JOINT PROBABILITY OF TWO OR MORE EVENTS is the probability of the intersection of the events.

EXAMPLE 4.13 (continued) (a) What is the probability that a randomly selected person is a man?

Solution: We can add the joint probabilities across the top row of cells. This is .4 + 0 + .2 = .6. Notice that we are here applying Equation 4.6 by adding the probabilities of mutually exclusive events. To show this more formally,

Let A_1 be the event "man."
Let A_2 be the event "woman."
Let B_1 be the event "working full-time."
Let B_2 be the event "working part-time."
Let B_3 be the event "not working."

We can then find the probability of a man, $P(A_1)$, by adding the probabilities of the three mutually exclusive kinds of men in the world, namely,

$A_1 \cap B_1$ = men working full-time
$A_1 \cap B_2$ = men working part-time
$A_1 \cap B_3$ = men not working

Thus, from Equation 4.6 and Table 4.2, we have

$$P(A_1) = P(A_1 \cap B_1) + P(A_1 \cap B_2) + P(A_1 \cap B_3) = .4 + 0 + .2 = .6$$

By similar reasoning, the probability of a woman being selected can be found by adding the probabilities of the three mutually exclusive events $A_2 \cap B_1$, $A_2 \cap B_2$, and $A_2 \cap B_3$. The probabilities of these three events add up to .4. Or we could have arrived at this figure by observing that the event A_2 is really A' (the nonoccurrence of A_1) and (from Equation 4.8)

$$P(A_2) = 1 - P(A') = 1 - .6 = .4$$

(b) What is the probability that a randomly selected person is working full-time?

Solution: There are two kinds of people working full-time, men working full-time and women working full-time—represented by $A_1 \cap B_1$ and $A_2 \cap B_1$. Since these are mutually exclusive, the probability of their sum is given by Equation 4.5.

$$P(B_1) = P(A_1 \cap B_1) + P(A_2 \cap B_1) = .4 + .1 = .5$$

Of course it is much quicker just to get this .5 figure by adding down the first column of Table 4.2, but it is important for you to see the logical reason for doing this.

(c) What is the probability that a person selected at random will be *either* a man *or* working full-time?

Solution: The question asks for $P(A_1 \cup B_1)$. Since A_1 and B_1 are not mutually exclusive events, we cannot use Equation 4.5 and simply add $P(A_1)$ to $P(B_1)$ to get a probability of 1.1. (You would know that this answer was wrong anyway, of course, since you cannot have a probability of more than 1.)

The problem with simply adding $P(A_1)$ to $P(B_1)$ is that you would be adding all the men *and* all those working full-time; that is, the sets A_1 and B_1. Of course that means that those who are men and are also working full-time get added in twice. To correct for this double adding, we have to subtract the probability of $A_1 \cap B_1$, so that every member of $A_1 \cap B_1$ will now have been counted just once. Therefore,

$$P(A_1 \cup B_1) = P(A_1) + P(B_1) - P(A_1 \cap B_1) \qquad \textbf{4.9}$$

and in terms of this example

$$P(A_1 \cup B_1) = .6 + .5 - .4 = .7$$

To confirm this result, go back to Table 4.1 and verify that 140 out of the 200 people (.7 or 70% of them) fit into either the category of "men or of those working full-time." ■

The Addition Theorem

Equation 4.9 give the probability of the union of the two events, A_1 and B_1. This equation can be stated in slightly more general form for any pair of events A and B, and is known as the Addition Theorem for Two Events.

Addition theorem for two events

$$P(A \cup B) = P(A) + P(B) - P(A \cap B) \tag{4.10}$$

Equation 4.10 holds for any pair of events, whether A and B are mutually exclusive events or not. If they are mutually exclusive, then $A \cap B = \phi$, the impossible event, and its probability is zero. Therefore Equation 4.10 says that $P(A \cup B) = P(A) + P(B) - 0 = P(A) + P(B)$ when A and B are mutually exclusive events. This is exactly what Equation 4.5 says that $P(A \cup B)$ is when A and B are mutually exclusive.

Conditional Probability

EXAMPLE 4.13 (continued) If someone is a man, what is the probability that he works full-time?

Solution: Look at Table 4.1. There are 120 men. Of these, 80 are working full-time. Therefore 80/120 = 2/3 of the men work full-time. The probability is therefore 2/3. The probability was found by dividing the number of men working full-time by the total number of men. Letting $P(B_1|A_1)$ be the probability of event B, given information that A_1 occurs, in this example we found that

$$P(B_1|A_1) = \frac{N(A_1 \cap B_1)}{N(A_1)} = \frac{8}{12}$$

where $N(A_1 \cap B_1)$ is the number in set $A_1 \cap B_1$ and $N(A_1)$ is the number in set A_1.

We can obtain the same results from the probabilities as well as from the numbers. Using Table 4.2 instead of Table 4.1, divide the *probability* of selecting a man working full-time, $P(A_1 \cap B_1)$, by the probability of selecting a man, $P(A_1)$. ■

> When information is given that an event A has occurred, the probability of another event B also occurring is known as the CONDITIONAL PROBABILITY of B given A, and is defined by
>
> $$P(B|A) = \frac{P(A \cap B)}{P(A)}$$

4.11

This definition of conditional probability is of fundamental importance in probability and statistics. It tells us the probability of an unknown event B when an event A is known to have occurred. If this conditional probability is substantially different from $P(B)$, the probability of B without any knowledge about A, then the information about A was of importance in revising the probability of B. In Example 4.13, the probability of B_1 (working full-time) is 1/2; but given information that event A_1 has occurred (that a person is a man) the conditional probability of B_1 has increased to 2/3. This can be shown by substituting the entries of Table 4.2 into Equation 4.11, the definition of conditional probability.

$$P(B_1|A_1) = \frac{P(A_1 \cap B_1)}{P(A_1)} = \frac{.4}{.6} = \frac{2}{3}$$

EXAMPLE 4.13 (continued) What is the probability that a selected individual is working part-time, given that the individual is a man?

Solution: The question asks for $P(B_2|A_1)$. From Equation 4.11,

$$P(B_2|A_1) = \frac{P(A_1 \cap B_2)}{P(A_1)} = \frac{0}{.6} = 0$$

Information on a person's sex has in this case told us that the person is not working part-time. ■

The essence of conditional probability is this concept of the "given." In order to spot whether a question is asking for a conditional probability, you should look for such phrases as:

"the probability of B *if* A occurs"
"the probability of B, given A"
"the probability of B conditional on A"
"the probability of B when A occurs"
"the proportion of A's that are B's"
"the probability of B, assuming A"
"the probability that an A will be a B"

The purpose of this list is not memorization; it is to give you more feeling of the concept. In a real-life situation, you should know when a conditional probability is called for, and you will not have a textbook to tell you what key word to look for.

Much statistical research is involved with assessing conditional probabilities and the way information on one event changes our knowledge and estimates about another event. For example, we may want to determine the probability that people will do well on a job from information on their scores on an aptitude test. The aptitude test is valuable if the probability of somebody's doing well on the job, given a high test score, is much higher than the probability would have been without information on the test.

EXAMPLE 4.14 When Mr. McArdle, the head of the Bureau of Motor Vehicles, comes to work he usually parks his maroon Imperial in the Number One Parking Space. Mr. Hope would like to talk to Mr. McArdle's secretary, but he wants to call her only if Mr. McArdle is not in the office. Mr. Hope determines whether or not Mr. McArdle is in his office by looking out the window to see whether a maroon Imperial is parked in Space Number One. The information on the car, though useful, is not totally reliable, as can be seen from the joint probabilities shown in Table 4.3.

(a) On any randomly selected occasion, what is the probability that the car is there?

Solution: $P(C_1) = P(M_1 \cap C_1) + P(M_1 \cap C_2)$
$= .4 + .2 = .6$

(b) What is the probability that the car is not there?

Solution: There are two ways to find this:

1. $P(C_2) = P(M_1 \cap C_2) + P(M_2 \cap C_2)$
$= .1 + .3 = .4$

2. $P(C_2) = 1 - P(C_1) = 1 - .6 = .4$

Table 4.3. *JOINT PROBABILITIES Mr. McArdle and His Car*

	C_1 THE CAR IS THERE	C_2 THE CAR IS NOT THERE
M_1 = McArdle is in his office	.4	.1
M_2 = McArdle is not in his office	.2	.3

(c) What is the probability that Mr. McArdle is in his office?

Solution: $P(M_1) = P(M_1 \cap C_1) + P(M_1 \cap C_2)$
$= .4 + .1 = .5$

(d) If the car is there, what is the probability that Mr. McArdle is in his office?

$$P(M_1|C_1) = \frac{P(M_1 \cap C_1)}{P(C_1)} = \frac{.4}{.6} = \frac{2}{3}$$

(e) If the car is not there, what is the probability that Mr. McArdle is in his office?

Solution: $$P(M_1|C_2) = \frac{P(M_1 \cap C_2)}{P(C_2)} = \frac{.1}{.4} = \frac{1}{4}$$

This example illustrates a problem that is widespread in almost all research and prediction. Apparently Mr. Hope's situation was such that he was unwilling or unable to get direct evidence on M_1 or M_2, so he had to use the indirect method of getting information on C_1 or C_2. This was of some use, because the conditional probabilities of M_1 were considerably revised from the original $P(M_1)$ when information on C_1 and C_2 was available. But on the other hand, information on the car was not a perfect predictor, so that Mr. Hope could still make the wrong decision. So it is with much research. Very often we are unable, unwilling, or not permitted, to get direct information on an event we are interested in, but we are able to get information on other related events. Information on these related events will seldom give us the direct answer to the question we are interested in, but it can help—by substantially increasing or decreasing the probability we had previously assigned to the event. ■

The Multiplication Theorem

Tables 4.2 and 4.3 give joint probabilities for pairs of events. Sometimes, however, the joint probability is not given and you want to find it from other information. Suppose, for example, that you knew that 60% of proposals to do research were approved by a Screening Committee and 30% of the proposals approved by the Screening Committee were approved by the Final Selection Committee. What is the probability that the proposal would be approved by both committees? Letting A be the event "the Screening Committee approves" and B be the event "the Final Selection Committee" approves, we have $P(A) = .6$ and the conditional probability $P(B|A) = .3$. We want to find the probability that both committees approve. This is $P(A \cap B)$.

We can determine $P(A \cap B)$ from Equation 4.11, the definition of conditional probability. Multiplying both sides of Equation 4.11 by $P(A)$ gives the multiplication theorem.

Multiplication theorem for two events

$$P(A \cap B) = P(A) \cdot P(B|A) \tag{4.12}$$

The probability of being approved by both committees is therefore seen to be

$$P(A) \cdot P(B|A) = (.6)(.3) = .18$$

By exchanging the A's and the B's, we can obtain from Equation 4.12 the equivalent expression

$$P(A \cap B) = P(B) \cdot P(A|B) \tag{4.13}$$

EXAMPLE 4.15 Seventy percent of the applicants to a law school are accepted into the program. Of those accepted, 40% complete the program. What is the probability that a randomly selected applicant will be both accepted into the program *and* complete the program?

Solution: Because the individual is randomly selected, we can assign equal probabilities to all applicants. Letting A be the event "accepted into the program" and C be the event "completed the program," we can assign

$$P(A) = .7 \qquad \text{and} \qquad P(C|A) = .4$$

The question asks for $P(A \cap C)$. By Equation 4.12, this is

$$P(A \cap C) = P(A) \cdot P(C|A) = (.7)(.4) = .28 \quad \blacksquare$$

The multiplication theorem can be extended to three or more events.

Multiplication theorem for three or more events

$$\begin{aligned} P(A \cap B \cap C \cap \ldots \cap K) \\ &= P(A) \cdot P(B|A) \cdot P(C|A \cap B) \cdot \ldots \\ &\quad \cdot P(K|A \cap B \cap C \cap \ldots \cap J) \end{aligned} \tag{4.14}$$

EXAMPLE 4.16 The School of Urban Affairs is looking for a chairperson for the Operations Research Department who meets the following qualifications. She or he

1. must be under 45 years of age
2. must have published at least one book and ten articles in the field

3. must have a substantial amount of administrative experience
4. must not be divorced
5. must be willing to travel out of town for periods of at least 10 days

Assuming that among prospective applicants

1. 50% are under 45 years of age
2. 40% of those meeting requirement 1 have the required publications
3. 20% of those meeting requirements 1 and 2 have the required administrative experience
4. 80% of those meeting requirements 1, 2, and 3 are not divorced
5. 30% of those meeting requirements 1, 2, 3, and 4 are willing to do the necessary traveling

What is the probability that a prospective applicant meets all the qualifications?

Solution:

Let A be the event "under 45 years old." $P(A) = .5$
Let PUB be the event "published the required number of books and articles." $P(PUB|A) = .4$
Let E be the event "has substantial administrative experience." $P(E|A \cap PUB) = .2$
Let D' be the event "not divorced." $P(D'|A \cap PUB \cap E) = .8$
Let T be the event "willing to travel." $P(T|A \cap PUB \cap E \cap D') = .3$

The question asks for $P(A \cap PUB \cap E \cap D' \cap T)$. By Equation 4.14 this is:

$$P(A) \cdot P(PUB|A) \cdot P(E|A \cap PUB) \cdot P(D'|A \cap PUB \cap E) \cdot P(T|A \cap PUB \cap E \cap D') = (.5)(.4)(.2)(.8)(.3) = .0096$$

Less than 1% of the prospective applicants meet the qualifications. ■

Independent Events

In Example 4.14, the probability that Mr. McArdle is in his office *depends* on whether his car is parked in Space Number One or not. In Example 4.13, the probability that a person is working full-time *depends* on the sex of that person.

As we have seen, information that the car is not parked in the space decreases the probability of Mr. McArdle's being in his office from .5 to .25—as can be seen by comparing

$$P(M_1) = .5 \qquad \text{and} \qquad P(M_1|C_2) = .25$$

When would the probability of one event B not depend on another event A? Only when the conditional probability of B given A is no different from the probability of B without any information on A. In this case we say that A and B are independent events.

> Two events A and B are INDEPENDENT EVENTS if and only if
>
> $$P(B|A) = P(B) \qquad \textbf{4.15}$$
>
> or
>
> $$P(A|B) = P(A) \qquad \textbf{4.16}$$

It can be shown that Equation 4.16 is the equivalent of Equation 4.15. Two events are therefore independent events if the probability of either of the events is unaffected by knowledge about the other.

EXAMPLE 4.17 In a local federal bureau, the probability that an employee with a GS-9 rating gets promoted to a GS-11 within a specified period of time is .4. The personnel director claims that "race is not a factor" in deciding who gets promoted. If the director's claim is correct, what is the probability that a randomly selected GS-9 who is a black will be promoted to GS-11 in the time period?

Solution: If "race is not a factor" we can consider that being promoted (A) and being black (B) are independent events. Therefore we can use Equation 4.16,

$$P(A|B) = P(A)$$

which says the probability of being promoted if you are black is the same as the probability of being promoted. Since $P(A) = .4$, then $P(A|B) = .4$, and the probability of a randomly selected black's being promoted is .4. ■

The assumption of the independence of two events enables us to express the multiplication theorem in a simpler form. If A and B are independent events, we can substitute $P(B)$ for $P(B|A)$ in Equation 4.12 so that

> **Multiplication theorem for two independent events**
>
> $$P(A \cap B) = P(A) \cdot P(B) \qquad \textbf{4.17}$$
>
> if and only if A and B are independent events.

Three or more events A, B, C, etc, are said to be MUTUALLY INDEPENDENT EVENTS if the probability of any one of them, $P(A)$, $P(B)$, $P(C)$ etc. is equal to the conditional probability of the event given any combination of the other events.

When three or more events are mutually independent, the joint probability of the occurrences of all of them is given by:

$$P(A \cap B \cap C \cap \ldots \cap K) = P(A) \cdot P(B) \cdot P(C) \cdot \ldots \cdot P(K) \tag{4.18}$$

if and only if $A, B, C, \ldots, K$ are mutually independent events.

EXAMPLE 4.18 A secretary is trying to find a time for a meeting of four busy executives. For any given hour of the day, the probability that executive A is free is .3, the probability that B is free is .2, the probability that C is free is .5, and the probability that D is free is .4. Assuming that the availability of any executive is independent of whether any of the others is available, what is the probability that for any given hour all the executives will be free to attend?

Solution:

Let A be the event "A is available."
Let B be the event "B is available."
Let C be the event "C is available."
Let D be the event "D is available."

The example asks for $P(A \cap B \cap C \cap D)$. The assumption of independence enables us to use Equation 4.18 to obtain

$$\begin{aligned} P(A \cap B \cap C \cap D) &= P(A) \cdot P(B) \cdot P(C) \cdot P(D) \\ &= (.3)(.2)(.5)(.4) \\ &= .012 \end{aligned}$$

The .012 probability may be a factor causing the high turnover rate of secretaries who have to set up meetings for busy administrators. ■

EXAMPLE 4.19 In a family with two children, what is the probability that both children will be girls?

Solution:

Let A be the event "first child is a girl."
Let B be the event "second child is a girl."

If both children are to be girls, we need $A \cap B$ to occur. We want to know $P(A \cap B)$.

Assume A and B are independent events, and, on the basis of empirical relative frequencies,

$$P(A) = P(B) = .5$$

Then, by Equation 4.17

$$P(A \cap B) = P(A) \cdot P(B) = (.5)(.5) = .25$$

The probability of both children being girls is .25. ■

EXAMPLE 4.20 If a pair of identical twins is born, what is the probability that they are girls?

Solution:

Let A be the event "first child is a girl."
Let B be the event "second child is a girl."

We want to find $P(A \cap B)$. In this example, we cannot assume that the events are independent, because identical twins are always of the same sex. Therefore to find $P(A \cap B)$ we cannot use Equation 4.17, which requires the assumption of independence. Instead, we must use the general multiplication theorem, Equation 4.12,

$$P(A \cap B) = P(A) \cdot P(B|A) = .5$$

Assume $P(A)$ is .5. Then $P(B|A)$ is 1, since if the first twin is a girl we know that the second will also be a girl.

Sets of identical twins have a .5 probability of being girls. ■

EXAMPLE 4.21 Consider the frequencies given in Table 4.4. Are being born under the sign of Scorpio and being successful independent events?

There are 6000 individuals. Assume that these 6000 individuals constitute the sample space and each individual has an equal probability

Table 4.4. *JOINT FREQUENCY TABLE*
Sign of Scorpio and Success

	SUCCESSFUL MANAGERS	UNSUCCESSFUL MANAGERS
Born under sign of Scorpio	300	200
Born under sign other than Scorpio	3300	2200

Table 4.5. *JOINT PROBABILITIES*
Sign of Scorpio and Success

	SUCCESSFUL	UNSUCCESSFUL	
Scorpio	.050	.033	.083
Not Scorpio	.550	.367	.917
	.600	.400	

of 1/6000 of being selected. Convert the table to a table of joint probabilities by dividing all frequencies by 6000, as shown in Table 4.5. Only if the events "Scorpio" and "Successful" are independent, will Equation 4.17 hold. The probability of "Scorpio" is .083. The probability of "Successful" is .600. Does

$$P(\text{Scorpio} \cap \text{Successful}) = P(\text{Scorpio}) \cdot P(\text{Successful})?$$

Check it. Does .050 = (.083)(.600)? Yes, except for a small error in rounding. The two events are independent. ■

Much statistical research is concerned with testing to see whether or not events are independent of each other. If they are *not* independent, then they are related somehow.

For example, in testing to find out whether a drug is harmful, a research team will look for a statistical relationship between use of the drug and some undesired symptom. If use of the drug and the occurrence of the symptom do not appear to be *independent*, then there is ground to suspect a causal relationship linking the two.

EXERCISES

4–17. Given the following classification of 100 employees by experience and receptivity to a new proposal:

	Receptive	Not Receptive
Experienced	28	12
Inexperienced	50	10

If *S*, one of the 100 employees, is selected at random,

a) what is the probability that *S* is receptive?
b) what is the probability that *S* is inexperienced?
c) If *S* is receptive, what is the probability that *S* is experienced?
d) If *S* is experienced, what is the probability that *S* is receptive?
e) What is the probability that *S* is receptive *or* experienced?

4–18. Last Monday 20% of the employees were sick and 30% celebrated a religious holiday; 6% were both sick *and* celebrated a religious holiday.

a) What is the probability that a randomly selected employee was *either* sick *or* celebrated a religious holiday?

b) If somebody celebrated a religious holiday, what is the probability the person was sick?

c) Are being sick and celebrating a religious holiday independent events?

***4–19.** Show that, if $P(B|A) = P(B)$, then $(A|B) = P(A)$. *Hint:* Use the definition of conditional probability.

4–20. Given $P(A) = .3$, $P(B) = .5$, and $P(A \cap B) = .1$, find:

a) $P(A \cup B)$

b) $P(A' \cap B)$

4–21. Given $P(A) = .7$, $P(B) = .8$, and $P(A \cup B) = .6$, find:

a) $P(A \cup B)$

b) $P(B|A)$

c) $P(B'|A)$

d) $P(B|A')$

4–22. Given $P(D') = .2$, $P(E|D) = .6$, and $P(E|D') = .3$,

a) draw a tree diagram showing the events D, D', E, and E'.

b) find $P(D \cap E)$.

c) find $P(E)$.

4–23. Forty percent of the trains are old and arrive late; sixty percent of the trains are old. If a train is old, what is the probability that it arrives late?

4–24. Given the following frequency table of a random sample of 300 pieces of heavy equipment:

	Maintenance Cost Under $1000	Maintenance Cost $1000 or More
Equipment Under 5 Years Old	120	60
Equipment 5 Years Old or More	80	40

does it appear from the frequencies that the events "under 5 years old" and "maintenance cost under $1000" are independent events? Why or why not?

***4–25.** Given that events A and B are independent, show:

a) that the events A and B' are independent

b) that the events A' and B are independent

c) that the events A' and B' are independent

4–26. Are the events "female" and "working part-time" as shown in Table 4.2 independent events? Why or why not?

4–27. a) Give an example in the real world of a pair of events that are independent.

b) Give an example of a pair of events that are mutually exclusive.

c) Distinguish between independent events and mutually exclusive events.

***4–28.** If two events, A and B, both have positive probabilities, show that if A and B are independent, they cannot be mutually exclusive.

*4.4 Bayes' Theorem

Conditional probability is the probability that an event B will occur if an event A has occurred. But suppose you have the information that B has occurred and you want to determine from this fact the probability that A has occurred. Or, more specifically, suppose there were two possible mutually exclusive prior events A_1 and A_2, either of which might have led to the occurrence of B. Given the observation of B, you want to know the probability that A_1 occurred rather than A_2. You want to know $P(A_1|B)$. If you know $P(A_1)$, the probability that you assigned to A_1 prior to observation of B; and $P(A_2)$, which will be $1 - P(A_1)$; $P(B|A_1)$, the conditional probability that B would occur if A_1 occurs; and $P(B|A_2)$, the conditional probability that B would occur if A_2 occurs, then it can be shown that

$$P(A_1|B) = \frac{P(A_1) \cdot P(B|A_1)}{P(A_1) \cdot P(B|A_1) + P(A_2) \cdot P(B|A_2)} \qquad \textbf{4.19}$$

which is BAYES' THEOREM for two mutually exclusive possible prior events A_1 and A_2.

Bayes' Theorem gives an "inverse probability," the probability of a prior event given information on a subsequent event. It provides an approach to determining the probability that a person has measles given that he has a certain symptom, the probability that a person committed a crime given a piece of circumstantial evidence, and the probability that a particular author wrote a disputed text given that the paper displays a particular stylistic pattern.

EXAMPLE 4.22 It is known that the users of a particular type of data processing equipment encounter trouble with the equipment about 2% of the time. Your office has recently purchased some of this equipment from a company with a very excellent reputation. The company claims that in not more than 1% of cases has there been any trouble with its equipment, and you have reason to believe this claim to the point where you would give a .95 probability to the statement. Yet your office has had trouble with the equipment. In view of your unfortunate experience, what is the probability that you would now assign to the company's claim?

Solution: Let A_1 be the event that there is trouble with the company's equipment only 1% of the time. Let A_2 be the event that the company's equipment causes trouble 2% of the time, the same as the industry average. Let B be the observed event that the equipment you purchased created trouble.

Before your unfortunate experience your assessment of the probability was

$$P(A_1) = .95, \qquad P(B|A_1) = .01, \qquad \text{and} \qquad P(B|A_2) = .02$$

You want to find $P(A_1|B)$. For this, you use Bayes' Theorem (Equation 4.19):

$$P(A_1|B) = \frac{(.95)(.01)}{(.95)(.01) + (.05)(.02)} = \frac{95}{105} = .905$$

In spite of your difficulties with the equipment, you would still give better than a 90% probability to the company's claim. ■

EXERCISES

***4–29.** The Municipal Building houses three departments: Health, Police, and Streets. Health employs 80% women, Police employs 20% women, and Streets employs no women. Each department selects one employee to receive its Outstanding Employee Award. Then the three outstanding employees compete to win the Grand Prize for Meritorious Service. The winner of the Grand Prize for Meritorious Service this year was a woman. Assuming that neither sex nor department was a factor in winning any award, what is the probability that the Grand Prize winner was employed by the Department of Health?

***4–30.** A person claims to have extrasensory perception (ESP). If you look at a randomly selected card from a deck of 52 cards, he claims that he can "read your mind" and tell what card you are looking at. You select the Jack of Clubs, and without seeing the card, he says "Jack of Clubs." Assuming you previously assigned a probability of .10 to his having ESP, what probability should you now assign on the basis of his correctly identifying your card?

Summary of Chapter

A situation involving uncertainty can be considered to be an experiment, the outcomes of which constitute the sample space. An event is a subset of the sample space. Combinations of events can be defined using the language of sets.

As a mathematical concept, probability is a number assigned to an event in a sample space. For practical purposes, we assign probabilities to events on the basis of either *a priori*, empirical, or subjective considerations.

Once probabilities are assigned to the elements of a sample space, it is possible to derive probabilities for events that are combinations of these elementary events. Conditional probabilities make use of information on the occurrence of one or more events to determine the probability of another event. If, on the other hand, the information about one event has no effect on the probability of another event, the events are independent events.

Bayes' Theorem is a method of using probabilities in a reverse direction. The Theorem makes use of the conditional probability of B given a prior event A to determine the probability of A given B.

REVIEW EXERCISES FOR CHAPTER 4

4–31. A volunteer worker is soliciting for the Community Chest. When he calls at a house, he wonders

(1) whether or not anyone will answer the door;
(2) if a person answers the door, whether or not the person will be willing to discuss contributing to the Community Chest;
(3) if the person is willing to discuss making a contribution, whether or not she or he will actually make a contribution.

Draw a tree diagram to represent the sample space of this experiment.

4–32. Suppose, in Exercise 4–31, that (1) The probability that someone will answer the door is .7. (2) If someone answers the door, the probability that that person will be willing to discuss contributing is .4. (3) If the person is willing to discuss contributing, the probability that she or he actually will contribute is .6. What is the probability:

a) that nobody answers the door?
b) that somebody answers the door but will be unwilling to discuss contributing?
c) that somebody answers the door, is willing to discuss contributing, but will not make a contribution?
d) that the volunteer will get a contribution?

4–33. Let A be the set of employed persons and B be the set of people receiving unemployment compensation. Describe, in words, the following sets.

a) A'
b) $A \cup B$
c) $A \cap B$
d) $A \cap B'$

4–34. The President is about to nominate a person to be Secretary of State. Let A be the event that the nominee is black, B be the event that the nominee is female, and C be the event that the nominee is confirmed by the Senate. Describe, in words, the following events.

a) A'
b) $A \cup B$
c) $A \cap B$
d) $A \cap B \cap C$

e) $A \cap B \cap C'$
f) $(A \cup B) \cup C$

4–35. You have been asked to determine the probability that an ICBM will hit within 1.6 kilometers of its target. What approaches to the concept of probability might you use to assess such a probability?

4–36. Given, the following joint probabilities:

	Job Rating		
	GS-13 or above	GS-9 to GS-12	Below GS-9
Completed less than 12 years of school	.05	.15	.10
Completed at least 12 years of school but less than 16 years	.20	.10	.10
Completed at least 16 years of school	.20	.05	.05

If a person is selected at random,

a) what is the probability of her/his being below GS-9?
b) what is the probability of her/his having completed at least 16 years of schooling?
c) what is the probability of her/his having completed at least 12 years of schooling?
d) If she (he) has completed at least 16 years of schooling, what is the probability that she (he) is GS-13 or higher?
e) If she (he) is a GS-13 or higher, what is the probability that she (he) has completed at least 16 years of schooling?

4–37. Refer to Exercise 4–36.

Let A_1 be the event "completed less than 12 years of schooling."
Let A_2 be the event "completed at least 12 years but less than 16 years of schooling."
Let A_3 be the event "completed at least 16 years of schooling."
Let B_1 be the event "GS-13 or above."
Let B_2 be the event "from GS-9 to GS-12."
Let B_3 be the event "below GS-9."

Express all the probabilities asked for in Exercise 4–36 in symbols.

4–38. Given $P(D) = .7$ and $P(E) = .3$, find $P(D \cap E)$ if

a) D and E are mutually exclusive events.
b) D and E are independent events.

4–39. In the city Transportation Department, 40% of the employees are black, 30% are females, and 10% are black females.

a) If a randomly selected employee is black, what is the probability that the person is female?

b) If a randomly selected employee is female, what is the probability that she is black?

c) Are being black and being female independent events? Why, or why not?

4–40. Depict, in a tree diagram, the possible actions that can be taken on a proposed piece of legislation originating in the Senate. To become law the bill must be passed by the Senate; passed by the House of Representatives; and either signed by the President, allowed by the President to become law without his signature, or vetoed by the President but passed by 2/3 majorities of both the Senate and the House over the presidential veto.

4–41. If 30% of people arrested on narcotics charges are under 20 years of age, and 25% of people arrested on narcotics charges are at least 20 years of age but less than 30 years of age, what is the probability that a person who is arrested on a narcotics charge is under 30 years of age? What probability rule did you apply to arrive at your answer?

4–42. Which of the following pairs of events would you expect to be independent events?

a) being over 35 years of age and earning over $10,000 a year

b) being a Roman Catholic and favoring a liberalized abortion bill

c) having an odd social security number and being employed at the local level of government

d) being female and singing in the bass section of the Bach Choir

e) working hard and getting promoted

f) smoking and contracting emphysema

g) smoking and contracting cirrhosis of the liver

4–43. Given the tree diagram shown below with probabilities where A_1 is "being overweight"; A_2 is "being not overweight"; B_1 is "earning at least $15,000"; and B_2 is "earning less than $15,000",

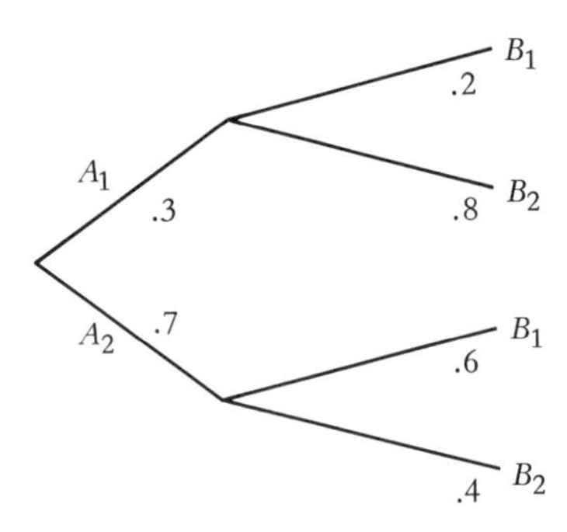

a) what is the probability of being overweight?

b) what is the probability of earning at least $15,000 if one is overweight?

c) what is the probability of being overweight and earning at least $15,000?

d) what is the probability of being not overweight and earning at least $15,000?

e) what is the probability of earning at least $15,000?

f) what is the probability of being overweight if one earns at least $15,000?

4–44. Many young readers may not recall the days when Christmas tree lights were connected "in series," so that if one bulb on the whole string was bad, none of the lights on the string would light up.

a) Let A_1 be the event "the first bulb on the string is good"; let A_2 be "second bulb on the string is good," etc. Depict symbolically the event that must occur for the bulbs on the string to light up.

b) Nowadays the new bulbs are strung "in parallel" so that a bulb will light up regardless of whether other bulbs on the string light up. Depict symbolically the event necessary for at least one of the bulbs to light up.

SUGGESTED READINGS

CLELLAND, RICHARD C., JOHN S. DeCANI, and FRANCIS E. BROWN. *Basic Statistics with Business Applications,* 2nd ed. New York: John Wiley and Sons, 1973.

GOLDBERG, S. *Probability: An Introduction.* Englewood Cliffs, N.J.: Prentice-Hall, Inc., 1960.

HAMBURG, MORRIS. *Statistical Analysis for Decision-Making,* 2nd ed. New York: Harcourt, Brace, Jovanovich, Inc., 1977. Chapter 2.

JOHNSON, ROBERT. *Elementary Statistics,* 2nd ed. North Scituate, Mass.: Duxbury Press, 1976. Chapter 5.

MOSTELLER, F., R. ROURKE, and G. THOMAS. *Probability with Statistical Applications.* Reading, Mass.: Addison-Wesley Publishing Co., Inc., 1961. Chapters 1–4; Appendix I.

FIVE

Discrete Probability Distributions

Outline of Chapter

5.1 The Nature of Random Variables and Probability Distributions

A random variable can take on different values with known probabilities.

5.2 The Binomial Probability Distribution

The binomial distribution is used to determine probabilities for the number of successes in a series of independent trials.

5.3 The Expected Value and Variance of a Discrete Random Variable

We can summarize or describe the behavior of random variables by measures of location and uncertainty.

Objectives for the Student

1. Become familiar with random variables and their importance in making inferences.
2. Learn to use the binomial table in determining probabilities.
3. Know the conditions under which the binomial distribution is the appropriate one to use.
4. Understand the concept and importance of the expected value of a random variable.

Chapter 1 introduced the concept of measured variables that *have taken on* certain values with observed frequencies. This chapter introduces random variables that *will take on* certain values with known or estimated probabilities. The frequencies with which measured variables took on certain values were observed. The probabilities with which random variables will take on certain values are either estimated from observed frequencies or deduced by means of mathematical logic. The recorded behavior of measured variables can be summarized by measures of central tendency and variability. The probable behavior of random variables can be summarized by measures of expectation and uncertainty.

Chapters 2 and 3 were devoted to describing how variables *have* varied. This chapter is devoted to estimating how random variables *will* vary.

5.1 The Nature of Random Variables and Probability Distributions

Jenny and Max Bishop have both applied for jobs; Jenny as a program evaluation specialist with the state department of health and Max as a teacher with the school system. Each has been told that she or he has a 50–50 chance of getting the job.

From reading Chapter 3 you should be able to identify the sample space for this "experiment," which consists of four outcomes. But Jenny and Max are not basically concerned about the whole sample space; they are basically concerned about how many of them get hired. If neither gets hired, they will be in deep economic trouble trying to feed their four children; if one of them gets hired; they will be able to make ends meet; if both of them get hired, they will be able to live very comfortably.

Jenny and Max are interested in the value of the random variable "the number who get hired."

> When we assign a unique number x to every element of a sample space, we call the set of these numbers the RANDOM VARIABLE X. The numbers that are so assigned are known as VALUES OF THE RANDOM VARIABLE.

The values of the random variables in this book will be designated by small letters, usually letters near the end of the alphabet, such as x, y, or z. The values of the random variable X will be known as x, the values of the random variable Y will be known as y, etc.

> If a random variable can take on only a finite number of possible values within any finite range of values, it is said to be a DISCRETE RANDOM VARIABLE.

Nearly all of the discrete random variables discussed in this book will take on only *integer* (whole number) values.

EXAMPLE 5.1 What are the values of the random variable "number hired" that Jenny and Max Bishop are interested in?

Discussion: Let X be the random variable "number hired." X can take on the values 0, 1, or 2. We can list the elements of the sample space as follows:

J_1 = "Jenny is hired"
J_2 = "Jenny is not hired"

M_1 = "Max is hired"
M_2 = "Max is not hired"

and assign values of x to these elements as follows:

Element of Sample Space	x
$J_1 \cap M_1$	2
$J_1 \cap M_2$	1
$J_2 \cap M_1$	1
$J_2 \cap M_2$	0

What are the probabilities of the values of the random variable X?

Solution: If we assign a probability of 1/2 to each person's being employed, and each one's being employed is independent of the other, we can assign probabilities to the elements of the sample space by the use of Equation 4.17.

$$P(J_1 \cap M_1) = P(J_1) \cdot P(M_1) = (1/2)(1/2) = 1/4$$
$$P(J_1 \cap M_2) = P(J_1) \cdot P(M_2) = (1/2)(1/2) = 1/4$$
$$P(J_2 \cap M_1) = P(J_2) \cdot P(M_1) = (1/2)(1/2) = 1/4$$
$$P(J_2 \cap M_2) = P(J_2) \cdot P(M_2) = (1/2)(1/2) = 1/4$$

Each element of the sample space therefore has a probability of 1/4. In order to assign probabilities to the values of the random variable X, add all the probabilities of the (mutually exclusive) elements of the sample space that are assigned to that value of X. Then

$$P(X = 2) = P(J_1 \cap M_1) = 1/4$$
$$P(X = 1) = P(J_1 \cap M_2) + P(J_2 \cap M_1) = 1/4 + 1/4 = 1/2$$
$$P(X = 0) = P(J_2 \cap M_2) = 1/4$$

We have developed a probability distribution for the discrete random variable X. ■

> A DISCRETE PROBABILITY DISTRIBUTION or a DISCRETE PROBABILITY FUNCTION is a rule that assigns a probability to every value of a discrete random variable.

The probabilities so assigned to the discrete random variable X will be denoted by $f(x)$, $g(x)$, or $h(x)$. $f(x)$ is read "f of x," meaning "the probability that a random variable takes on the value x," or $P(X = x)$.

The assignment of these probabilities must be in accordance with the mathematical rules for assigning probability given by Equations 4.3, 4.4, and 4.5. The following specifications for a discrete probability function are in accordance with those rules:

> A discrete probability function assigns probabilities $f(x)$ to each value x of the random variable X in a way that:
>
> $f(x) \geq 0$ for every real number x **5.1**
>
> $f(x) > 0$ for only a finite number of values within any finite interval of x values **5.2**
>
> $\Sigma f(x) = 1$ **5.3**

Equation 5.1 says that $f(x)$ is defined for every real number (all the positive and nonpositive numbers, including fractions and irrational numbers*) and that $f(x)$ cannot be negative.

Equation 5.2 says that within any limited range of x values there is only a finite number of them with positive probabilities. There are "spaces" between x values with positive probabilities.

Equation 5.3 says that all of the probabilities add up to 1.

A common source of confusion is the distinction between the *name* of the random variable, which we will always designate by a capital letter, and the *value* that a random variable takes on, which we will designate by the corresponding small letter. It may help to picture X as a man who enters a restaurant with n tables, numbered 1 to n, and x as the number of the table that he chooses to sit at on a particular evening. From past experience, we can assign a probability $f(x)$ for each table x. Mr. X is the random variable whose table-choosing behavior conforms to the known probability distribution $f(x)$. The actual table that Mr. X sits at on any given evening is the value that the random variable takes on. $P(X = x)$ means "the probability that Mr. X sits at any table x" where x can be 1, 2, 3, or any number up to n. $P(X = 1)$ or $f(1)$ is the probability that Mr. X sits at table 1, $f(2)$ is the probability he sits at table 2, etc.

Another person, Ms. Y, enters the restaurant. We designate the table she sits at as y, and the probability distribution of the table she chooses to sit at as $g(y)$, where y can be any table from 1 to n. Because Ms. Y has a different pattern of table-choosing behavior from Mr. X, $g(y)$ will in general have a different form from $f(x)$.

EXAMPLE 5.2 Let the random variable X be the number of complaints received in the Ombudsman's office in a day. X can take on the values 0, 1, 2, etc. If we designate the probability as $f(x)$, which of the following statements contradict conditions 5.1, 5.2, or 5.3?

*An irrational number is a number that cannot be expressed as an integer or a fraction.

(a) $f(2) = .2$
(b) $f(1.3) = .3$
(c) $f(3) = .4$, $f(4) = .5$, and $f(5) = .2$
(d) $f(0) = -1$
(e) $f(-1) = 0$

Discussion: Statement (a) is all right. It assigns a probability to the event "2" which is in accordance with the conditions.

Statement (b) does not violate conditions 5.1, 5.2, or 5.3, but it does violate the definition of the random variable X *in this problem*. According to this definition, the only x values that can have positive probabilities are the integers 0, 1, 2, etc.

The three statements taken together in (c) violate the conditions, since the probabilities add up to more than 1.

Statement (d) contradicts the conditions, because $f(x)$ is not allowed to be negative.

Statement (e) is all right. Although -1 is not a possible value of X, the probability function is defined for all values of X. For all values except the integers 0, 1, 2, etc. this probability is zero. ■

Probability functions can be defined either by a listing of all the probabilities, as

x	$f(x)$
-1	.2
11	.5
3	.3

or by a mathematical equation that instructs you how to compute the probabilities, such as

$$g(x) = 1/3 \qquad x = -1, 0, 1$$

or

$$h(y) = y/6 \qquad y = 1, 2, 3$$

As an exercise, check to see whether $g(x)$ and $h(y)$ fulfill the conditions required for a probability distribution.

EXAMPLE 5.3 Which of the following are discrete probability distributions?

(a)

x	$h(x)$
-1	.1
0	.3
7	.4

(b) $g(y) = \dfrac{2 + y}{3} \qquad y = -1, 0$

Solution: (a) is not a probability distribution, since the probabilities do not add to 1. (b) is a discrete probability distribution. Computing the probabilities yields

$$g(-1) = \frac{2 + (-1)}{3} = \frac{1}{3}$$
$$g(0) = \frac{2 + 0}{3} = \frac{2}{3}$$

The probabilities are nonnegative and add to 1. ■

METHODS FOR DETERMINING REALISTIC PROBABILITY DISTRIBUTIONS

You can assign probabilities to values of random variables pretty much arbitrarily as long as you conform to the mathematical requirements. But you also want the probabilities to be realistic—to conform to your expectations in the real world.

One method of assigning probabilities to values of random variables is the *a priori* method. For example, the random variable "number of spots showing on the roll of a fair die" can take on the values 1, 2, 3, 4, 5, or 6; and we may *a priori* assign a probability of 1/6 to each of these events.

Another method is to assign probabilities on the basis of relative frequency. For example, the chief resident of a medical unit may determine the probability distribution of Y, the number of patients admitted to Ward C in a day, as

y	$f(y)$
0	.25
1	.35
2	.20
3	.15
4	.05

$$f(y) = P(Y = y)$$

on the basis of his past observations taken over a long period of time.

Another method, and one you may often have to fall back on when the conditions necessary for *a priori* or relative frequency do not hold, is to use subjective probabilities. On the basis of information that an epidemic of mumps has broken out, the chief resident may revise his probability distribution of Y to

y	$f(y)$
0	.05
1	.10
2	.25
3	.30
4	.15
5	.10
6	.05

In most cases, the determination of the probabilities of values of a random variable can make use of the mathematics of probability. Many random variables in the real world behave to a large extent in accordance with mathematical theorems that are based on assumptions that approximate a true-to-life situation. The assumptions seldom fit the real-life situation perfectly; nevertheless, use of the theorems leads to probabilities that are far more realistic than one might intuitively have assigned without the use of analysis. Sections 5.2 and 5.3 will discuss two probability distributions that are mathematically derived from sets of assumptions that fit some common real-world situations.

EXAMPLE 5.4 Describe how you might assess the probability distribution of X, the number of visitors who will come to the state penitentiary tomorrow during visiting hours.

Discussion: If conditions have been essentially the same over a considerable period of time, you might look at any records that have been kept over this period to determine the number of days that 0, 1, 2, etc. visitors showed up. From these you could develop a probability distribution that is based on relative frequency. For example, on the basis of records over a 200-day period, you could determine the probability distribution as follows:

x	*Recorded Frequency of x*	$f(x)$
0	0	0
1	0	0
2	3	.015
3	6	.030
4	11	.055

If however, conditions have drastically changed, such as a recent transfer of a large class of prisoners into another prison, then you may do better to scrap these probabilities and use a more subjective evaluation. ∎

Cumulative Functions

In many situations we are interested not so much in the probability of the occurrence of a particular value of a random variable as in the occurrence of "at most" a certain value, or "at least" a certain value. To find the probabilities of "at most" or "at least," we can use a cumulative function. A cumulative function adds all the probabilities of the values of the random variable that are more than or equal to (at least), or less than or equal to (at most) a certain value. This book will define the cumulative function only as the "less than or equal to" variety.

> The CUMULATIVE DISTRIBUTION, or CUMULATIVE FUNCTION OF THE RANDOM VARIABLE X gives the probability that the random variable is less than or equal to any value x, or
>
> $$F(x) = P(X \leqslant x)$$

5.4

The cumulative probability of a value x can be found by adding up all the probabilities of the probability distribution of X, for all values up to and including the value x. This is given by

$$F(x) = \Sigma f(x) \qquad X \leqslant x \tag{5.5}$$

The cumulative probability of x will be denoted by a capital letter, such as F or G, corresponding to the small letter denoting the probability of x (such as f or g).

Graphical Representation

We can draw a graph to represent a discrete probability distribution by letting the horizontal axis indicate values of the random variable and by letting the vertical axis indicate the probabilities. Since there are spaces between the x values that have positive probabilities, those probabilities can be shown by a series of discrete points, as in Figure 5.1 where the vertical stems from each point to the horizontal axis are merely visual aids. Figure 5.1a is an example of the *uniform* discrete probability function, in which all of the positive probabilities are equal. Figure 5.1b shows the probabilities that are of concern to Jenny and Max Bishop, the probabilities that 0, 1, or 2 of them will get the jobs they are seeking.

Since the probabilities must be defined for *all* real numbers x, then we should indicate that $f(x)$ is zero for all values of x other than the discrete points where $f(x)$ is positive; we do this by drawing a solid line along the x axis to indicate those zero probabilities.

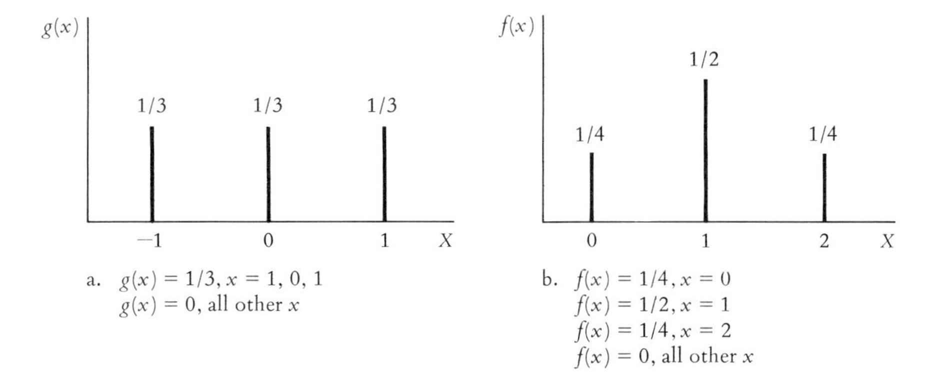

Figure 5.1. *GRAPHS OF TWO DISCRETE PROBABILITY DISTRIBUTIONS*

In order to draw a graph of a cumulative probability distribution, it is best to start by setting up a table of cumulative probabilities. To do this for the Bishops' distribution shown in Figure 5.1b, ask yourself the following questions:

1. If x is less than 0, what is the probability of being $\leqslant x$? Answer: 0
2. If x is at least 0 but less than 1, what is the probability of being $\leqslant x$? Answer: 1/4
3. If x is at least 1 but less than 2, what is the probability of being $\leqslant x$? Answer: 3/4
4. If x is at least 2, what is the probability of being $\leqslant x$? Answer: 1

The answers to these four questions can be translated into the following cumulative distribution.

1. $F(x) = 0$, if $X < 0$.
2. $F(x) = 1/4$, if $0 \leqslant X < 1$
3. $F(x) = 3/4$, if $1 \leqslant X < 2$
4. $F(x) = 1$, if $X \geqslant 2$

This cumulative function can be graphed as shown in Figure 5.2.

Cumulative functions of discrete random variables have the staircase appearance of Figure 5.2. As you move left to right, you "accumulate" probabilities; the cumulative probability can never go down; it can only stay at the same level or go up.

If you are given a cumulative distribution $F(x)$, you can find the probability that the random variable X takes on a particular value x, by

$$f(x) = F(x) - F(x - 1) \qquad \textbf{5.6}$$

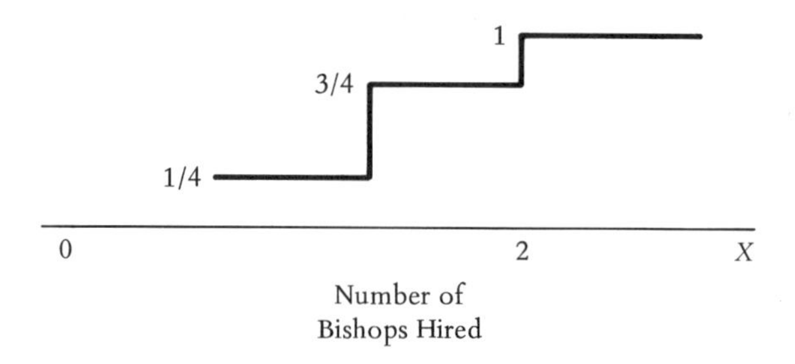

Figure 5.2. *GRAPHING OF CUMULATIVE DISTRIBUTION*
Discrete Probability Distribution Shown in Figure 5.1b

EXERCISES

5–1. A rock festival is being planned for next Tuesday in the city of Oxford. The Chief of Police believes that the number of people attending will depend primarily on the temperature.

If the temperature is very hot, he believes 5000 will attend.
If the temperature is hot, he believes 8000 will attend.
If the temperature is warm, he believes 12,000 will attend.
If the temperature is medium, he believes 12,000 will attend.
If the temperature is cool, he believes 8000 will attend.
If the temperature is cold, he believes 5000 will attend.
If the temperature is very cold, he believes 2000 will attend.

He assigns the following probability distribution to the temperatures.

$P(\text{very hot}) = .1$
$P(\text{hot}) = .2$
$P(\text{warm}) = .3$
$P(\text{medium}) = .2$
$P(\text{cool}) = .1$
$P(\text{cold}) = .08$
$P(\text{very cold}) = .02$

The Chief of Police is primarily interested in how many people will attend the rally.

a) What is the name of the random variable X in which the Chief of Police is interested?
b) What are the values x that the random variable X can take on?
c) What is the probability distribution of X according to the subjective probabilities of the Chief of Police?
d) What is the cumulative distribution of X?
e) Draw a graph of the probability distribution and the cumulative distribution of X.
f) The Chief of Police believes that he will have enough police officers available as long as the number of people attending the festival is not

over 10,000. What is the probability that he will have enough police for the festival?

5–2. Which of the following qualify as discrete probability distributions? In each case, give a reason for your answer.

a)

x	$f(x)$
0	.3
.1	.3
.2	.4

b)

y	$f(y)$
−2	1/4
0	1/4
2	1/4
4	1/4

c)

Z	$g(Z)$
0	.5
1	.6
2	−.1

d)

y	$g(y)$
10	.4
20	.4
30	.3

e) $g(x) = \frac{x}{10}$, $x = 1, 2, 3, 4$

f) $f(y) = \frac{1}{4}$, $y = -2, 0, 2, 4$

g) $h(x) = \frac{x + 2}{6}$, $x = -1, 0, 1$

h) $f(z) = \frac{2 - Z}{2}$, $Z = 0, 1, 2, 3$

5–3. The times consumed by typists to type a test are recorded to the nearest minute. The following is the probability distribution of typing times:

t	$f(t)$	
10	.05	
11	.10	
12	.25	
13	.20	
14	.15	$f(t) = P(T = t)$
15	.10	
16	.10	
17	.05	

a) Find the cumulative distribution of the random variable T.
b) What is the probability that a randomly selected typist will take no more than 15 minutes?
c) What is $F(13)$?
d) What is the probability that a typist will take between 12 and 15 minutes inclusive?
e) What is the probability that a typist will take either 13 or 14 minutes?
f) What probability theorem from Chapter 4 is necessary in answering part (e)?

5–4. Mr. Cravath has promised to call his secretary between the hours of 10:00 and 11:00 on Thursday morning. If the time he calls is given to the nearest minute, and if you assign equal probabilities to each minute between 10:00 and 11:00 inclusive, what is the probability distribution of the time at which Mr. Cravath calls his secretary?

5–5. Given the following probability distribution of the random variable Y:

$$f(y) = \frac{y - 1}{15} \qquad y = 2, 3, 4, 5, 6$$

a) Find the cumulative distribution $F(y)$.
Hint: It will be easier to list the probabilities rather than to attempt a general formula.
b) What is the probability that $Y = 4$?
c) What is the probability that $Y \leq 4$?
d) What is the probability that $Y = 2.5$?
e) What is the probability that $Y \leq 2.5$?

5–6. Distinguish between X, the random variable, and x, the value of the random variable.

5–7. A time-honored method of drafting young people into military service involves the drawing of capsules at random from a large container of capsules. One such method is to mix 366 capsules in the container, each capsule containing a different date of the year. Whoever has a birthday on the date contained in the selected capsule is selected for induction. Let X be a number from 1 to 366 corresponding to the date in the selected capsule.

a) If the selection is random, what is the probability distribution of X?
b) What is the probability that X is 13? (The 13th day of the year, or January 13).
c) What is the probability that X is less than 30?
d) Give a general formula for the probability that X is less than any value x.

5.2 The Binomial Probability Distribution

If you have invested your time or money in two ventures, each one of which has a 50–50 chance of turning out successfully, then you might reasonably want to assess the probability that both will be successful, or the probability that at least one of them will be successful. To analyze this problem, consider an analogous situation: you flip two coins and want to calculate the probabilities of 0, 1, and 2 heads.

This situation can be illustrated by the tree diagram in Figure 5.3. The two possible outcomes of the first coin are represented by the branches emanating from the left-hand node. The possible outcomes of the second coin are represented by the second set of nodes. The 4 paths in the tree (from top to bottom) represent the 4 possibilities. These are summarized in Table 5.1.

The probability of 2 heads is .25. We can determine this from the fact that the probability of heads on the 2nd coin is .5, and we can multiply .5 by .5 (using Equation 4.17) because the outcomes of the 2 coins are *independent* of each other. By similar reasoning, the probability of each of the other paths in also .25.

Figure 5.3. *TREE DIAGRAM*
The Possible Outcomes of the Tosses of Two Coins

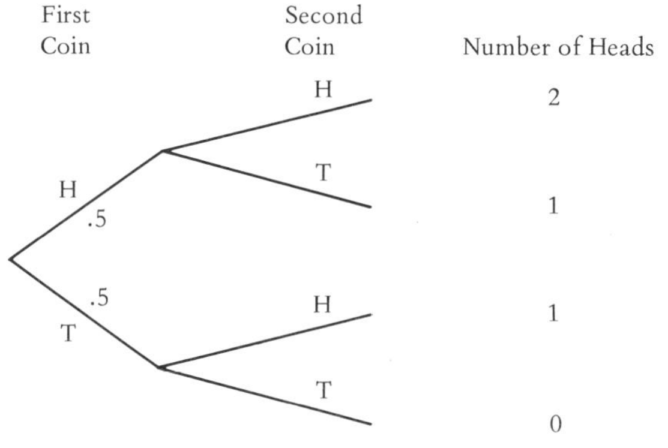

First Coin
Second Coin
Number of Heads
H
2
T
H
1
.5
.5
H
1
T
T
0

Table 5.1. *FOUR POSSIBLE OUTCOMES OF THE TOSSES OF TWO COINS*

PATH	DESCRIPTION	NO. OF HEADS	PROBABILITY
1	Head, Head	2	.25
2	Head, Tail	1	.25
3	Tail, Head	1	.25
4	Tail, Tail	0	.25

Associated with each path is a value of the random variable "number of heads."

One of the paths results in "2 heads." Since the path has a probability of .25, the probability of 2 heads is .25.

Two of the paths result in "1 head." Since each of these two paths has a probability of .25, the probability of 1 head is $2 \times .25 = .50$.

One of the paths results in "0 heads." Since that path has a probability of .25, the probability of 0 heads is .25.

From this analysis we can present the probability distribution of the random variable X = number of heads in Table 5.2. The results in Table 5.2 are applicable to the determination of the probability that 0, 1, or 2 investments will turn out to be successful.

The coin problem is an application of the binomial probability distribution.

The BINOMIAL PROBABILITY DISTRIBUTION or FUNCTION is defined by

$$f(x) = \frac{n!}{x!(n-x)!} p^x (1-p)^{n-x} \qquad 5.7$$

where n is a positive integer
$0 \leq p \leq 1$
$x = 0, 1, \cdots, n$

In Equation 5.7,

n is a fixed number of trials
p is the probability of a success on a trial
x is a value of the random variable "number of successes"

$$n! = n(n-1)(n-2) \cdots (2)(1)$$
$$x! = x(x-1)(x-2) \cdots (2)(1)$$
$$(n-x)! = (n-x)(n-x-1)(n-x-2) \cdots (2)(1)$$

The expression $n!$ is read "n factorial." The factorial of any integer n is the product of all the integers from n down to 1. Therefore,

Table 5.2. *PROBABILITY DISTRIBUTION OF THE RANDOM VARIABLE X*
X = Number of Heads in the Toss of Two Coins

x	$f(x)$
0	.25
1	.50
2	.25

$$3! = 3 \times 2 \times 1 = 6$$
$$4! = 4 \times 3 \times 2 \times 1 = 24$$
$$\text{etc.}$$

In writing out a factorial, you can continue writing factors from n down as far as you want, and then write a factorial sign, !, after the last number you wrote. Therefore you may write

$$4! = 4 \times 3!$$
$$37! = 37 \times 36 \times 35 \times 34!$$
$$\text{etc.}$$

This is a very useful device for computing the solution to division problems involving factorials. For example, if you want to evaluate the expression $9!/7!$ you need not compute the numerator and the denominator. You can simply rewrite the numerator in a form that readily cancels with the denominator. Therefore,

$$\frac{9!}{7!} = \frac{9 \times 8 \times 7!}{7!} = 9 \times 8 = 72$$

For any integer n,

$$n! = n(n - 1)!$$

From this expression we get

$$\frac{n!}{n} = (n - 1)!$$

If we let $n = 1$, then we get

$$\frac{1!}{1} = (1 - 1)! \qquad \text{so that} \qquad 0! = 1$$

Although your intuitions may balk at saying that $0! = 1$, this value is important to remember in working with the binomial distribution.

The conditions necessary and sufficient for using the binomial probability distribution are as follows.

CONDITIONS FOR USE OF THE BINOMIAL PROBABILITY DISTRIBUTION

1. There is a fixed number of trials, n.
2. On each trial there are 2 possible outcomes, a "Success" and a "Failure."
3. The probability of a success on any trial is a constant, designated by p.
4. A success or failure on any trial is independent of a success or failure on any other trial.

Equation 5.7 is not easy to use when n gets above 5 or 6. However, you can use Table I in Appendix A to find the binomial probabilities for any number n up to 20. Table I gives cumulative probabilities. For any value x, the table gives the probability that the random variable will be *less than or equal to* x. For example, when $n = 12$ and $p = .2$, and you want to find the probability that X is less than or equal to 4, then enter the table at $n = 12$, go across to the column headed .2, and down to the row marked 4. This probability is .9274. This is $F(4)$, or $P(X \leq 4)$, the probability that the random variable takes on a value less than or equal to 4. If you want to find the probability that X is *exactly* 4, then use Equation 5.6 to find

$$\begin{aligned} f(4) &= F(4) - F(3) \\ &= .9274 - .7946 = .1328 \end{aligned}$$

In order to become familiar with Table I, try using it for a problem that you can easily do using Equation 5.7 and compare the results. For example, to find the probability of getting 1 head on 2 tosses of a coin, Equation 5.7 gives

$$f(1) = \frac{2!}{1!1!}(.5)^1(.5)^1 = 2(.25) = .50$$

Table I (for $n = 2$, $p = .5$) gives

$$F(1) = .75 \qquad \text{and} \qquad F(0) = .25$$

then, by Equation 5.6,

$$f(1) = F(1) - F(0) = .75 - .25 = .50$$

The values that are "set" or held constant are known as the PARAMETERS of the distribution.

A binomial distribution can be specified only by setting values for the two parameters n and p. The parameter n can be any positive integer. The parameter p can be any real number between 0 and 1.

EXAMPLE 5.5 If a consulting firm submits 5 proposals to do research for government agencies, and each proposal has a probability of .1 of getting approved, what is the probability distribution of the number of successful proposals?

Solution: If we assume that approvals on the different proposals are independent of each other (which is more likely if the proposals all go to different agencies) then we can use the binomial distribution, because we have:

i. 5 trials ($n = 5$)
ii. 2 possible outcomes per trial
iii. constant probability of a success, which is .1
iv. independent trials

Using Table I, at $n = 5$, $p = .1$, read the cumulative probabilities as:

x	$F(x)$
0	.5905
1	.9185
2	.9914
3	.9995
4	.9999
5	1.0000

Use Equation 5.6 to find

$$f(0) = .5905$$
$$f(1) = .9185 - .5905 = .3280$$
$$f(2) = .9914 - .9185 = .0729$$
$$f(3) = .9995 - .9914 = .0081$$
$$f(4) = .9999 - .9995 = .0005$$
$$f(5) = 1.0000 - .9999 = .0001$$

These results could be of considerable interest to the firm. There is more than an even chance of getting no approvals. In addition to the concern about not getting any approvals, the firm might be equally wary about getting too many. If, for example, three or more proposals were approved, the firm might not have the resources to perform all the research they undertook to do. Such an eventuality would take place only if $X = 3$, 4, or 5. The probability of this happening would be (from Equation 4.5)

$$P(3 \cup 4 \cup 5) = f(3) + f(4) + f(5)$$
$$= .0081 + .00045 + .00001$$
$$= .00856$$

The chances are less than 1 in 100 of becoming overcommitted in this way. ■

EXAMPLE 5.6 The Inter City Transit Authority (ICTA) has 6 seats available for next Wednesday's 2:14 train. From past experience it is known that only 40% of those reserving seats actually show up. For this reason ICTA is tempted to reserve more than 6 seats in order to decrease the probable number of unsold seats. There is a risk in reserving too many, however, because of the possibility that more than 40% (possibly even all) of those reserving seats will show up this time.

Suppose ICTA reserves 9 seats. What is the probability that no more than 6 people will show up and claim their seats?

Discussion: The binomial probability distribution would be appropriate if we assume independence, that is, that the probability of one person's showing up does not depend on whether anyone else shows up. This assumption will tend to be closer to the truth if the reservations are made separately rather than for groups. Assuming the events are independent, we have the binomial parameters

$$n = 9 \qquad \text{and} \qquad p = .4$$

and the probability that X is no more than 6 is given from Table I as .9750. ■

When to use the Binomial Distribution

The assumptions necessary for using the binomial function have been presented, but students often have difficulty deciding when a true-to-life problem can be solved by use of this distribution. It may help to consider the prototype of all binomial problems—the flipping of a coin. On each flip there are two possible outcomes (call a head a "success" and a tail a "failure") the probability of a success p is given by the physical characteristics of the coin—we do not have to assume p is .5. The model would fit the case of warped or unbalanced coins which land heads with probabilities such as .2, .6, or .9 as well as the case of "fair" coins. The method of tossing a coin guarantees that the trials are independent. Therefore, if we flip a coin n times which has a probability p of landing heads each time, we can calculate the probability of getting exactly x heads by Equation 5.7.

Why illustrate the binomial distribution by the apparently irrelevant example of flipping coins? Because the flipping of coins can serve as a model with which you can compare some real managerial situations. You can ask yourself: is this situation like that of tossing a coin and counting heads?

EXAMPLE 5.7 Among all the workers in a large government department, it is known that:

40% are married
20% are single
15% are widowed
20% are divorced
5% are legally separated.

In a random sample of 4 employees, what is the appropriate probability distribution to use to find the probabilities of X, the number of divorced persons?

Discussion: Although there are more than 2 possible outcomes on a trial, we can make this a binomial problem by defining 2 possible outcomes: a success is "divorced" and a failure is "not divorced." If p is the probability of being divorced, then $p = .2$ and $(1 - p) = .8$. The random selection of employees enables us to assume the trials are independent. We therefore have the assumptions necessary for using a binomial distribution with parameters $n = 4$ and $p = .2$. The situation is like that of tossing a coin, which has a probability of .2 of landing heads 4 times, and counting the number of heads.

As an exercise, set up the distribution and compute the probabilities. ■

EXAMPLE 5.7 (continued) Suppose in this example we are interested in the number who are either single, widowed, or divorced (the union of 3 mutually exclusive events). We can use the binomial distribution to calculate the probability of any number of people falling in this category. The parameter p is found by summing the probabilities of the three mutually exclusive events: single, widowed, and divorced. This is $.20 + .15 + .20 = .55$. This gives us a binomial distribution with $n = 4$ and $p = .55$. ■

EXAMPLE 5.8 Two cards are drawn from a bridge deck consisting of 52 cards. Four of the cards in the deck are queens. Would the probability distribution of X, the number of queens in the sample of 2, conform to the binomial probability distribution? Why or why not?

Discussion: Unless you replaced the first card you drew and reshuffled the deck before drawing the second card, the probability of a queen on the second draw would depend on whether or not you got a queen on the first draw. The probability of a queen on the first draw is

4/52. If you got a queen on the first draw, the conditional probability of a queen on the second draw is only 3/51, but if you did not get a queen on the first draw the conditional probability of a queen on the second is 4/51.

The two events are not independent, and the binomial distribution is not appropriate. ■

EXAMPLE 5.9 If you take a random sample of two voters from the city of Tallahassee, what is the probability distribution of the random variable "number of Democrats"?

Discussion: This question is in principle like the one about the queens in the deck of cards. Unless you sample with replacement, the probability of the second person's being a Democrat *depends* on whether or not the first person was a Democrat. As a practical matter, however, the dependency is negligible. If a sample of 2 is taken from a city of 72,000 people, the probability of getting a Democrat on the second trial is for all practical purposes unaffected by whether or not you got a Democrat on the first trial. Therefore you can use the binomial distribution with parameters $n = 2$ and $p =$ the proportion of voters in Tallahassee who are Democrats. In general, when taking a small random sample from a large population, you can ignore the fact that you are not replacing, so that the trials are for all practical purposes independent. ■

EXAMPLE 5.10 In July the temperature in Slippery Rock gets above 30 degrees Centigrade on 10% of the days. Would the binomial distribution be an appropriate model for computing the probabilities of number of "30-degree-plus" days next July?

Discussion: You should not use the binomial distribution unless the assumption of independent events is approximately true. Daily weather conditions generally do not meet this assumption. Weather tends to come in spells, and a high pressure area can create a long series of hot (or cold) days. The conditional probability that tomorrow will be hot, given that today was hot, is generally going to be higher than the unconditional probability that tomorrow will be hot. In this example, the use of the binomial distribution would be at best only a crude approximation. ■

EXERCISES

5–8. Assume that 25% of all people who are qualified for a particular job classification are black. An agency has hired 4 people in the past year to fill jobs in this classification. If the agency filled those positions without regard to color, what is the probability distribution of the number of blacks hired?

5–9. Assume that a binomial probability distribution has parameters $n = 6$ and $p = .5$. Find the probability distribution of "number of successes" using Equation 5.7. Check to see that these results agree with those you find in Appendix Table I.

5–10. It has been estimated that 5% of the books acquired by a Public Library will be lost within two years. If the library purchased 16 books, what is the probability that in two years:

a) none of the 16 books will have been lost?
b) not more than one book will have been lost?
c) exactly one book will have been lost?
d) at least two books will have been lost?

What assumptions must you make in order to find these probabilities?

5–11. Forty percent of the criminal cases coming before Judge Crosman result in convictions. The judge will hear 5 cases tomorrow. If these 5 are considered a random selection of cases coming before Judge Crosman, what is the probability of

a) exactly two convictions?
b) not more than two convictions?
c) at least two convictions?
d) no convictions?
e) more than 3 convictions?

5–12. From the records of an agency in the Space Program that has awarded a large number of contracts to manufacturers it has been found that

5% of contracts have had no cost overruns.
10% of contracts have had overruns between 0 and 10%.
25% of contracts have had overruns of more than 10% but less than 40%.
30% of contracts have had overruns of at least 40% but less than 100%.
30% of contracts have had overruns of at least 100%.

If 15 contracts are selected at random, what is the probability distribution of

a) the number of contracts that have had cost overruns of less than 40%?
b) the number of contracts that have had cost overruns of up to 10%?

5–13. A dolphin has been trained to distinguish between the colors of two buttons, so that if she touches the "right" one she will be rewarded with fish, but if she chooses the "wrong" one she will get no reward. On each trial the positions of the two buttons are randomly set by a human experimenter. After a series of successful trials, she gets 17 out of 20 wrong.

a) What is the probability that she would get at least 17 wrong if she had no ability to discriminate between the colors?
b) On the basis of your answer to (a) what is your conclusion about the dolphin's ability to distinguish colors?

(It may be relevant to your answer to consider the fact that dolphins have a higher ratio of brain weight to body weight than do humans.)

5–14. In his drive to work Mr. Coveleski must go through 3 traffic lights. Each one has a probability of .4 of being green, and since there is no synchronization of lights, "hitting a green light" is independent of "hitting any other light" when it is green. What is the probability that, on a given day, when Mr. Coveleski drives to work,

a) all three of the lights will be green as he approaches them?
b) two lights will be green and one red or yellow?
c) one light will be green and two red or yellow?
d) all the lights will be red or yellow?

5.3 The Expected Value and Variance of a Discrete Random Variable

Would you be willing to pay $1.00 for a lottery ticket if there were a $25,000 grand prize and 4 $10,000 runner-up prizes, and there were 100,000 tickets sold? Although any individual's answer to this question might reflect personal tastes, optimism, views on gambling, needs, etc., your answer might also be related to the answer to another question: How much does a ticket pay off on the average?

The mean payoff per ticket is given by

$$\overline{X} = \frac{\Sigma X}{n} = \frac{\$25{,}000 + \$10{,}000 + \$10{,}000 + \$10{,}000 + \$10{,}000}{100{,}000}$$
$$= \$.65$$

The mean winnings are only 65 cents, so that on the average those who paid $1.00 for tickets will lose 35 cents a ticket, and the promoter of the lottery would have made an average (mean) profit of 35 cents a ticket, or a total of $35,000.

This average or mean payoff per ticket, $\overline{X}$, is the mean value *after the fact,* after you have observed the number of tickets sold and the amount paid out in prizes. In this example, however, we are concerned with a mean computed before the fact, in anticipation of or in expectation of a certain number of tickets sold and a certain amount paid out in prizes. A mean so based on anticipations or expectations is known as an expected value.

> The EXPECTED VALUE OF A DISCRETE RANDOM VARIABLE X is defined to be
>
> $$\mu = E(X) = \Sigma x P(X = x)$$
>
> or
>
> $$= \Sigma x f(x) \qquad \textbf{5.8}$$
>
> where the summation is taken over all values of X.

The Greek letter μ (pronounced "mew") is a shorter notation for $E(X)$.

The expected value is found by taking each value x of the random variable, multiplying it by its probability $f(x)$, and summing these products. In the case of the lottery tickets the expected value of the amount of winnings per ticket is calculated in Table 5.3.

An expected value is very similar to a weighted mean, in which the weights of the values are the probabilities. The primary difference is that a weighted mean is based on recorded data, while the expected value is based on the future—the "average" we can expect.

EXAMPLE 5.11 On the basis of extensive past records, the probability distribution of the number of reported gas leaks in an hour (symbolized here by Y) for the United Gas Works has been computed to be

y	$f(y)$
0	.00
1	.01
2	.08
3	.25
4	.30
5	.22
6	.10
7	.03
8	.01

What is $E(Y)$, the expected number of reported leaks in an hour?
Solution: Use Equation 5.8 to compute

$$\begin{aligned} E(Y) &= (1)(.01) + 2(.08) + 3(.25) + 4(.30) + 5(.22) \\ &\quad + 6(.10) + 7(.03) + 8(.01) \\ &= 4.11 \text{ reported leaks} \end{aligned}$$ ■

Paradoxically, the "expected value" is not usually the value you most expect, as we ordinarily use the word "expect." We do not expect in

Table 5.3. *CALCULATION OF E(X) FOR WINNINGS OF A LOTTERY TICKET*

x	$f(x)$	$xf(x)$
0	$\frac{99{,}995}{100{,}000}$	0
10,000	$\frac{4}{100{,}000}$	.40
25,000	$\frac{1}{100{,}000}$	.25
		.65 $= \Sigma xf(x)$

any hour to receive 4.11 calls. But the expected value has an important "long-run" significance. If you play a game many times, or if you undertake similar ventures many times, in the long run your mean gain or loss will get very close to the "expected value" of the outcome of the game or venture.

EXAMPLE 5.12 Wildcat, Inc., is a very speculative and risky business enterprise. In fact, it loses money in more months than it makes money. It depends for its survival on occasional very profitable months. The probability distribution of Z, the monthly profits of Wildcat, is given by

z	$g(z)$
−$10,000	.60
0	.10
$10,000	.10
$50,000	.10
$100,000	.10

Would you say that Wildcat is a profitable company?

Discussion: The expected value of Wildcat's monthly profits are calculated in Table 5.4. From the table, we see that over a long period of time Wildcat will earn an average of $10,000 a month, in spite of the fact that both the mode and the median of its monthly profit over a long period of time would be negative. The expected value is the more relevant measure in the long run. ■

THE VARIANCE AND STANDARD DEVIATION OF A DISCRETE RANDOM VARIABLE

Just as the expected value of a random variable is a measure of the average that we expect in the long run, the variance of a random variable

Table 5.4. *CALCULATION OF EXPECTED PROFIT Wildcat, Inc.*

z	$g(z)$	$zg(z)$
−10,000	.6	−6,000
0	.1	0
10,000	.1	1,000
50,000	.1	5,000
100,000	.1	10,000
		10,000 = $E(Z)$

is a measure of how much we expect the random variable to deviate from that expected value.

> The VARIANCE OF A DISCRETE RANDOM VARIABLE X is written σ_X^2 or σ^2 and is defined to be
>
> $$\sigma^2 = \Sigma(x - \mu)^2 f(x)$$

5.9

When we are dealing with only one random variable, so that there is no confusion as to which variable we are defining the variance for, we generally leave off the name of the variable and write the variance as σ^2.

It can be shown that Equation 5.9 is equivalent to

$$\sigma^2 = \Sigma x^2 f(x) - \mu^2 \qquad \textbf{5.10}$$

Equation 5.10 is generally easier to use for calculation than is Equation 5.9, especially when the expected value μ is not an integer.

EXAMPLE 5.13 The Surething Corporation has the following probability distribution of its monthly profit Y.

y	$f(y)$
\$4000	.1
5000	.8
6000	.1

Compute the expected value and variance of Surething's profit.

Solution: These are calculated in Table 5.5.

How does the expected profit and variance of Surething compare with the expected profit and variance of Wildcat?

Solution: The expected profit of Surething is $E(Y) = \$5000$. The expected profit of Wildcat is $E(Z) = \$10{,}000$. The variance of

Table 5.5. *CALCULATION OF μ AND σ^2*
Profits of Surething, Inc.

y	$f(y)$	$yf(y)$	$y - \mu$	$(y - \mu)^2$	$(y - \mu)^2 f(y)$
4000	.1	400	−1000	1,000,000	100,000
5000	.8	4000	0	0	0
6000	.1	600	1000	1,000,000	100,000
	1	5000			200,000

$$E(Y) = \Sigma y f(y) = 5000$$
$$\sigma_Y^2 \Sigma(y - \mu)^2 f(y) = 200{,}000$$

Surething's profit is 200,000. The variance of Wildcat's profit is calculated in Table 5.6.

The variance of Surething's monthly earnings is 200,000; the variance of Wildcat's is 1,220,000,000, or over 6000 times as much! Although it seems obvious that Wildcat's earnings vary more than Surething's it seems incredible that they vary 6000 times as much. The exaggeration in the difference is due to the fact that the variance is based on *squared* deviations, and squaring tends to magnify differences. If you take the *square root* of the variance you obtain a measure that eliminates the distortion due to squaring. This measure is a better one for comparison of the variability of the earnings.

> The STANDARD DEVIATION OF A DISCRETE RANDOM VARIABLE X [denoted by σ_X or simply σ] is the square root of the variance of X, or
>
> $$\sigma = \sqrt{\sigma^2} = \sqrt{\Sigma(x - \mu)^2 f(x)}$$

5.11

The standard deviation of Surething's earnings is

$$\sigma_Y = \sqrt{200{,}000} = \$447.21$$

The standard deviation of Wildcat's earnings is

$$\sigma_Z = \sqrt{1{,}220{,}000{,}000} = \$34{,}928.50$$

which is about 78 times as much as Surething's.

The variance and standard deviation of a random variable are measures of variability, but they can also be looked upon as measures of risk or uncertainty. When the variance and standard deviation are low, the values of the random variable tend to be close to their expected value, so that we have a high degree of confidence that the random variable will take on a value near the expected value. On the other hand, if the vari-

Table 5.6. *CALCULATION OF VARIANCE*
Monthly Earnings of Wildcat, Inc.

z	$f(z)$	$z - \mu$	$(z - \mu)^2$	$(z - \mu)^2 f(z)$
−10,000	.6	−20,000	400,000,000	240,000,000
0	.1	−10,000	100,000,000	10,000,000
10,000	.1	0	0	0
50,000	.1	40,000	1,600,000,000	160,000,000
100,000	.1	90,000	8,100,000,000	810,000,000
				1,220,000,000

ance and standard deviation are high, then we know that the values of the random variable tend to deviate more from their expected value, so that the expected value will be a very unreliable indicator of the value that the random variable will take on in any single trial.

Consider again the cases of the two companies Wildcat and Surething. If you had your choice of a fixed share of the profits of either company, which would you choose?

Wildcat has the higher expected profit ($10,000 per month vs. $5000) but because of the high standard deviation, the $10,000 expected profit does not give a reliable picture of what you would gain or lose in any month. This value does not reveal the fact that 60% of the months you would be losing money, and that you may be bankrupt before you had any chance of realizing the long-run expected profit. The high risk of Wildcat may lead you to prefer Surething, even though in the long run your gains would be only half as much. However, a wealthy person or large investment company might be willing to go with Wildcat. Such a person or company can afford to take a long-range view, and ultimately gain more money with Wildcat.

The expected value (μ) and the variance (σ^2) are summary measures of a probability distribution of a random variable. For many kinds of decisions, it is not necessary to consider all possible values of the random variable and their probabilities. Knowledge of the general order of magnitude and the variability is often sufficient.

EXAMPLE 5.14 Find the expected value, variance, and standard deviation of the following probability distribution

x	$f(x)$
0	.2
1	.4
2	.3
3	.1

Solution: The calculations are shown in Table 5.7.

EXAMPLE 5.15 When a probability distribution is defined by a mathematical rule or formula, it is sometimes possible to calculate the expected value of the random variable from the formula.

Suppose that trains stop at a station every 30 minutes. A man unfamiliar with the train schedule arrives at the station. The probability distribution of the time he has to wait for the train (all times rounded to the next whole minute) is a uniform probability distribution given by

$$f(x) = \frac{1}{30} \qquad x = 1, 2, 3, \cdots, 30$$

Table 5.7 *CALCULATION OF EXPECTED VALUE, VARIANCE, AND STANDARD DEVIATION*

x	$f(x)$	$xf(x)$	x^2	$x^2f(x)$	$x - \mu$	$x - \mu^2$	$(x - \mu)^2f(x)$
0	.2	0	0	0	−1.3	1.69	.338
1	.4	.4	1	.4	−.3	.09	.036
2	.3	.6	4	1.2	.7	.49	.147
3	.1	.3	9	.9	1.7	2.89	.289
		1.3		2.5			.810

$$\mu = 1.3$$

(5.10) $\sigma^2 = \Sigma x^2 f(x) - \mu^2 = 2.5 - (1.3)^2 = .81$

(5.9) $\sigma^2 = \Sigma(x - \mu)^2 f(x) = .81$

$$\sigma = \sqrt{.81} = .9$$

Find the expected value of X.

Solution: The general formula for the expected value (Equation 5.8) gives

$$E(X) = \sum_{i=1}^{30} x_i f(x_i) = \sum_{i=1}^{30} \frac{1}{30} x_i$$

Any constant can be factored out of a summation sign. Doing this with the constant 1/30 yields

$$E(X) = \frac{1}{30} \sum_{i=1}^{30} x_i$$

There is a theorem that says that the sum of all the integers from 1 to n is given by $n(n + 1)/2$, so that

$$\sum_{i=1}^{30} x_i = \frac{30(31)}{2} = 465$$

Therefore

$$E(X) = \frac{1}{30}(465) = 15.5 \text{ minutes}$$ ■

The Expected Value of a Sum of Random Variables

Suppose you are in charge of a road-building project that consists of 3 tasks:

1. getting approval for the project
2. clearing the land
3. constructing the road

Task 3 cannot start until task 2 has been completed; task 2 cannot start until 1 has been completed.

Let X_1 be the time to complete task 1.
Let X_2 be the time to complete task 2.
Let X_3 be the time to complete task 3.
Let $E(X_1) = 5$ weeks.
Let $E(X_2) = 12$ weeks.
Let $E(X_3) = 20$ weeks.

You are interested in how long you expect to take to perform the whole project.

Let Z be the time to do the whole project. Then

$$Z = X_1 + X_2 + X_3$$

There is a theorem that the expected value of a sum of random variables equals the sum of the expected values of the random variables, or

$$E(X_1 + X_2 + \cdots + X_n) = E(X_1) + E(X_2) + \cdots + E(X_n) \qquad \textbf{5.12}$$

so that $E(Z) = 5 + 12 + 20 = 37$ weeks.

The Expected Value and Variance of the Binomial Distribution

The expected value of a random variable having a binomial probability distribution is given by

$$E(X) = np \qquad \textbf{5.13}$$

Equation 5.13 says that if you flip a fair coin 10 times, the expected number of heads is $10 \times .5 = 5$; or that if you write 16 proposals, each with a probability of .2 of getting funded, and if each one's being funded is independent of any other being funded, then the expected value of the number of proposals that get funded is $16 \times .2 = 3.2$.

The variance of a random variable that has a binomial probability distribution is given by

$$\sigma^2 = np(1 - p) \qquad \textbf{5.14}$$

If you write 16 proposals with a probability of .2 of getting funded, and you assume independence, then the variance of the number of successes is

$$\sigma^2 = 16 \times .2 \times .8 = 2.56$$

If you write twice as many proposals, then the variance of the number of successes will double to

$$\sigma^2 = 32 \times .2 \times .8 = 5.12$$

Suppose, however, you are interested in the *proportion* of the proposals that get funded. The proportion is given by x/n, where x is the number of successes. If you get 4 successes out of 16, then your proportion of successes is 4/16 = .25.

The expected value of the proportion is given by

$$E\left(\frac{X}{n}\right) = \frac{1}{n} \times E(X) = \frac{1}{n} \times np = p \qquad \textbf{5.15}$$

Equation 5.15 gives the nonsurprising result that the expected value of the proportion of successes is simply p, the probability of a success on a single trial.

The variance of the proportion is given by

$$\sigma^2_{X/n} = \frac{1}{n^2}\sigma^2 = \frac{np(1-p)}{n^2} = \frac{p(1-p)}{n} \qquad \textbf{5.16}$$

The variance of the *number* of successes is $np(1 - p)$. The variance of the *proportion* of successes is $p(1 - p)/n$. Notice that in the first case n is in the numerator, and in the second case n is in the denominator. This means that the *number* of successes tends to vary more as the sample size increases, while the *proportion* of successes tends to vary less as the sample size increases.

EXAMPLE 5.16 A random sample is taken of 10 people who have been on a treatment for heroin addiction for 6 months. If the probability of being cured by the treatment is .4, what is the expected value of the number cured and the expected value of the proportion cured?

Solution: Assuming the cases are independent, we have

$$E(X) = np = (10)(.4) = 4$$

$$E\left(\frac{X}{n}\right) = p = .4$$

What are the variance and standard deviation of the number cured and the variance and standard deviation of the proportion cured?

Solution: The variance of the number cured is given (from Equation 5.13) as

$$\sigma^2 = np(1 - p) = 10(.4)(.6) = 2.4$$

The standard deviation is given by the square root

$$\sigma = \sqrt{np(1 - p)} = \sqrt{2.4} = 1.55$$

The variance of the proportion cured is given as

$$\sigma^2_{X/n} = \frac{p(1 - p)}{n} = \frac{(.4)(.6)}{10} = .024$$

The standard deviation is given by the square root

$$\sigma_{X/n} = \sqrt{\frac{p(1 - p)}{n}} = .024 = .155$$

Notice in particular how large is the standard deviation of the proportion. If we assume that the probability is .95 of being within 2 standard deviations of the expected value, then we would have about 95% confidence that the proportion would lie within 2 × .155 = .310 of the expected value of .4. That is not very close, since .4 minus or plus .310 goes all the way down to .09 and up to .710. Although this is only an approximation to the true binomial probabilities, which you can check from Table I, the standard deviation of .155 gives you an idea of the tremendous range of values the proportion in a small sample of 10 can take on. ■

EXAMPLE 5.17 From the same population as in Example 5.16, take a sample of 100 instead of 10. Let Y be the number cured.

Find the expected value of Y and the expected value of the proportion of those cured, $Y/100$.

Solution:

$$E(Y) = np = 100 \times .4 = 40$$
$$E\left(\frac{Y}{100}\right) = p = .4$$

Find the variance and standard deviation of Y and the variance and standard deviation of $Y/100$, the proportion cured.

Solution:

$$\sigma^2_Y = np(1 - p) = 100 \times .4 \times .6 = 24$$
$$\sigma_Y = \sqrt{np(1 - p)} = \sqrt{24} = 4.9$$
$$\sigma^2_{Y/100} = \frac{p(1 - p)}{n} = \frac{(.4)(.6)}{100} = .0024$$
$$\sigma_{Y/100} = \sqrt{\frac{p(1 - p)}{n}} = \sqrt{.0024} = .049$$ ■

In comparing these two examples, notice what has changed as the sample size increased from 10 to 100

The expected number of successes has increased by a factor of 10.
The variance of the number of successes has increased by a factor of 10.
The standard deviation of the number of successes has increased by a factor of $\sqrt{10}$.
The expected proportion of successes has remained constant.
The variance of the proportion of successes has been reduced to 1/10 of before.
The standard deviation of the proportion of successes has been reduced to $1/\sqrt{10}$ of before.

These examples illustrate the benefit of larger samples. If someone were taking a sample to estimate the rate of cure, or the probability of success, then the expected value of the sample proportion would be the universe parameter p, no matter what the sample size. But the standard deviation of the proportion in the sample you take from the parameter p gets smaller as n gets larger. This means that with a larger sample you can generally expect your sample proportion to be closer to p than with a small sample. In Example 5.17 the standard deviation of the proportion from a sample of 100 was .049. Two standard deviations would be .098, or 9.8 percentage points. With a probability of about .95 the sample proportion will be within 9.8 percentage points of the parameter p. You might still consider this too big a range, but it is a vast improvement over the range for a sample of size 10.

Chapters 6 and 7 will consider more thoroughly the questions of taking samples from a binomial distribution. The subject was introduced at this point to illustrate an important use of the standard deviation, that of telling you how much trust you can have in your sample.

EXERCISES

5–15. Mr. Waddell operates the only ambulance in the town of Leftbank. The probability distribution for the number of calls (y) for Mr. Waddell's ambulance in a day is

y	$g(y)$
0	.7
1	.1
2	.1
3	.1

Find the expected value, variance, and standard deviation of the number of calls.

5–16. Given a situation in which you have a 50–50 chance of selecting a man or a woman, the probability distribution of X, the number of women selected if you selected one person, would be

x	$f(x)$
0	.5
1	.5

a) Find the expected value of X. Can you give this value a meaningful interpretation?

b) Find the variance and standard deviation of X.

5–17. A state lottery awards the following prizes each week.

1st Prize of $50,000
5 prizes of $10,000 each
50 prizes of $1000 each
500 prizes of $100 each
1000 prizes of $50 each
5000 prizes of $10 each

Assume that there are 500,000 tickets sold each week.

a) What is the expected value of a ticket?

b) If tickets are sold for $1.00 apiece, what is the expected value *to the state* of a ticket that is sold?

c) How much net revenue will the state receive each week from the lottery?

5–18. A bus is scheduled to make the round trip from 69th Street Terminal to Mount Airy and back in 84 minutes. From the recorded times of a large number of trips, the following probability distribution of the time (t) for a trip was computed.

t	$f(t)$
80	.1
84	.6
88	.1
92	.1
96	.1

a) Find the expected value and the variance of the time to make a round trip.

b) If the bus is scheduled to make 5 round trips a day, with no delay scheduled between trips, do you think the bus will be on schedule by the end of the day? Why or why not?

5–19. If you roll a fair die, the probability distribution of the number of spots that appear is a uniform discrete probability distribution given by

$$f(x) = \frac{1}{6} \qquad x = 1, 2, 3, 4, 5, 6$$

Calculate the expected value and variance of X.

5–20. Large-scale projects, such as construction, research and development, etc., are often planned by use of a PERT (Program Evaluation and Review Technique) network. This technique looks at the paths through the network, a path consisting of a series of time-consuming activities that follow one another in a required order.

Suppose a path consists of 4 activities, as follows:

Activity	*Expected Time*
Do basic research	10 weeks
Write 1st draft	5 weeks
Write 2nd draft	4 weeks
Type final copy	1 week

Assume the activities must be done in the order shown, and that the times to complete them are independent of each order. Find the expected value of the time to do all 4 activities.

5–21. One hundred people get on an airplane after having checked their luggage. Let X = the weight of a person (including her/his luggage) and $E(X) = 200$ pounds. Assuming that the weights of different individuals are independent of each other, what is the expected value of the total weight of the 100 people?

5–22. We will assume that the probability that a prisoner, after being granted a parole, will return to prison within a year is .4.

a) What is the expected value of the number that will return within a year out of 16 prisoners paroled this month?
b) What are the variance and standard deviation of the number of prisoners returning?
c) What are the variance and standard deviation of the proportion of prisoners returning?

Summary of Chapter

In many practical situations a person is likely to be concerned not so much with all the possible outcomes of an experiment as with the values of a random variable that might occur. The concept of the probability distribution of a random variable was discussed.

Of particular importance is the binomial distribution. It is important to know not only how to apply the binomial distribution but also to be aware of the conditions necessary for its use.

A probability distribution can be described by the expected value of a random variable and the variance of a random variable. The expected value is a "long-run" expectation and not necessarily the outcome one would expect on a particular trial. The variance and standard deviation are measures of the degree of uncertainty or risk.

REVIEW EXERCISES FOR CHAPTER 5

5–23. Which of the following fulfill the mathematical requirements for a discrete probability function?

a) $f(x) = \frac{x}{10}$, $x = 1, 2, 3, 4$.

b)

y	$f(y)$
1	.1
$\sqrt{2}$	.35
2	.1
$2\sqrt{2}$	.35
4	.1

c) $f(t) = t - 1$, $t = 0, 1, 2$.

d) $g(y) = \frac{3!}{y!(3-y)}(.4)^y(.6)^{3-y}$, $y = 0, 1, 2, 3$.

e) $h(z) = \frac{2!}{(2-z)!}(.5)^2 \frac{1}{z!}$, $z = 0, 1, 2$.

f) $X = 1$ if you get paid this month.
$X = 0$ if you do not get paid this month.

g) $f(x) = \frac{1}{10}$, $x = 0, 1, 2, \cdots, 10$.

5–24. Representatives from 12 Federal Reserve Districts are holding a meeting at 10:00 A.M. in Washington tomorrow. All are arriving in the morning on separate planes. If the probability of each plane's being more than half an hour late is .2, and the lateness of each of the 12 planes is independent of the others, what is the probability that

a) none of the twelve planes will be late?
b) not more than 2 planes will be late?
c) not more than 5 planes will be late?
d) more than 5 planes will be late?
e) exactly two planes will be late?

5–25. A baseball player has a .300 batting average. This means that she gets a hit on the average 30% of the times she comes to bat. Use the binomial probability distribution to determine the probability that in the next 6 times at bat she will get

a) 0 hits?
b) 1 or 2 hits?
c) more than 3 hits?
d) 6 hits?

5–26. Given the following probability distribution of X

$$f(x) = \frac{1}{5} \qquad x = 1, 2, 3, 4, 5$$

Find the cumulative function $F(x)$.

5–27. An economist assigns the following probability distribution to r, the rate of inflation one year hence.

r (%)	$f(r)$
−2	.05
0	.05
2	.10
4	.10
6	.15
8	.15
10	.15
12	.15
14	.10

Find the expected value and variance of r.

5–28. The frequencies of the ages of Miss America contest winners from 1959 through 1976 are:

Age (X)	*Frequency*
18	3
19	3
20	3
21	6
22	1
23	2
	18

Using the relative frequencies as the probability distribution $f(x)$ for the ages of future Miss Americas, find the following:

a) the probability that $X = 18$
b) the probability that $X = 20$
c) $f(24)$
d) $F(19)$
e) $P(19 \leqslant X \leqslant 21)$

f) $F(24)$
g) the expected value of X

Discuss the merits and faults in using a set of 18 observations as the basis for determining a probability distribution.

5–29. If 10% of the population will get influenza this winter, and the economic cost per case of influenza is $60, what is the expected economic cost of the influenza epidemic in a city of 250,000 people who constitute a random sample of the population?

5–30. Suppose a vaccine was developed that reduced the rate of influenza this winter from 10% to 6%. What would now be the expected economic cost of the influenza epidemic in the same city mentioned in Exercise 5–29 if everyone in the town were vaccinated? On the basis of a comparison of your answers to this and problem 5–29, what is the economic benefit of the vaccine?

SUGGESTED READINGS

HAMBURG, MORRIS. *Statistical Analysis for Decision-Making,* 2nd ed. New York, N.Y.: Harcourt, Brace, Jovanovich, Inc., 1977. Chapter 3.

HARNETT, DONALD L. *Introduction to Statistical Methods,* 2nd ed. Reading, Mass.: Addison-Wesley Publishing Company, 1975. Chapters 3 and 4.

JOHNSON, ROBERT. *Elementary Statistics,* 2nd ed. North Scituate, Mass.: Duxbury Press, 1976. Chapter 6.

MOSTELLER, F., R. ROURKE, and G. THOMAS. *Probability with Statistical Applications,* 2nd ed. Reading, Mass.: Addison-Wesley Publishing Co., Inc., 1970. Chapter 5.

SIX

The Normal Probability Distribution

Outline of Chapter

6.1 Continuous Random Variables and Their Probability Distributions

Measures of quantity, length, and time are usually continuous random variables. Their probability distributions have features similar to those of discrete random variables, but they differ in certain respects.

6.2 The Normal Distribution and the Table of Normal Areas

The bell-shaped normal distribution is a two-parameter distribution that approximately fits a wide variety of situations. By converting a normally distributed random variable to a standard normal variable, you can use the table of normal areas to determine probability.

6.3 The Normal Distribution as an Approximation to the Binomial Distribution

When the number of trials get too large to use a binomial table, you can use the table of normal areas as an approximation.

Objectives for the Student

1. Distinguish between discrete random variables and continuous random variables.
2. Be able to transform a value of a normally distributed random variable to the corresponding value of the standard normal variable.
3. Be able to use the table of normal areas to find probabilities that a random variable will take on a value in a specified interval.
4. Be able to use the table of normal areas to find approximate probabilities for the binomial distribution when n is large.

Chapter 5 discussed discrete random variables, the values of which can usually be determined by counting. This chapter will discuss continuous random variables, the values of which are usually found by measuring. Like discrete random variables, continuous random variables can be distributed according to a variety of probability models. The most important of these is the normal distribution. Although the formula for the normal distribution is very difficult to apply, by use of a simple transformation you can use a table to determine probabilities for any normally distributed random variable.

Chapter 5 introduced the binomial distribution and a table for finding probabilities for small values of n. When the number of trials is too large to use the Table of Binomial Probabilities, the normal distribution is a good enough approximation to enable you to use it to estimate binomial probabilities.

6.1 Continuous Random Variables and Their Probability Distributions

The United States population by age group, according to the 1970 Census, is given in Table 6.1.

How could you use such data to determine probabilities such as the probability that a randomly selected person was 43 years old on her/his last birthday, the probability that a person is below 3 years of age, etc.? To answer questions like these you need to make certain assumptions about the distribution of the ages within age classes. One assumption that you might make is that the ages of individuals are uniformly distributed throughout each age group. This assumption is reflected if we draw a histogram of the data as in Figure 6.1.

The Y axis of Figure 6.1 is a frequency *density*. We can call it the "frequency per 5-year interval." The X axis represents the continuous variable X = age; and the probability of being between any two ages x_1

Table 6.1. *POPULATION OF U.S. BY AGE GROUPS, 1970*

Age Group	Number of Persons (thousands)
Under 5	17,154
5 to 9	19,956
10 to 14	20,788
15 to 19	19,070
20 to 24	16,371
25 to 29	13,477
30 to 34	11,430
35 to 39	11,107
40 to 44	11,981
45 to 49	12,116
50 to 54	11,104
55 to 59	9,973
60 to 64	8,617
65 to 69	6,992
70 to 74	5,444
75 to 79	3,835
80 to 84	2,284
85 and over	1,511

Source: U.S. Census Bureau

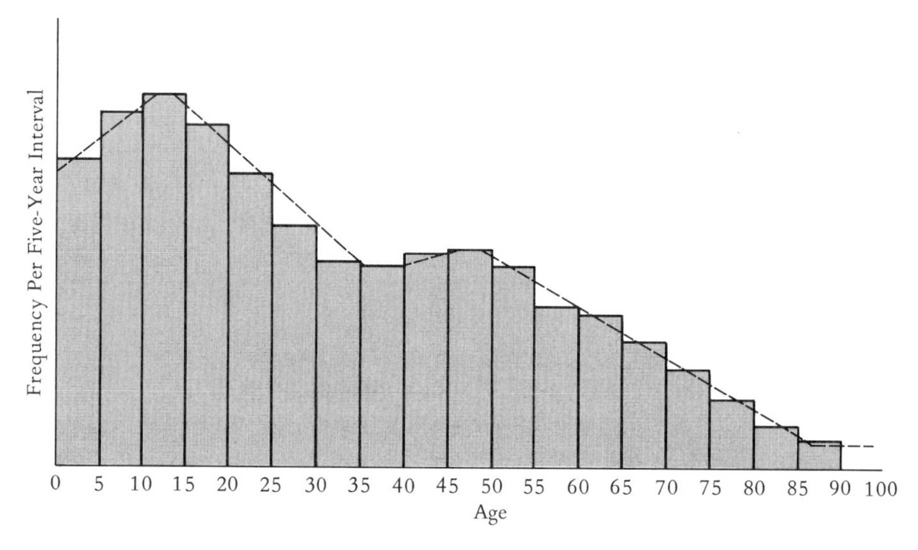

Figure 6.1. *HISTOGRAM: U.S. POPULATION BY AGE GROUPS, 1970*

and x_2 is the ratio of the *area* between the two points representing the ages and the total area of the rectangles, or

$$P(x_1 < X < x_2) = \frac{\text{area of rectangles between } x_1 \text{ and } x_2}{\text{total area of all rectangles}}$$

This probability is, of course, based on the assumption of uniform frequency density within each interval, an assumption that is only a crude approximation to reality. It is not reasonable to assume that the density remains constant over each five-year interval and then takes an abrupt jump or drop at the beginning of the next interval. It is reasonable to expect that the probabilities will be better approximated if we plot a smooth curve which rises and falls more gradually than the heights of the rectangles. Such a smooth curve is drawn with a dotted line in Figure 6.1. Although this curve is only a free-hand drawing representing an estimate, it is probably more realistic than the histogram in estimating the true probability.

If we randomly select an individual from this population, the smooth curve will help us to determine the probability that the individual will be between any two given ages. If we define the area under the entire curve as equal to one, then the area under the curve between any two points will be the probability that a value is between those two points. In Figure 6.2, the shaded area between 12.5 and 26 represents the probability that an individual randomly selected from the population given in Table 6.1 will be between 12.5 and 26 years of age.

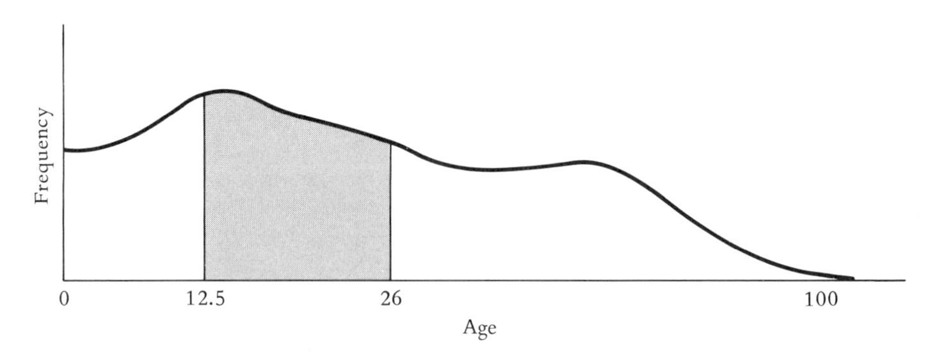

Figure 6.2. *PROBABILITY THAT A RANDOMLY SELECTED INDIVIDUAL IS BETWEEN 12.5 AND 26 YEARS OF AGE*
Probability = Shaded Area ÷ Entire Area

A solid curve such as that shown in Figure 6.2 is the general method of displaying a probability function of a continuous random variable.

> A CONTINUOUS RANDOM VARIABLE is a random variable that can take on infinitely many values in an interval. There is therefore an infinite number of possible values that a continuous random variable can take on in the interval.

To determine whether a random variable is discrete or continuous, ask yourself: How many possible values can the random variable take on between two designated points? If the number is finite, the random variable is discrete; if the number is infinite, the random variable is continuous.

EXAMPLE 6.1 Is the amount of time that a person requires to travel by public transportation between two points in a city a discrete or a continuous random variable?

Discussion: In concept, time is a continuous random variable. There are an infinite number of possible times, even within a fixed interval such as 30 to 40 minutes. But if we have operationally defined time as a recording "to the nearest minute," this variable (call it "recorded time") is a discrete random variable. There are only 11 possible recorded times between 30 and 40 minutes inclusive. ■

If a random variable is discrete, but the units are very small, it is often more convenient to treat the random variable as continuous. For example, income is technically a discrete variable, since there is only a finite number of incomes between any two values, assuming income cannot be expressed in smaller units than cents. But when the units are as small as that, most analyses involving income are greatly facilitated if we treat the variable as continuous. It is a little like sand—it is made up of discrete particles, but for most practical purposes you treat it as a continuous substance.

There is a serious problem in representing a probability distribution of a continuous random variable by a free-hand curve. The probabilities can be determined only be measuring the areas under this curve, and measurement of such areas is not always easy. It would be much more satisfactory if we could assume a mathematical equation for the curve, so that we could more readily determine the probabilities.

> The probability function of a continuous random variable is known as a PROBABILITY DENSITY FUNCTION. The probability density function $f(x)$ can be represented by a curve, and in many cases can be determined by a mathematical formula. The area under this curve is defined to be equal to 1.

The value of a probability density function is *not* a probability. The area under the curve between any two values of the random variable is the probability that the random variable will fall between these two values.

EXAMPLE 6.2 A fire drill is scheduled to take place between 11:00 and 12:00 on Thursday morning. If any moment during this hour is equally probable for the drill to start, the continuous random variable T, the time of the start of the fire drill, is said to have a *uniform probability density*, or a *uniform probability distribution*.

Figure 6.3 illustrates the uniform distribution of probabilities. Letting the entire curve (which is in this case a rectangle) have an area of one, the probabilities of the drill coming between any two selected points of time, such as 11:20 and 11:27, is given by the area of the curve between those two points.

The curve in this example has the formula

$$f(t) = 1/60 \qquad 0 \leqslant t \leqslant 60$$

where t is time in minutes past 11:00.

Find the probability that the fire drill will be between 11:20 and 11:27.

Solution: The shaded area in Figure 6.3 represents the probability.

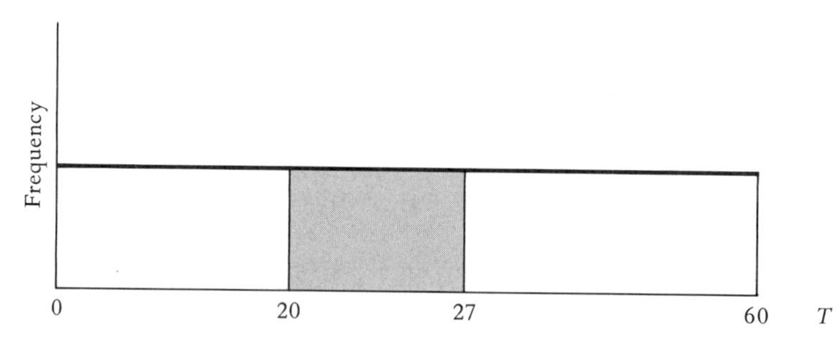

Figure 6.3. *MEANS OF FINDING PROBABILITIES FOR A UNIFORM DISTRIBUTION Probability That $20 \leq T \leq 27$ = Shaded Area ÷ Total Area*

Since the "curve" is a rectangle, the area between these points is a rectangle whose length is 7 and whose height is 1/60. The probability is $7 \times 1/60 = 7/60$.

Find the probability that the time for the fire alarm is before 11:50.

Solution: In this case we want the cumulative probability $F(50)$, or $P(T \leq 50)$. This is the area of the rectangle to the left of 50 in Figure 6.3. It has a length of 50, a height of 1/60, and, hence, an area of 5/6. The probability is 5/6. ■

EXERCISES

6–1. A summer drought has destroyed a large proportion of the tomato crop over a section of the country. A survey of the crop damage by farm led to the following estimates.

Damage to Crop (%)	*Number of Farms*
At least 0 but less than 20	12,000
At least 20 but less than 40	35,000
At least 40 but less than 60	25,000
At least 60 but less than 80	21,000
At least 80	7,000
	100,000

a) Plot a histogram showing these estimates.
b) On the basis of the histogram, find the probability that a randomly selected farm sustained between 30 and 50% damage.
c) What is the probability that a randomly selected farm sustained more than 90% damage?

d) Fit a smooth curve to the histogram that gives what seems to you to be a more realistic estimate of the probability distribution.
e) On the basis of the curve that you fitted in part (d), estimate the probability that a randomly selected farm sustained more than 90% damage. Compare this answer to your answer to part (c).

6–2. A housing project houses only families with incomes between \$5000 and \$7000 a year. Assume that any income within that range is equally probable. Draw a graph of the probability density of the incomes of families living in this project, and find the probability that a randomly selected family has an income between \$5300 and \$5700 a year.

6–3. Which of the following random variables would generally be treated as discrete, and which as continuous random variables?

a) number of persons per family
b) the time someone arrives at work as measured by a timeclock
c) the diastolic blood pressure of volunteers for an experiment
d) the proportion of a set of 12 subjects who score over 60% in a test
e) the proportion of a set of 142,738 school pupils who score over 60% in a test.

6.2 The Normal Distribution and the Table of Normal Areas

You may have observed that most adult females are between 5 and 6 feet tall, with the greatest concentration of heights a little below midway between these two values. The most common height is fairly close to 5′5″; most women are within 3 or 4 inches of that height on either side, while very few are more than 6 inches below that height, and an approximately equal number are more than 6 inches above that height. If you drew a curve to represent the probability density of the heights of adult women in the United States, you would make the curve high in the neighborhood of 5′5″ and let it taper off somewhat equally in both directions as you get farther from 5′5″. The result would be a bell-shaped curve, such as the one shown in Figure 6.4.

The normal probability distribution has been found to be a reasonable approximation to the behavior of such phenomena as heights and linear dimensions of human beings and animals, the times required to perform certain tasks, etc. But the feature giving the normal distribution its greatest importance is the fact that many measures taken from large samples are approximately normally distributed, even when the population from which the sample was drawn is far from normal. Researchers working with large or moderately large samples are likely to be dealing constantly with the normal distribution.

The symmetrical bell-shaped curve, such as that shown in Figure

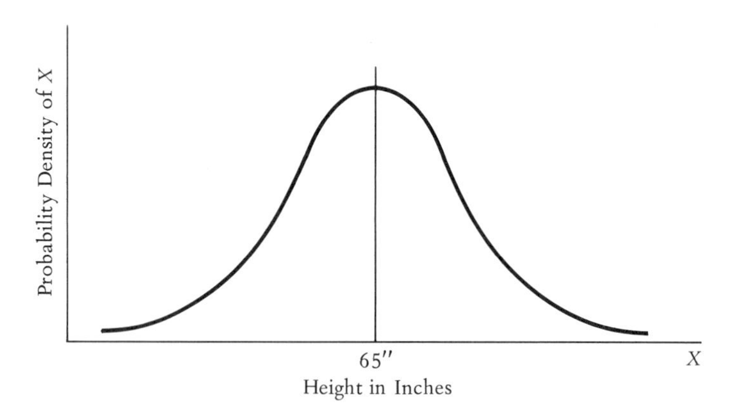

Figure 6.4. *PROBABILITY DENSITY*
Heights of Adult Women in U.S.

6.4, is a fairly good representation of the probability densities of a number of random variables with which you may be familiar, such as weights and strengths of products of manufacturing processes, aptitude test scores, as well as heights. A number of mathematical functions can also be plotted as bell-shaped curves. A particularly important one is the normal probability function, or the normal probability distribution.

There is a big advantage in being able to express a probability density with a mathematical formula. In Section 6.1, I mentioned that finding the probability that a continuous random variable takes on a value between 2 points involves finding the area under the curve between these 2 points. When the curve expresses a mathematical formula, then this area can be found by the process of taking a *definite integral* between the two points. This is often a laborious process and requires a knowledge of integral calculus. However, you will be spared this chore because in the case of the normal distribution this work has been done for you and been recorded in Table II in Appendix A. There is still one little problem, however, and that is that the normal distribution has two parameters, μ and σ. It would require voluminous sets of tables to give areas under the curve for many values of x for many different values of μ and for many different values of σ. Table II gives areas for only the normally distributed random variable whose expected value is 0 and whose standard deviation is 1.

> The STANDARD NORMAL VARIABLE (designated by z) is the normally distributed random variable whose expected value (μ) is 0 and whose standard deviation (σ) is 1.

It is possible to transform the value of any random variable X (no matter what its parameters μ and σ are) by the following transformation.

> The formula for transforming any normally distributed random variable into the standard normal variable is
>
> $$z = \frac{x - \mu}{\sigma}$$

6.1

EXAMPLE 6.3 Assume the height of randomly selected women is a normally distributed random variable with an expected value of 65 inches and a standard deviation of 3 inches. Express the height of a woman who is 68 inches tall in standard units.

Solution: Applying Formula 6.1, we have

$$z = \frac{68 - 65}{3} = +1$$

Express the height of a woman who is 60 inches tall as a standard normal variable.

Solution:

$$z = \frac{60 - 65}{3} = -1.67$$

The standard normal variable z can be interpreted as the number of standard deviations a value is above or below its expected value. A woman who is 68 inches tall is 1 standard deviation above the expected value; a woman who is 60 inches tall is 1.67 standard deviations below the expected value. ■

Finding Probabilities Using the Table of Normal Areas

Table II tells you the probability that the standard normal variable will be between 0 and any value you look up. This probability is indicated by the shaded area in the figure at the top of the table.

EXAMPLE 6.4 Find the probability that z is between 0 and 1.

Solution: The z values (to one decimal place) are shown in the left margin. Go down to $z = 1.0$. The second decimal place of z is given across the top. Go to the first column (headed by 0.00), since the second decimal place value of 1.00 is 0. The probability in the table is .3413. This is $P(0 \leqslant z \leqslant 1)$, or the probability that z is between 0 and 1. ■

EXAMPLE 6.5 Find $P(0 \leqslant z \leqslant 1.54)$.

Solution: Go down the left margin to $z = 1.50$ and over to the column headed by 0.04. The probability is 0.4382. ■

EXAMPLE 6.6 Find $P(-1.75 \leqslant z \leqslant 0)$.

Solution: Since the normal distribution is symmetrical, the probability that z is between any negative value and 0 is equal to the probability that it is between 0 and the corresponding positive value. Therefore

$$P(-1.75 \leqslant z \leqslant 0) = P(0 \leqslant z \leqslant 1.75)$$

Go down to row $z = 1.70$, across to column 0.05, and read the probability as .4599. ■

EXAMPLE 6.7 Find $P(z \leqslant 0)$.

Solution: Since z is symmetrically distributed around 0, the probability of its being less than or equal to 0, or $F(0) = .5000$. ■

EXAMPLE 6.8 Find $P(z \leqslant 1.54)$.

Solution: This is the probability of the union of two mutually exclusive events covered in Examples 6.5 and 6.7.

$$P(z \leqslant 1.54) = P(z \leqslant 0) + P(0 \leqslant z \leqslant 1.54) = .5000 + .4382 = .9382$$ ■

EXAMPLE 6.9 Find $P(-1.96 \leqslant z \leqslant 1.96)$.

Solution: Break the event $(-1.96 \leqslant z \leqslant 1.96)$ into two mutually exclusive events, and add the probabilities, according to Equation 4.10. We have

$$P(-1.96 \leqslant z \leqslant 1.96) = P(-1.96 \leqslant z < 0) + P(0 \leqslant z \leqslant 1.96)$$

Look up 1.96 in Table II. The probability is .4750. Therefore,

$$P(-1.96 \leqslant z \leqslant 1.96) = .4750 + .4750 = .9500$$

This result is a well-known one. A normally distributed random variable has a .95 probability of being within 1.96 standard deviations of its expected value. ■

EXAMPLE 6.10 Find $P(1.00 \leqslant z \leqslant 2.00)$.

Solution:

$$P(0 \leqslant z \leqslant 2.00) = .4772$$
$$P(0 \leqslant z < 1.00) = .3413$$

and

$$P(1.00 \leqslant z \leqslant 2.00) = P(0 \leqslant z \leqslant 2.00) - P(0 \leqslant z < 1.00)$$
$$= .4772 - .3413 = .1359 \quad ■$$

EXAMPLE 6.11 What z value is exceeded with a probability of .01?

Solution: This question is in the reverse order of the previous questions. It gives the probability and asks for the z values.

Let $z_{.01}$ be the value of z that we want. We are given $P(z > z_{.01}) = .0100$. The table gives $P(0 \leqslant z \leqslant z_{.01})$ for any z_1. We can find $P(0 \leqslant z \leqslant z_{.01})$ by

$$P(0 \leqslant z \leqslant z_{.01}) = P(z \geqslant 0) - P(z \geqslant z_{.01}) = .5000 - .0100 = .4900$$

Look up the probability .4900 in the body of the table. The closest value to it is .4901, which is associated with a z of 2.33. We have $P(0 \leqslant z \leqslant 2.33) = .4901$. Therefore $z_{.01} = 2.33$. ■

EXAMPLE 6.12 The length of time that it takes to drive to Friendship Airport from Ms. Didrickson's house in Baltimore is a normally distributed random variable with parameters $\mu = 42$ minutes and $\sigma = 5$ minutes.

If Ms. Didrickson starts to drive from her house to the airport now, what is the probability that it will take her more than 45 minutes to reach the airport?

Solution: If X is the time to drive to the airport, we want to know $P(X \geqslant 45)$. Using Equation 6.1, we have

$$P(X \geqslant 45) = P\left(z \geqslant \frac{45 - 42}{5}\right) = P(z \geqslant .6)$$

and since

$$P(z \geqslant .6) = P(z \geqslant 0) - P(z \leqslant .6) = .5000 - .2257 = .2743$$
$$P(X \geqslant 45) = .2743$$

If Ms. Didrickson has to be at the airport in 45 minutes she is running a risk of over 27% of being late.

Suppose Ms. Didrickson decides she does not want to run a risk of more than .05 of being late for her plane. How much time should she allow for her drive?

Solution: Let x_1 be the amount of time she should allow. Ms. Didrickson wants $P(X \geqslant x_{.05})$ to be no more than .05. If $P(X \geqslant x_{.05}) = .05$, then we can find $x_{.05}$ from Table II by transforming the X values to z values, and finding $z_{.05}$, the z value that will be exceeded with a probability of .05.

$$P(X \geqslant x_{.05}) = P(z \geqslant z_{.05}) = .0500$$

But

$$P(z \geqslant z_{.05}) = P(z \geqslant 0) - P(0 \leqslant z \leqslant z_{.05}) = .05$$
$$\text{and } .5000 - P(0 \leqslant z \leqslant z_{.05}) = .05$$

so that

$$P(0 \leqslant z \leqslant z_{.05}) = .4500$$

According to Table II, a z value of 1.64 gives a probability of .4495, and a z value of 1.65 gives a probability of .4505. Since you want the z value for a probability of .4500, split the difference, so that $z_{.05} =$ 1.645.

From Equation 6.1 it follows that

$$x_{.05} = \sigma z_{.01} + \mu = (5)(1.645) + 42 = 50.225 \qquad \textbf{6.2}$$

Ms. Didrickson should allow 50.225 minutes to go to the airport. If she prefers whole numbers, she had better round *upwards* to the next number and allow 51 minutes to make sure that the risk of being late is not more than .05. ■

EXAMPLE 6.13 In casting for the lead part in a play, the director is seeking a woman whose height is between 64 inches and 70 inches. What is the probability that a randomly selected woman will meet the height requirements?

Solution: Assume that women's heights are normally distributed with $\mu = 65$ inches and $\sigma = 3$ inches. Letting X be the height of a randomly selected woman, we want to know $P(64 \leqslant X \leqslant 70)$.

Using Equation 6.1, transform each member of this inequality into z values, so that

$$P(64 \leqslant X \leqslant 70) = P\left(\frac{64 - 65}{3} \leqslant z \leqslant \frac{70 - 65}{3}\right) = P(-.33 \leqslant z \leqslant 1.67)$$

Using the reasoning of Example 6.9, find

$$P(-.33 \leqslant z \leqslant 1.67) = P(-.33 \leqslant z \leqslant 0) + P(0 < z \leqslant 1.67)$$
$$= .1293 + .4525 = .5818$$

About 58% of women meet the height requirement. ■

EXAMPLE 6.14 The Commissioner of Parks is planning to move into the new county building as soon as construction is completed. The building is scheduled to be finished in 6 months, but the commissioner, having learned to think in terms of probability, considers the time to complete construction to be a random variable. Assume that the time to complete the building is a normally distributed random variable whose expected value is 6 months and whose standard deviation is 3 months; what is the probability that the commissioner will be able to move into the new building within a year's time?

Solution: The commissioner wants to know the probability that X is less than or equal to 12 months. If $\mu = 6$ and $\sigma = 3$, then

$$P(X \leqslant 12) = P\left(z \leqslant \frac{12 - 6}{3}\right) = P(z \leqslant 2)$$

Look up $P(0 \leqslant z \leqslant 2)$ in Table II. This is .4772.

$$P(z \leqslant 2) = P(z \leqslant 0) + P(0 < z \leqslant 2) = .5000 + .4772 = .9772$$

The commissioner has 97.72% confidence of moving within a year. ■

EXAMPLE 6.15 What are the probabilities that

(a) a normally distributed random variable will be within 1 standard deviation of its expected value?
(b) a normally distributed random variable will be within 2 standard deviations of its expected value?
(c) a normally distributed random variable will be within 3 standard deviations of its expected value?

Solution: (a) The question asks for $P(\mu - \sigma \leqslant X \leqslant \mu + \sigma)$. This can be transformed into standard form by subtracting μ from each member of the inequality and dividing by σ. We therefore want

$$P\left(\frac{\mu - \sigma - \mu}{\sigma} \leqslant \frac{X - \mu}{\sigma} \leqslant \frac{\mu + \sigma - \mu}{\sigma}\right) = P(-1 \leqslant z \leqslant 1)$$
$$= .3413 + .3413 = .6826$$

where .3413 is found from Table II for $P(0 \leqslant z \leqslant 1)$.

(b) By similar reasoning,

$$P(\mu - 2\sigma \leqslant X \leqslant \mu + 2\sigma) = P(-2 \leqslant z \leqslant 2) = .4772 + .4772 = .9544$$

(c) By the same line of reasoning,

$$P(\mu - 3\sigma \leqslant X \leqslant \mu + 3\sigma) = P(-3 \leqslant z \leqslant 3) = .49865 + .49865 = .99730$$

The solutions to this example form the basis for the statements that a normally distributed random variable (a) has more than a ⅔ probability of being within one standard deviation of its expected value; (b) has more than a .95 probability of being within 2 standard deviations of its expected value; and, (c) is almost certain to be within 3 standard deviations of its expected value. ■

You may have noticed that in the normal distribution, the probability that a random variable *is less than or equal to* ($\leq$) some value x is identical with the probability that it is less than ($<$) x. The addition of the equal sign has no effect on probabilities. This is because the probability that a continuous random variable takes on a particular value x is zero. This follows from the fact that probabilities for continuous random variables are represented by areas under a curve. The area over a single point will always be zero.

Although I have emphasized the importance of the normal distribution in statistics, I must warn you against overemphasizing it. Contrary to its name, it is not the "normal" way for most variables to behave, and in fact very few variables in the natural or social world have distributions very close to the normal. Only persons inexperienced in statistical work or unfamiliar with probability theory will assume that a variable is normal unless there is some empirical or theoretical justification for making this assumption.

The great importance of the normal distribution is in the applications discussed in the next section and in Chapter 7. If you take a large sample, the probability distribution of such statistics as the sample mean or sample proportion has a normal probability distribution, even though the distribution of individual values may be far from normal. In general, the larger the sample the less you have to worry about whether the distribution of individual values is normal. For those who use mostly large samples in their statistical work, the normal distribution is indeed the "normal" one to use.

EXERCISES

6–4. The amount of time that a patient spends in the recovery room after a certain type of surgery is a normally distributed random variable with parameters $\mu = 45$ minutes and $\sigma = 10$ minutes. Find the probability that a randomly selected patient will be in the recovery room

a) less than 35 minutes.
b) between 30 and 60 minutes.
c) more than 70 minutes.

6–5. Find the probability that the standard normal variable z will be

a) between 0 and 2, inclusive.
b) between 0 and 2, exclusive.
c) between -1.5 and $+2.5$.
d) between 1.2 and 2.2.
e) either less than -1.64 or more than 1.64.

6–6. Jenny Bishop is 5 feet 11 inches tall and her husband Max is 6 feet tall. Jenny claims that in terms of the standard normal variable she is taller than Max.

a) In what sense is Jenny right?

b) Assume that men's heights X are normally distributed with $E(X) =$ 69 inches and $\sigma_X = 4$ inches. Assume, as in other examples, that women's heights Y are normally distributed with $E(Y) = 65$ inches and $\sigma_Y = 3$ inches. Express Jenny's and Max's heights in terms of the standard normal variable.

***6–7.** What value of the standard normal variable z will be exceeded with a probability of .05?

***6–8.** A street is blocked off while a sewer line is being replaced. The time required to replace the sewer line is a normally distributed random variable with $\mu = 18.5$ days and $\sigma = 3$ days.

Mrs. Tyus is planning to install a new heating system in Building H. She does not plan to start the installation until she is "95% sure" that the sewer-line replacement has been completed. In how many days should Mrs. Tyus plan to start installation?

***6–9.** Between what two z values do the middle 99% of all z values fall?

6–10. The number of people coming into the Horsechester post office during each morning hour is an approximately normally distributed random variable with $\mu = 36$ and $\sigma = 6$. What is the probability that more than 45 people will enter the post office in a randomly selected morning hour?

6–11. The number of crimes in a month in South Horsechester is an approximately normally distributed random variable with $\mu = 250$ and $\sigma = 16$.

(a) What is the probability that the number of crimes in a randomly selected month will be between 255 and 270 inclusive?

(b) Since "number of crimes" is a discrete random variable, what justification, if any, is there for treating it as a continuous random variable?

6–12. The daily electric power consumption in a city is a normally distributed random variable with $\mu = 7{,}500{,}000$ kilowatt hours and $\sigma = 150{,}000$ kilowatt hours. What is the probability that consumption will be between 7,400,000 and 7,730,000 kilowatt hours on a given day?

6–13. A professor spends her lunch hour running 1500 meters on the university indoor track. Her time is a normally distributed random variable with μ equal to 4 minutes 50 seconds and σ equal to 5 seconds. What is the probability that in a randomly selected lunch hour her time will be

(a) less than 4 minutes 40 seconds?

(b) less than 5 minutes?

6–14. The price of bananas each day in the GIA Market is a normally distributed random variable with $\mu = 27.4$ cents per pound and $\sigma = 4$ cents. What is the probability that bananas will sell for between 19 and 22.5 cents a pound on a randomly selected day?

6–15. Miss Fitzgerald makes it a policy not to buy bananas unless the price is 20 cents a pound or less. If the price of bananas is distributed as in Exercise 6–14, what proportion of time will Miss Fitzgerald buy bananas?

***6–16.** Suppose that Miss Fitzgerald decides that she needs to buy bananas only 20% of the time, and only when "the price is right." What is the maximum price per pound she should pay?

***6–17.** The speeds of cars on Maple Shade Drive is a normally distributed random variable with μ equal to 66 miles per hour and σ equal to 4 miles per hour. The police have a policy of giving speeding tickets to 1% of the drivers. If only the speediest 1% of the cars are selected for ticketing, how fast can you go without getting a ticket?

6–18. The amount of time an ambulance takes to get from the Exit Number 4 of the Penn-Washington Highway to Aaron Burr Hospital is a normally distributed random variable with μ equal to 12 minutes and σ equal to 2 minutes. If a patient must get to the hospital in no more than 13 minutes in order to survive, what is the probability that the ambulance gets there in time?

6.3 The Normal Distribution as an Approximation to the Binomial Distribution

In Section 5.2 you studied the binomial probability distribution. Because the calculation of binomial probabilities using Equation 5.11 becomes very difficult as n gets above 6 or 7, you can use Table I to calculate binomial probabilities for n up to 20. But tables of binomial probabilities get very cumbersome as n gets large, so that for values of n greater than 20, once again you are going to need a new method of computing (or approximating) binomial probabilities.

Fortunately, the normal probability distribution comes to the rescue. It can be shown that

> As n gets large, the binomial probability distribution is closely approximated by the normal distribution whose expected value is np and whose standard deviation is $\sqrt{np(1 - p)}$.

The closeness of the fit can be illustrated by examining the probabilities of a binomial variable with parameters $n = 16$ and $p = .5$. Suppose, for example, that one half of all those arriving at an accident clinic require surgery. If you assume independence, what is the probability distribution of the number requiring surgery if 16 patients arrive at the clinic in an evening?

From Table I you can find the cumulative probabilities $F(x)$ for $n = 16$ and $p = .5$, and from these you can calculate $f(x)$. These probabilities and calculations are shown in Table 6.2.

Now compare these probabilities with those of a normally distributed random variable Y whose parameters are $\mu = np = 16(.5) = 8$ and σ

Table 6.2. *CALCULATION OF $f(x)$ FROM $F(x)$*

x	$F(x)$	$f(x)$
0	.0000	.0000
1	.0003	.0003
2	.0021	.0018
3	.0106	.0085
4	.0384	.0278
5	.1051	.0667
6	.2272	.1221
7	.4018	.1746
8	.5982	.1864
9	.7728	.1746
10	.8949	.1221
11	.9616	.0667
12	.9894	.0278
13	.9979	.0085
14	.9997	.0018
15	1.0000	.0003
16	1.0000	.0000

$$f(x) = F(x) - F(x - 1)$$

or

$$f(x) = \frac{16!}{x!(16 - x)!} (.5)^x (.5)^{16-x}$$

$= \sqrt{np(1 - p)} = \sqrt{(16)(.5)(.5)} = 2$. The following probabilities can be computed by the methods of Section 6.2.

Compare the probabilities in the last column of Table 6.2 with those in the last column of Table 6.3. For example, compare the binomial probability that $X = 6(.122)$ with the probability that the normal variable Y is between 5.5 and 6.5(.1210). Some of the corresponding probabilities are even closer. Notice that in comparing probabilities of the discrete random variable X to those of the continuous random variable Y, that the interval of Y consists of those y values that are closer to the corresponding x value than to any other x value. The y values in the interval $(5.5 \leq y \leq 6.5)$ are closer to 6 than to any other x.

Not all of the probabilities correspond as closely as this. However, if n were larger, the normal approximation would be a better fit.

The probability distributions of the discrete random variable X and the continuous random variable Y are superimposed in Figure 6.5. The probabilities $f(x)$ are shown as the tops of the discrete "stems." The probability density $g(y)$ is shown as the smooth curve.

Table 6.3. *CALCULATION OF PROBABILITIES FOR A NORMALLY DISTRIBUTED RANDOM VARIABLE ($\mu = 8$, $\sigma = 2$)*

Y VALUES	z VALUES (= (x − 8)/2)	PROBABILITIES FROM TABLE III
$Y < 2.5$	$z < -2.75$	$.5000 - .4970 = .0030$
$2.5 \leq Y < 3.5$	$-2.75 \leq z < -2.25$	$.4970 - .4878 = .0092$
$3.5 \leq Y < 4.5$	$-2.25 \leq z < -1.75$	$.4878 - .4599 = .0279$
$4.5 \leq Y < 5.5$	$-1.75 \leq z < -1.25$	$.4599 - .3944 = .0655$
$5.5 \leq Y < 6.5$	$-1.25 \leq z < -.75$	$.3944 - .2734 = .1210$
$6.5 \leq Y < 7.5$	$-.75 \leq z < -.25$	$.2734 - .0987 = .1747$
$7.5 \leq Y < 8.5$	$-.25 \leq z < .25$	$.0987 + .0987 = .1974$
$8.5 \leq Y < 9.5$	$.25 \leq z < .75$	$.2734 - .0987 = .1747$
$9.5 \leq Y < 10.5$	$.75 \leq z < 1.25$	$.3944 - .2734 = .1210$
$10.5 \leq Y < 11.5$	$1.25 \leq z < 1.75$	$.4599 - .3944 = .0655$
$11.5 \leq Y < 12.5$	$1.75 \leq z < 2.25$	$.4878 - .4599 = .0279$
$12.5 \leq Y < 13.5$	$2.25 \leq z < 2.75$	$.4970 - .4878 = .0092$
$13.5 \leq Y$	$2.75 \leq z$	$.5000 - .4970 = .0030$

This tendency of the binomial distribution to approach the normal enables us to use the table of normal areas to approximate binomial probabilities when the sample size n gets too large to use in a table of binomial probabilities. You must be careful, however, to take into consideration the fact that you are approximating a discrete distribution with a continuous distribution. If X is a discrete random variable, and you want to approximate its probabilities with a normal variable whose distribution is $g(y)$, then corresponding to any integer x you would want to find

Figure 6.5. *PROBABILITY CURVE*
Binomial Probability of X where $n = 16$, $p = .5$
Normal Probability Density of Y where $\mu = 8$, $\sigma = 2$

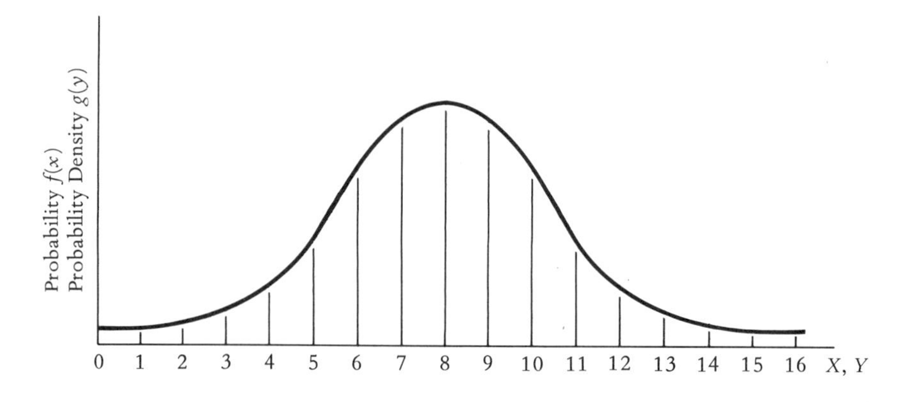

the probability that Y would be in the interval of y values that are closer to x than to any other integer. To approximate $f(6)$, for example, we would find $P(5.5 \leqslant y < 6.5)$; to approximate $f(100)$, we would find $P(99.5 \leqslant y < 100.5)$, etc.

EXAMPLE 6.16 If you want to approximate the probability that a binomial random variable X takes on the values 3, 4, or 5, for what corresponding values of the normal random variable Y would you find probabilities?

Solution: You are seeking the probability of the union of 3 mutually exclusive events. Therefore you add the probabilities $f(3) + f(4) + f(5)$.

$f(3)$ is approximated by $P(2.5 \leqslant Y < 3.5)$
$f(4)$ is approximated by $P(3.5 \leqslant Y < 4.5)$
$f(5)$ is approximated by $P(4.5 \leqslant Y < 5.5)$

Adding the 3 probabilities for both X and Y gives

$$f(3 \cup 4 \cup 5) = P(2.5 \leqslant Y < 5.5) \quad \blacksquare$$

EXAMPLE 6.17 If 10% of all claims for unemployment compensation contain at least one false statement, what is the probability that in a random sample of 16 claims for unemployment compensation either 3, 4, or 5 of them will contain at least one false statement? Find the probability first by using Table I, and then by using Table II to find the normal approximation.

Solution: We want probabilities for the same values of X and Y asked for in Example 6.16. We can find $f(3 \cup 4 \cup 5) = f(3) + f(4) + f(5)$ by subtracting $F(2)$ from $F(5)$. From Table I ($n = 16$, $p = .1$), the required probability is

$$F(5) - F(2) = .9967 - .7892 = .2075$$

For the normal approximation, we want $P(2.5 \leqslant Y < 5.5)$. Letting $\mu = np = 16(.1) = 1.6$ and $\sigma = \sqrt{np(1-p)} = \sqrt{(16)(.1)(.9)} = 1.2$, we obtain from Table II,

$$\begin{aligned} P(2.5 \leqslant Y \leqslant 5.5) &= P\left(\frac{2.5 - 1.6}{1.2} \leqslant z < \frac{5.5 - 1.6}{1.2}\right) \\ &= P(.75 \leqslant z < 3.25) = P(.75 \leqslant z) \\ &= .5000 - .2734 \\ &= .2266 \end{aligned}$$

The correct probability is .2075; the normal approximation is .2266. The approximation is not very close because of a combination of two reasons; the smallness of n and the fact that p is far from .5. Recall that the normal distribution is symmetrical. The binomial distribution is symmetrical when p is .5 and departs from symmetry as p departs from .5. For very high or very low values of p, a larger value of n is required to make the binomial symmetrical enough to be approximated by the normal distribution. The closeness of the normal approximation can be estimated by the minimum of either np or $n(1 - p)$. If this is below 3 (as in this example) the normal approximation is going to be a very crude one. But if this minimum is above 5, the approximation is going to be good enough for most purposes. ■

EXAMPLE 6.18 It has been determined that a person who has been arrested for drunken driving has a .2 probability of being involved in an accident resulting in bodily injury within a year, if permitted to drive. Of 64 persons who have been arrested for drunken driving, what is the probability that within a year of being permitted to drive at least 10 will be involved in accidents resulting in bodily injury?

Solution: "At least 10" calls for $f(10) + f(11) + \ldots + f(64)$. Approximate the binomial variable X with Y, a normally distributed random variable with $\mu = (64)(.2) = 12.8$ and $\sigma = \sqrt{(64)(.2)(.8)} = 3.2$. By the reasoning of Example 6.16, the normal probability corresponding to the binomial probability is

$$P(9.5 \leqslant Y < 10.5) + P(10.5 \leqslant Y < 11.5) + P(11.5 \leqslant Y < 12.5) + \ldots$$
$$= P(9.5 \leqslant Y) = P(Y \geqslant 9.5)$$

By Equation 6.1, this is equivalent to

$$P\left(z \geqslant \frac{9.5 - 12.8}{3.2}\right) = P(z \geqslant -1.03) = .3485 + .5000 = .8485$$

There is a probability of about .85 that at least 10 of the 64 will be involved in accidents resulting in bodily injury.

What is the probability that *more than* 10 of the 64 people will be involved in accidents that result in bodily injury?

Solution: This question differs from the previous one in that "more than 10" does not include 10. The binomial probability is $f(11) + f(12) + \ldots + f(64)$. The normal probability that approximates this is

$$P(10.5) \leqslant Y < 11.5) + P(11.5 \leqslant Y < 12.5) + \ldots = P(10.5 \leqslant Y)$$
$$= P(Y \geqslant 10.5)$$

This probability is equal to

$$P\left(z \geqslant \frac{10.5 - 12.8}{3.2}\right) = P(z \geqslant -.72)$$

Using Table II, we can find this probability to be .5000 + .2642 = .7642.

Notice that, as in any problem using a discrete binomial variable, the probability is affected by whether or not the equality sign is included or not. ■

EXAMPLE 6.19 If a political candidate is favored by 50% of the voters, what is the probability that in a random sample of 100 voters, between 45 and 55 will favor the candidate?

Solution: If "between 45 and 55" is interpreted to include the values 45 and 55, then the sum of the probabilities $f(45) + f(46) + \ldots + f(55)$ can be approximated by the sum of the normal probabilities $P(44.5 \leqslant Y < 45.5) + P(45.5 \leqslant Y < 46.5) + \ldots + P(54.5 \leqslant Y < 55.5)$, or $P(44.5 \leqslant Y < 55.5)$, where Y is a normally distributed random variable with $\mu = (100)(.5) = 50$ and $\sigma = \sqrt{(100)(.5)(.5)} = 5$.

$$P(44.5 \leqslant Y < 55.5) = P\left(\frac{44.5 - 50}{5} \leqslant z \leqslant \frac{55.5 - 50}{5}\right)$$
$$= P(-1.1 \leqslant z < 1.1) = .3643 + .3643 = .7286$$

What value of X, the number responding favorably to the candidate, would be exceeded with a probability of only .01?

Solution: We want x such that $P(X \geqslant x) = .01$. Using the normal approximation, we want to find a value $y_{.01}$, so that $P(Y \geqslant y_{.01}) = .01$. To do this, find a value $z_{.01}$ of the standard normal variable z, so that $P(z \geqslant z_{.01}) = .01$. If $P(z \geqslant z_{.01}) = .01$, then $P(0 \leqslant z < z_{.01}) = .4900$. Look up the probability .4900 in the body of Table II. The nearest probability is .4901, and the corresponding z value is 2.33. Therefore $z_{.01} = 2.33$, and $P(z \geqslant 2.33) = .01$. Now transform the 2.33 back to $y_{.01}$ by Equation 6.2,

$$y_{.01} = 5(2.33) + 50 = 61.65$$

What binomial value x corresponds to $y_{.01} = 61.65$? The closest value is 62. Since 62 is a little higher than $y_{.01}$, the probability of X exceeding 62 is going to be a little *less than* .01. ■

EXERCISES

6–19. If 40% of the employees in a large organization favor going on strike, what is the probability that at least 4 out of a randomly selected committee of 6 will be in favor of going on strike? Use the binomial distribution and the normal approximation to the binomial. Compare your answers to determine how good an approximation the normal distribution is.

6–20. If 25% of the people are silent, what is the probability that in a sample of 20 people you will find a silent majority?

6–21. If the President's Council on Physical Fitness correctly claims that only 30% of America's young men and women are physically "fit," what is the probability that in a randomly selected group of 84 of America's young men and women not more than 30 will be physically fit?

6–22. In a randomly selected 9th grade class of 30 pupils, what is the probability that more than 20 will be girls? Assume that the probability distribution of "number of girls" is a binomial distribution with $p = .5$.

On the basis of your answer to this question, what conclusion might you draw if you found 21 girls in a randomly selected class of 30 pupils? This is the kind of question with which we will be concerned when we test statistical hypotheses.

6–23. What is the probability that in a randomly selected school of 1000 pupils there will be more than 550 girls?

6–24. An agency is planning to increase its staff by 10 employees. From past experience it is known that only 60% of new employees are still working with the agency a year after they were hired. Assuming that every employee staying at least a year is independent of what any other employee does, find the probability that more than 10 new employees will be working for the agency a year hence if 15 new employees are hired now.

6–25. In a study of the therapeutic benefit of one's experience in a state prison, it was found that

10% of prisoners are helped by the prison experience;
60% of prisoners are harmed by the prison experience;
30% of prisoners are unaffected by the prison experience.

Out of 100 prisoners, what is the probability that

a) at least 20 are helped by the prison experience?
b) at least 50 are harmed by the prison experience?

6–26. A random sample of 150 voters was asked which candidate they favored. If 60% of all voters favor Miss Evert, between what two sample values would you expect, with a probability of .95, the number in the sample who favor Miss Evert to be?

6–27. If 40% of the heads of planning agencies have gray hair, what is the probability that out of 22 heads of agencies not more than 10 will have gray hair?

Summary of Chapter

The concept of a continuous random variable was discussed. Some random variables, even though they are discrete, can vary by very small increments. It is often useful to treat such random variables as continuous in statistical analysis.

The most widely used continuous probability distribution is the normal distribution. Probabilities for the standard normal distribution can readily be found by use of the Table of Normal Areas. Since any normally distributed random variable can be transformed into the standard normal variable, the Table of Normal Areas can be used to find probabilities for any normal variable.

The binomial probability distribution closely resembles the normal distribution when n is fairly large. For this reason it is possible to find approximate probabilities for the binomial distribution by use of the Table of Normal Areas. A somewhat tricky problem in approximating probabilities of a discrete distribution with a continuous distribution lies in setting the lower and upper limits on the relevant values of the continuous random variable.

REVIEW EXERCISES FOR CHAPTER 6

6–28. Tours of a state capitol building start every 30 minutes. For a person arriving at the capitol at a randomly selected time, the probability distribution of time until the next tour is

$$f(t) = 1/30 \qquad 0 \leqslant t \leqslant 30$$

What is the probability that if you arrive at a randomly selected time, your wait for the next tour will be

a) 10 minutes or less?
b) less than 10 minutes?
c) more than 27½ minutes?

6–29. The income of the attendants in the Hermann Goering Nursing Home is a normally distributed random variable with parameters μ = \$7500 and σ = \$1000. What is the probability that a randomly selected attendant has an income

a) of between \$7500 and \$8500?
b) of over \$8500?
c) that deviates from \$7500 by more than \$1500?

6–30. What is the probability that the standard normal variable z will be

a) between its expected value and 1 standard deviation above its expected value?
b) more than 1 standard deviation below its expected value?
c) within 1.8 standard deviations of its expected value?
d) more than 1.8 standard deviations from its expected value?

***6–31.** Within what two z values will the random variable z be with a probability of .90? Select an interval (z_1, z_2) so that the probability that z will lie below z_1 will be equal to the probability that z will lie above z_2.

6–32. In a newly installed data-processing system, the number of machine hours X used by the Personnel and Payroll Division is a normally distributed random variable with $\mu = 1.4$ hours per day and $\sigma = .5$ hours per day. Within what X values will the number of hours used by P and P on a randomly selected day be with a probability of .90? Select an interval (x_1, x_2) so that the probability that X will lie below x_1 is equal to the probability that X will lie above x_2.

6–33. If 20% of young people are arrested before they reach the age of 18, use the binomial distribution to find the probability that in a family of 5 children at least 2 children will be arrested before reaching the age of 18.

Discuss the appropriateness of the binomial model for this problem.

6–34. In Magee Hospital, newly born boys are wrapped in blue blankets and newly born girls are wrapped in pink blankets. A consulting psychologist says that it would be very damaging psychologically for any baby to be wrapped in a blanket of the wrong color. There are 12 blue blankets and 12 pink blankets available for tomorrow's babies. If 20 babies will arrive tomorrow, what is the probability that there will be enough blankets of each color to prevent psychological damage?

6–35. The time to cut hair in the senatorial Barber Shop is a normally distributed random variable with μ equal to 35 minutes and σ equal to 10 minutes. As Senator Dome enters the shop, 1 customer is just getting into the chair and 3 others are waiting. Assuming that there is only 1 barber, and Senator Dome decides to wait until all the customers in front of him have received haircuts,

a) what is the probability distribution of the time that Senator Dome will have to wait until he gets into the chair?

b) what is the probability distribution of the time until Senator Dome is finished getting his hair cut?

***6–36.** From the information given in Exercise 6–35, find the amount of waiting time until Senator Dome gets into the chair that will be exceeded with a probability of .05.

6–37. An educational psychologist believes that 20% of the second grade pupils who are labeled as "slow learners" have visual problems that could be corrected with glasses. Of a group of 26 "slow learners," what is the probability that at least 5 of them have visual problems that could be corrected with glasses?

6–38. There are two ways to drive across Eesahallat Falls. Route A is short and generally fast, but traffic conditions vary considerably. Route B is longer and generally takes a little longer, but there is little variability in traffic. Assume that the time to drive across Eesahallat Falls by Route A is a normally distributed random variable with $\mu = 20$ minutes and $\sigma = 5$ minutes, and that the time to drive by Route B is a normally distributed random variable with $\mu = 25$ minutes and $\sigma = 2$ minutes.

a) Which route offers a higher probability of getting across Eesahallat Falls in 25 minutes or less?

b) Which route offers a higher probability of getting across Eesahallat Falls in 30 minutes or less?

SUGGESTED READINGS

JOHNSON, ROBERT. *Elementary Statistics,* 2nd ed. North Scituate, Mass.: Duxbury Press, 1976. Chapter 7.

MENDENHALL, WILLIAM, and LYMAN OTT. *Understanding Statistics,* 2nd ed. North Scituate, Mass.: Duxbury Press, 1976. Chapter 6.

MORONEY, M. J. *Facts from Figures,* 3rd ed. Harmondsworth, Middlesex: Penguin Books, Ltd., 1956. Chapter 9.

SEVEN

Taking Samples

Outline of Chapter

7.1 Sampling Distributions of Statistics

A sample statistic has a probability distribution, just like any other random variable.

7.2 Estimation of Parameters

A sample statistic is used to estimate a parameter. Among the qualities to look for in a good estimator are lack of bias and a small standard error.

7.3 Confidence Intervals

You cannot trust a sample statistic to estimate a parameter "on the nose," but an interval around the statistic has a good chance of containing the parameter.

7.4 Sampling Designs

Sometimes you can use some prior knowledge about the population to get more information from the sample.

Objectives for the Student

1. Know what a random sample is.
2. Understand the concept of sampling distribution, both for a mean and for a proportion.
3. Understand the concept of standard error, both for a mean and for a proportion.
4. Understand the meaning of independent random sampling.
5. Understand the meaning and usefulness of the Central Limit Theorem.
6. Become familiar with the conditions necessary for using the normal distribution to approximate probabilities for sample proportions.
7. Know what unbiased estimators and efficient estimators are.
8. Be able to set up and interpret a confidence interval for a mean and for a proportion.
9. Be able to determine the sample size in order to keep the error within the limits you want.
10. Distinguish between simple random sampling, systematic sampling, stratified sampling, and cluster sampling.
11. Understand the advantages and disadvantages of random sampling methods *vis a vis* nonrandom methods, such as quota sampling.

You now have some knowledge of what statistical data look like, what they mean, and how you can describe the data with compact measures. You also know something about the laws of probability, or chance. In the rest of the book you will combine your knowledge of data with that of probability in order to draw inferences from the data to the population you are interested in.

7.1 Sampling Distributions of Statistics

Betty Bishop returned to the family bridge table after a trip to the kitchen to get some ginger ale. She picked up the hand that had been dealt to her in her absence and discovered to her astonishment that it contained 13 diamonds. Betty tried for a minute to keep a straight face, until she saw that the others were looking at her and starting to laugh. "It is impossible!" Betty burst out.

"What's impossible?" asked her brother Ben in feigned innocence. "The deal was rigged!" she cried. "The odds against getting 13 diamonds are something tremendous!"

"About 635 billion to one," her mother Jenny suggested. "But, on the other hand, the same odds are against any other specific hand," she added.

If you are not a bridge addict yourself, ask somebody who is how often he or she was dealt a hand containing 13 diamonds. Unless there were some rigged deals, the answer would almost inevitably be never. Even 11 or 12 cards of the same suit would be very unlikely, even over a bridge player's lifetime (and bridge players tend to live longer than poker players). Occasionally you will find people who have been dealt hands with 9 or 10 of a suit. The *expected* number of diamonds is $13(1/4) = 3.25$, but any bridge player can tell you how often the actual number of diamonds (or any other suit) departs considerably from this expected value. Six diamonds is not at all unusual, and zero diamonds (a void in diamonds) is also not rare. Bridge players, perhaps more than anyone else, can tell you how much and how little faith you can have in a sample of size 13.

If you want to know the expected value of the ages of welfare recipients, you can obtain information on 13 recipients and calculate $\overline{X}$, the mean age of the 13. But before you jump to the conclusion that the $\overline{X}$ you calculated is very close to the mean age of the entire population, you should stop to consider what would happen if someone else took another sample of 13 and calculated the mean $\overline{X}_2$ from that sample. The two samples could vary from each other as much as two bridge hands can.

The Sampling Distribution of the Mean

Although the probability distribution of the number of diamonds in a bridge hand is an example familiar to many readers, it is too complicated to serve as a good example for detailed analysis. To illustrate how sample means can vary, consider an example of a small universe of 6 elements.

EXAMPLE 7.1 There are 6 fathers in the Parents' Club of Kosciusko County. Somebody is curious to find out the mean number of children these fathers have. Of course the surest way to answer this question is to interview all six fathers. But suppose the curious person, a statistical methodologist at heart, is interested in evaluating how good a sample of size 1 is, and how much better a sample of size 2 is. In order to compare the two sample sizes, let us assume that (unknown to the sampler) the number of children of the 6 fathers were, respectively, 1, 2, 3, 4, 5, and 6, so that the true expected value is

$$(1)(1/6) + (2)(1/6) + (3)(1/6) + (4)(1/6) + (5)(1/6) + (6)(1/6) = 3.5 \quad \blacksquare$$

A sample of size 1 can have 6 possible results. The probability distribution of these sample results is shown in Table 7.1 along with the difference between each result and the population expected value of 3.5. The deviation of each value from 3.5 is known as the "error." This is how much we would be off if we used the sample result to estimate the population expected value.

> The probability distribution of the statistic $\overline{X}$ is known as the SAMPLING DISTRIBUTION OF THE MEAN.

A graph of this sampling distribution of $\overline{X}$ is shown in Figure 7.1

A sample of size 2 can be obtained in 15 ways. This is shown in Table 7.2, which can be condensed by lumping all samples with the same $\overline{X}$ together and adding their probabilties. For example there are 2 possible samples with a mean of 2.5, each with a probability of 1/15, so that the probability of a sample with an $\overline{X}$ of 2.5 is 2/15. The probabilities of the

Table 7.1. *PROBABILITY DISTRIBUTION OF X AND DEVIATIONS FROM E(X) X = Number of Children of Six Fathers*

NUMBER OF CHILDREN X	PROBABILITY $f(x)$	ERROR $(x - \mu)$
1	1/6	−2.5
2	1/6	−1.5
3	1/6	−0.5
4	1/6	+0.5
5	1/6	+1.5
6	1/6	+2.5

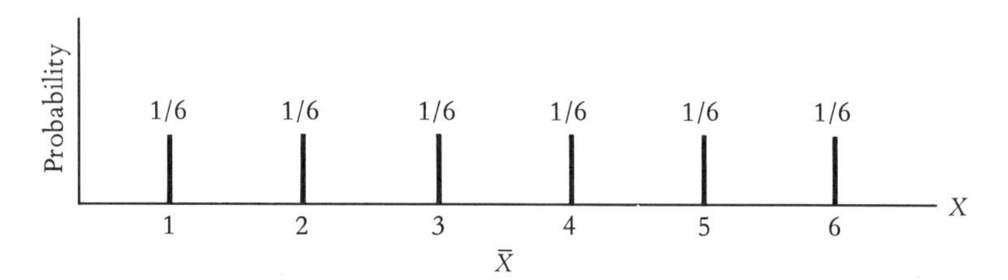

Figure 7.1. *SAMPLING DISTRIBUTION OF X FOR n = 1*
Population from Table 7.1

other sample means are shown in Table 7.3. The graph of this distribution is shown in Figure 7.2.

Several things are worth noting in comparing the distributions of $\overline{X}$ for the two sample sizes. First, the range of the means of a sample of size 2 is smaller than the range of the individual values we would get from a sample of size 1.

Second, and more fundamentally, the probabilities of being close to the population expected value increases considerably as n increases from 1 to 2. The expected value of both distributions is 3.5. It would be desirable if the sample means were close to 3.5, such as between 3 and 4. In the case of $n = 1$, the probability of 3 or 4 is $1/6 + 1/6 = 1/3$. In the case

Table 7.2. *POSSIBLE SAMPLES OF SIZE TWO*
Data from Table 7.1

Sample	Sample Mean $\overline{X}$	Probability	Error $(\overline{X} - \mu)$
1,2	1.5	1/15	−2.0
1,3	2.0	1/15	−1.5
1,4	2.5	1/15	−1.0
1,5	3.0	1/15	−0.5
1,6	3.5	1/15	0.0
2,3	2.5	1/15	−1.0
2,4	3.0	1/15	−0.5
2,5	3.5	1/15	0.0
2,6	4.0	1/15	0.5
3,4	3.5	1/15	0.0
3,5	4.0	1/15	0.5
3,6	4.5	1/15	1.0
4,5	4.5	1/15	1.5
4,6	5.0	1/15	1.5
5,6	5.5	1/15	2.0

Table 7.3. *SAMPLING DISTRIBUTION OF $\overline{X}$ FOR $n = 2$*
Data from Table 7.1

$\overline{X}$	PROBABILITY	ERROR
1.5	1/15	−2.0
2.0	1/15	−1.5
2.5	2/15	−1.0
3.0	2/15	−0.5
3.5	3/15	0.0
4.0	2/15	0.5
4.5	2/15	1.0
5.0	1/15	1.5
5.5	1/15	2.0

of $n = 2$, the probability of either 3, 3.5, or 4 is $2/15 + 3/15 + 2/15 = 7/15$, which is considerably higher than 1/3.

Third, the graph of the distribution for $n = 2$ shows that the probabilities are higher for values of $\overline{X}$ near the middle than they are for values of $\overline{X}$ near the ends, even though the probabilities of the individual x values were equal throughout.

The errors from a sample of size 2 appear to be generally smaller than the errors from a sample of size 1. In order to measure the size of the errors, you can compute the means and standard deviation of the probability distributions of X and $\overline{X}$ from a sample of size 2. If you do this as an exercise you should get the following results:

$$\mu_X = 3.5 \qquad \sigma_X = 1.7078 \qquad n = 1$$

and

$$\mu_{\overline{X}} = 3.5 \qquad \sigma_{\overline{X}} = 1.0801 \qquad n = 2$$

Figure 7.2. *SAMPLING DISTRIBUTION OF $\overline{X}$ FOR $n = 2$*
Data from Table 7.1

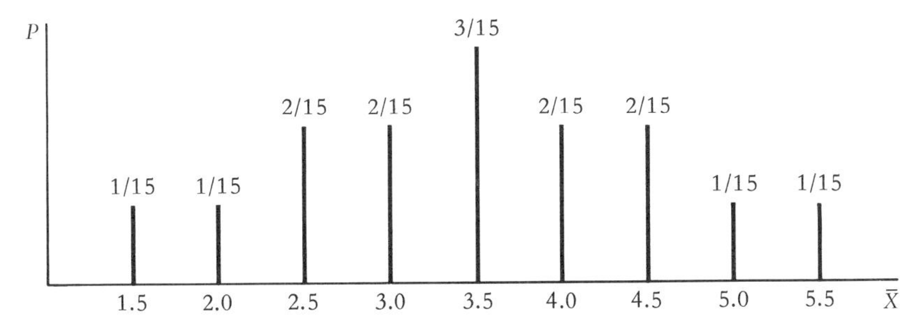

Both distributions have the same expected value, and it is nice to know that this is the expected value of the population. But the standard deviation of $\overline{X}$ from a sample of size 2 is considerably smaller than the standard deviation of the individual x values.

The probability distribution of the individual x values is the probability distribution of the random variable X. The x values can be considered to be the results of a random sample of size 1.

One method that will help to conceptualize the sampling distribution of the mean is to imagine a large number of experimenters repeatedly taking samples of size n from the same population. Naturally, different samples will have different sample means. If you tabulate all these sample means in a frequency table, the relative frequencies will tend to approximate the probabilities given by the sampling distribution of the mean.

In discussing the sampling distribution of the mean, it is important that the sampling procedure be a random one.

> A RANDOM SAMPLE is a sample that is obtained in such a way that every element in the population or universe from which it was drawn has an equal probability of being selected.

It is necessary to take great care in obtaining a random sample. Haphazard methods or methods using intuition or judgment may seem to be random, but in fact such methods will almost inevitably tend to "favor" some members of the population over others.

For example, if you want to sample the population of those who watched a presidential speech on TV, you should not sample people walking along the street. Such a sampling method would give too low a probability for invalids and stay-at-homes to be selected. Nor should you take a sample from the names in the telephone directory; those with unlisted numbers and those without telephones would be underrepresented.

The sampling distribution of the mean, like any probability distribution, has an expected value and a standard deviation. The expected value of the mean is simply the expected value of the population from which the random variable was drawn. It is important to be aware of the danger of drawing a sample from a population different from the one you are interested in. For instance, if a doctor makes generalizations on the basis of observations of his patients, he should generalize to the population of patients who would come to him, not to the population of all people.

If a random sample of size n is drawn from a population of size N, the expected value of the sample mean $\overline{X}$ is given by **7.1**

$$E(\overline{X}) = E(X) = \mu$$

If you look at Figure 7.2, you can see that the sample mean can vary all the way from a low value of 1.5 up to a high value of 5.5. That means that two experimenters could each take a sample of 2 from the same population and come up with sample means far apart from each other, purely on the basis of chance. This variability is a major problem in sampling, especially when the sample size is small. It means that if somebody generalizes on the basis of a mean of a small sample, watch out!

One of the major tasks of statistics is to keep this variability as small as possible, or at least tolerably small. The smaller the variability of the mean, the more confidence a person can have that the sample mean $\overline{X}$ is close to the true mean.

We measure the variability in the mean by the standard deviation of the statistic $\overline{X}$, which we call the standard error of the mean.

If a random sample of size n is drawn from a population of size N, the STANDARD ERROR OF THE MEAN, denoted as $\sigma_{\overline{X}}$, is defined to be

$$\sigma_{\overline{X}} = \frac{\sigma}{\sqrt{n}}\sqrt{\frac{N-n}{N-1}} \qquad \textbf{7.2}$$

where σ is the population standard deviation
n is the number in the sample
N is the number in the population

The factor $\sqrt{(N-n)/(N-1)}$ is known as the *finite population correction factor.* It is the reduction of the standard error of the mean caused by the fact that you are "using up" part of the universe and reducing the possible amount of variation as you increase your sample size. This factor is of importance only when the sample size n is fairly substantial in proportion to the number in the universe N. The factor is close to 1 when n is small compared to N, as is the case when the population is very large or infinite. In such a case, this factor can be ignored and Equation 7.2 simplifies to

$$\sigma_{\overline{X}} = \frac{\sigma}{\sqrt{n}} \qquad 7.3$$

standard error of the mean when n is small compared to N, or if N is indefinitely large.

A rough rule of thumb is that you can use Equation 7.3 when the sample size n is not more than 1/10 of N. Since most statistical work of a scientific nature is concerned with generalizations to indefinitely large populations, Equation 7.3 is very widely used for the standard error.

EXAMPLE 7.1 (continued) Use Equation 7.2 to compute the standard error of the number of children of the 6 fathers in a sample of size 2.

Solution: Since σ_X can be computed to be 1.7078, $\sigma_{\overline{X}}$ is computed as

$$\sigma_{\overline{X}} = \frac{\sigma}{\sqrt{n}}\sqrt{\frac{N-n}{N-1}} = \frac{1.7078}{\sqrt{2}}\sqrt{\frac{6-2}{6-1}} = 1.0801$$

This agrees with the earlier calculation. ■

We noted in taking a sample of size 2 that the probability distribution of the sample mean $\overline{X}$ showed that values of X near $E(X)$ had higher probabilities than values of X far away from $E(X)$. This tendency toward higher probabilities near the expected value would be evident even if we had sampled the fathers with replacement.

INDEPENDENT RANDOM SAMPLING is a sampling procedure in which the result of one observation is not related to the result of another observation. You can achieve independent random sampling by sampling at random from an infinitely large universe, or from a finite universe if you sample with replacement, that is, if each element is returned to the universe before another element is drawn.

In independent random sampling, it can be shown that the sampling distribution of the sample mean $\overline{X}$ approaches the normal probability distribution as the sample size increases, no matter what the probability distribution of the random variable X. This property of the sample mean, plus Equations 7.1 and 7.3, give us the Central Limit Theorem, which is fundamental to the theory of large samples.

> **Central Limit Theorem**
>
> If X is a random variable with expected value μ and standard deviation σ, the sampling distribution of $\overline{X}$, the mean of a sample of size n from the population X, approaches the *normal* probability distribution, the expected value of which is μ, and the standard deviation of which is $\sigma/\sqrt{n}$.

It can be shown from the Central Limit Theorem that the statistic

$$z = \frac{\overline{X} - \mu}{\sigma/\sqrt{n}}$$

approaches the *standard* normal distribution as n gets large.

The Central Limit Theorem is of great use because it is so general. No matter what the shape of the underlying probability distribution of the random variable X is, the mean from a large sample will have a normal probability distribution.

EXAMPLE 7.2 The probability distribution of incomes in almost any conceivable society will be far from normal. In almost every case it will be skewed, or lopsided. Suppose the probability distribution of incomes in the village of Skew Gardens looked like that of Figure 7.3, with an expected value of $10,000 and a standard deviation of $6,000.

If a random sample of 100 people is taken from Skew Gardens, what will be the sampling distribution of $\overline{X}$, the sample mean?

Solution: According to the Central Limit Theorem, the probability distribution (or sampling distribution) will be a normal probability distribution with

Figure 7.3. *PROBABILITY DISTRIBUTION OF PERSONAL INCOMES IN SKEW GARDENS*

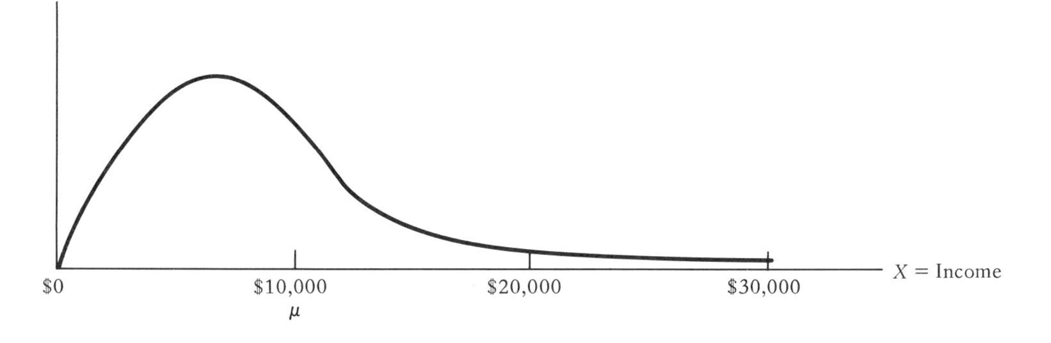

$$E(\overline{X}) = \mu = \$10{,}000$$

$$\sigma_{\overline{X}} = \frac{\sigma}{\sqrt{n}} = \frac{\$6{,}000}{\sqrt{100}} = \$600$$

This sampling distribution is shown in Figure 7.4. ■

Having determined the probability distribution of $\overline{X}$, we can determine the probability that $\overline{X}$ will be greater than \$11,000, greater than \$12,000, more than \$1,000 away from the expected value, and many other probabilities that may be of interest. These probabilities are very useful in evaluating methods of sampling. If there is too low a probability that $\overline{X}$ is within \$1,000 of μ, then we should consider taking a larger sample.

The Sampling Distribution of a Sample Proportion

Public opinion polls frequently publish conclusions about proportions or percentages. For example, you may read that 40% of the voters approve a particular policy of the President. If this conclusion is based on a sample of, say, 1600, how much faith can you have in an inference about the proportion of 100 million voters who approve of the policy? The question is one involving the sampling distribution of a proportion, or how much the proportion of "successes" in a sample of a given size is likely to vary from the proportion of "successes" in the entire population.

Let $\hat{p}$ be the statistic, "the proportion of successes x in a sample of size n," or $\hat{p} = x/n$.

Figure 7.4. *SAMPLING DISTRIBUTION OF $\overline{X}$ FROM SAMPLE OF SIZE 100 Personal Incomes in Skew Gardens*

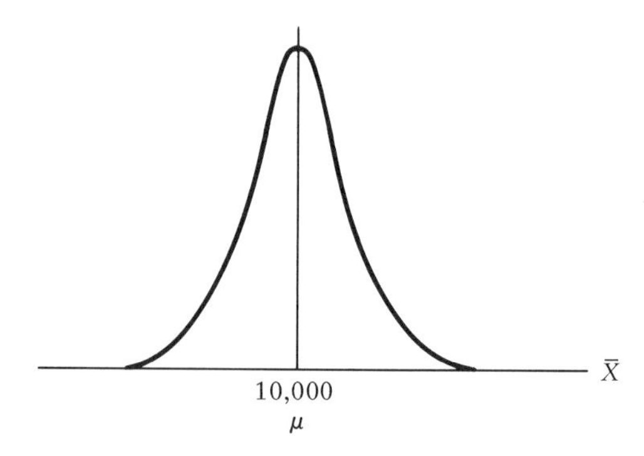

> The SAMPLING DISTRIBUTION OF A PROPORTION $\hat{p}$ is the probability distribution of the proportion of successes in n independent trials from a population in which the probability of a success is p.

The sampling distribution of the proportion $\hat{p}$ approaches the normal probability distribution as n gets large.

> The expected value of $\hat{p}$, $E(\hat{p}) = p$, and the standard deviation of $\hat{p}$ (known as the STANDARD ERROR OF THE PROPORTION) is given by
>
> $$\sigma_{\hat{p}} = \sqrt{\frac{p(1-p)}{n}}$$

EXAMPLE 7.3 Suppose that 60% of the population favor a proposal permitting a woman to have an abortion on demand in the first trimester (3-month period) of her pregnancy. What is the sampling distribution of the proportion who would favor the proposal in a sample of 96 randomly selected people? What is the probability that fewer than 50% in the sample would favor the proposal?

Solution: The sampling distribution of the proportion $\hat{p}$ would be approximately a normal probability distribution with $E(\hat{p}) = p = .60$ and the standard error

$$\sigma_{\hat{p}} = \sqrt{\frac{p(1-p)}{n}} = \sqrt{\frac{(.6)(.4)}{96}} = .05$$

The probability that $\hat{p}$ is less than .50 is approximated by a normal random variable X, where $\mu_X = .50$ and $\sigma = .05$.

$$\begin{aligned} P(\hat{p} < .50) &= P(X \leqslant 49.5) \\ &= P\left(z \leqslant \frac{.495 - .600}{.05}\right) = P(z \leqslant -2.10) \\ &= .5000 - .4821 = .0179 \end{aligned}$$

The probability of getting a sample proportion below .50 is .0179. ■

WHEN IS n LARGE ENOUGH TO USE THE NORMAL APPROXIMATION?

It is very nice to know that the sampling distribution of $\overline{X}$ or $\hat{p}$ approaches the normal distribution as n gets large, but how large does n have to get in order for the normal distribution to be close enough?

In the case of the mean, the answer to this question depends on how close to the normal distribution the probability distribution of the random variable X is. If X is itself normally distributed (as in the case of heights of an adult population) then the means of a sample of any size (including 1) will be normally distributed. If the distribution of X is reasonably symmetrical around its mode, such as in Figure 7.5a, then a sample of size 10 should be ample for using the normal distribution.

If the distribution of X is very skewed, as in Figure 7.5b, or U shaped, as in Figure 7.5c, then you can make use of the Central Limit Theorem only if the sample size is at least around 20. No matter what the underlying shape of the distribution, however, a sample size of 30 is nearly always considered large enough to use the normal distribution.

In the case of a sample proportion, the size of the sample you need in order to use the normal approximation depends on the value of the parameter p. If p is .5 or within .1 of .5 (that is, from .4 to .6), then samples of size 10 or 15 give a probability distribution of $\hat{p}$ reasonably close to the normal; and for a sample of size 20, the distribution of $\hat{p}$ is very close to normal. For very large or very small values of p (over .95 or under .05), you may require a sample of 30 or more before you can safely use the normal as an approximation.

EXERCISES

7–1. Suppose the ages of the persons in Doerr County who receive Medicare benefits is a random variable with expected value $\mu = 72$ years and $\sigma = 6.0$ years.

a) What is the sampling distribution of the mean age of a sample of 36 randomly selected recipients of Medicare benefits in Doerr County?

b) What is the probability that the sample mean is greater than 75?

7–2. The ages of 5 employees are 30, 35, 40, 45, and 50.

a) Calculate the expected value and standard deviation of the 5 ages.

b) List all possible samples of size 2 that you could take from the population and calculate the mean $\overline{X}$ of each sample.

c) Find the expected value and standard deviation of the sample means you calculated in part (b).

Figure 7.5. *POSSIBLE SHAPES OF PROBABILITY DISTRIBUTION f(x)*

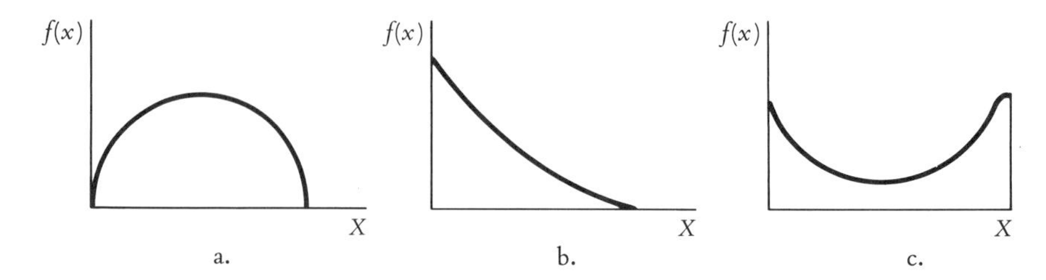

d) Find the expected value and standard deviation of X from a sample of size 2, using Equations 7.1 and 7.2. Compare your answers with those in part (c).

7–3. Suppose for the employees in Exercise 7–2 you took a sample of 2 with replacement (that is, you could pick the same employee twice).

a) List all possible samples and calculate $\overline{X}$ for each.
b) Find the expected value and standard deviation of the sample means you calculated in part (a).
c) Find the expected value and standard deviation of $\overline{X}$ from a sample of size 2 with replacement, using Equations 7.1 and 7.3. Compare your answer with that in part (b).
d) In sampling with replacement why do you use Equation 7.3 instead of Equation 7.2 to calculate $\sigma_{\overline{X}}$?

7–4. If the probability distribution of the time that students spend studying for a statistics course each week has an expected value of 12.5 hours and a standard deviation of 3 hours,

a) what is the sampling distribution of the mean number of hours studied in a random sample of 16 students?
b) what is the probability that the mean will be at least 13 hours? Assume that the probability distribution of study times is fairly symmetrically distributed around its mode.

7–5. If 70% of drivers are buckled in with their seat belts when crossing the Tri-Boro Bridge, what is the sampling distribution of the proportion of "drivers who are buckled in" in a random sample of 84 drivers taken as they start to cross the Tri-Boro Bridge?

7–6. What is the approximate value of the finite population correction factor, $\sqrt{(N - n)/(N - 1)}$, if

a) $N = 100$ and $n = 20$?
b) $N = 10$ and $n = 5$?
c) $N = 10{,}000$ and $n = 200$?
d) N is large and n is 5% of N?
e) the population is infinitely large?
f) the sampling is done with replacement?

7.2 Estimation of Parameters

A random sample of 60 city employees who were over 40 years old showed a mean loss of 6.4 days of work during the past year because of illness. What does this result tell us about how many days on the average the 19,369 employees of this city who are over 40 years old lose in a year?

There are two extreme and equally fallacious conclusions that can be drawn from this information. One is the naive conclusion that the average rate of days missed for all employees must be exactly 6.4 days per year, a conclusion that is based on the assumption that what's true for the sample is true for the universe. The other extreme is the skeptical conclusion that we can conclude nothing from the data, since a sample of 60 employees tells us nothing about the other 19,309 employees over 40 years of age.

Both conclusions, one based on unbridled faith in samples and the other based on total skepticism, are unwarranted. The faith is not justified; the skepticism is not necessary. From the previous section you know that sample means are subject to chance variation, and hence cannot be simply accepted as population means. But in that section it was also pointed out that this variation can be measured, and this measurement enables us to say things about the population with a specified degree of confidence.

Parameters and Statistics

An astronaut has just brought a randomly selected lunar widget back from the moon; it weighs 16.8 kilograms. From this information, estimate μ, the expected value of the weight of *all* lunar widgets.

Although the information is very skimpy, being based on a sample of size 1 we have at least *something* to base our estimate on. It seems reasonable, if we have to pick some value to estimate μ, to estimate μ to be 16.8 kilograms, even though we are almost certain not to be perfectly correct in our estimate. There is nothing in using your sample result that predisposes you either to underestimate or to overestimate.

The problem is a case of estimating a parameter on the basis of information about a statistic.

> A PARAMETER is a characteristic of a population or a probability distribution. A STATISTIC is a value or measurement that one can take from a sample.

Among the parameters that you will encounter are: μ, the expected value of a probability distribution or population; p, the proportion in a population having a specified characteristic; σ^2, the variance of the population or probability distribution; and σ, the standard deviation. Among the statistics you have encountered and will encounter are $\overline{X}$, the sample mean; $\hat{p}$, the sample proportion; s^2, the sample variance; and z, the standard normal variable computed from a sample. You can also construct statistics that are compounded from other statistics, for example, by taking the difference between two sample means, and thereby constructing the statistic $\overline{X}_1 - \overline{X}_2$. You can also multiply or divide a statistic by a constant or assumed value to get another statistic, as when you compute the statistic z from the statistic $\overline{X}$ and the known or assumed parameters μ and σ.

As long as your knowledge about a population is confined to information derived from samples (other than 100% samples, which are known as censuses), you will never know exactly the value of a parameter. You can only estimate the value of the parameter from some statistic, which is known as an *estimator* of the parameter. It is often a matter of some ingenuity to determine what statistic to use as an estimator of the parameter you want.

The problem of having to rely on a statistic to estimate a parameter is that the statistic is subject to chance or random error. In the previous section you studied sampling distributions, which describe just how much a statistic can vary because of chance. A statistic that has a small standard error will generally give you more reliable information about a parameter than a statistic that can jump all over the place.

Unbiased Estimators

But even a statistic with a small standard error is not going to help you much if it does not reflect some parameter that you are interested in. For example, the statistic $\overline{X}$ will be of little use in estimating σ, the population standard deviation. But $\overline{X}$ will be a very useful estimator of μ, the expected value of X, because sample $\overline{X}$'s will on the average be μ, or

$$E(\overline{X}) = \mu$$

> When the expected value of a statistic Y is equal to a parameter (the Greek letter "theta" is commonly used to represent an unspecified parameter), so that
>
> $$E(Y) = \theta$$
>
> then the statistic Y is said to be an UNBIASED ESTIMATOR of θ.

When you use the weight of the one widget in order to estimate μ, the estimator is unbiased, since it is a sample mean (albeit of a sample of only size 1).

Efficient Estimators

In general we like an estimator to be unbiased, but that is not the only quality we look for in a good estimator. Of two unbiased estimators, the one with the smaller standard error is going to be preferred. And, while a smaller standard error will result from increasing the sample size, it can also involve selecting an estimator that is more efficient in giving you information for any given sample size.

If you wanted to estimate μ from a sample of n lunar widgets, you could use either the sample mean $\overline{X}$ or the sample median Md as the estimator. It can be shown that

$$E(Md) = \mu$$

so that the sample median, like $\overline{X}$, is also an unbiased estimator of μ. But the variance of the median is given by

$$\sigma^2_{Md} = \frac{1.57\sigma^2}{n}$$

which is 57% larger than the variance of the mean $\overline{X}$, for any given n. Put another way, if you use the sample median rather than $\overline{X}$ to estimate μ, you would need to take a sample that is 57% larger to reduce the variance or standard error of the median to that of the mean. The mean is therefore a more efficient estimator of μ than is the median. For any given sample size, you can have more confidence in its being within any specified limits of μ.

In order to estimate the variance of a population σ^2, we use the sample variance $s^2 = \Sigma(X - \overline{X})^2/(n - 1)$. It can be shown that

$$E(s^2) = \sigma^2$$

or that s^2 is an unbiased estimator of σ^2. Notice that when $n = 1$, s^2 is undefined (since division by zero is undefined). Therefore when you have a sample of size 1, you do not have any estimator of the variance of the population.

In order to estimate p, the proportion of "successes" in a two-valued population, you can use the sample proportion $\hat{p}$, which is an unbiased estimator of p.

EXAMPLE 7.4 Suppose you want to estimate the mean number of units of sulphur dioxide per cubic foot of air in Fume City at 8:00 A.M. Readings are taken on 100 randomly selected mornings. What statistics could be used to estimate μ? What would be their expected value and standard deviation?

Discussion: Two statistics mentioned in the section that could be used are the sample mean $\overline{X}$ and the sample median Md.

The expected value of the mean is $E(\overline{X}) = \mu$, and its standard error is

$$\sigma_{\overline{X}} = \frac{\sigma}{\sqrt{n}} = \frac{\sigma}{10}$$

where σ is the standard deviation of sulphur dioxide counts at 8:00 in the morning.

The expected value of the median, $E(Md)$ is also μ. Its standard error is

$$\sqrt{\frac{1.57\sigma^2}{n}} = \frac{1.25\sigma}{\sqrt{n}} = \frac{\sigma}{8}$$

Suppose that μ is 13.5 and $\sigma = 4$. What would be the probability that the estimate would be within 1 unit of the expected value (a) if we use $\overline{X}$ as an estimator, and (b) if we use Md as the estimator?

Solution: Using $\overline{X}$, we have a standard error of

$$\frac{\sigma}{\sqrt{n}} = \frac{4}{\sqrt{100}} = .4$$

The question asks for $P(12.5 \leqslant \overline{X} \leqslant 14.5)$. Because of the large sample size, we can use the Central Limit Theorem. The probability can therefore be expressed in terms of the standard normal variable.

$$P\left(\frac{12.5 - 13.5}{.4} \leqslant z \leqslant \frac{14.5 - 13.5}{.4}\right) = P(-2.5 \leqslant z \leqslant 2.5)$$
$$= .4946 + .4946 = .9892$$

Using the median, we have a standard error of

$$\frac{1.25\sigma}{\sqrt{n}} = \frac{5}{10} = .5$$

The question asks for $P(12.5 \leqslant Md \leqslant 14.5)$. Using an extension of the Central Limit Theorem to sample medians, this probability is equivalent to

$$P\left(\frac{12.5 - 13.5}{.5} \leqslant z \leqslant \frac{14.5 - 13.5}{.5}\right) = P(-2 \leqslant z \leqslant 2)$$
$$= .4772 + .4772 = .9544$$

This example illustrates the advantage of using the more efficient estimator. By using the sample mean there is a probability of .9892 of the estimator's being within 1 unit of the expected value μ; by using the median, there is a probability of only .9544 of being within one unit of μ. This difference becomes more striking if we look at the probability of *not* being within one unit of μ.

Using $\overline{X}$, this is $1 - .9892 = .0108$.
Using Md, this is $1 - .9544 = .0456$.

If our goal is to estimate with an error of no more than one unit, then we have more than 4 times as much risk of missing our goal if we use the median rather than the mean. ■

EXAMPLE 7.5 Suppose you wanted to estimate what proportion of the population of Left Chester were male. If you took a sample of size 1, would the proportion of males in the sample be an unbiased estimator of the proportion of males in Left Chester? Do you think it would be a good estimator?

Discussion: Although such a sampling method might seem absurd, it does have the advantage of being unbiased. The expected value of the sample proportion is p. The problem with this estimator, of course, is that it comes from such a ridiculously small sample size. The estimator would have to predict either that 0% or 100% of the population of Left Chester are male. The standard error of the proportion $\sigma_{\hat{p}} = \sqrt{\frac{p(1-p)}{n}}$, which in this case is $\sqrt{p(1-p)}$

If the true proportion, for example, were .5, then the standard error would be $\sqrt{(.5)(.5)} = .50$, or 50%, which is a rather large standard error. On the other hand, if p were 0 or 1 (an all-female or an all-male population) then this standard error would be 0 and the estimator would give us perfect information. ■

EXAMPLE 7.6 Suppose a Turkish bath had separate rooms for ladies and for gentlemen. If you were unsure of which room you were in, how large a sample would you need to take in order to have a precise estimate of p, the proportion of males in the room?

Discussion: Because of the segregation policy, you have reason to believe that p (the proportion of males) is either 0 or 1. Then the sample proportion $\hat{p}$ should be a perfect estimator of p even if the sample size were only one, since

$$E(\hat{p}) = p$$

and

$$\sigma_{\hat{p}} = \sqrt{\frac{p(1-p)}{n}} = 0 \qquad \text{if } p = 0$$
$$= 0 \qquad \text{if } p = 1$$

Since the expected value of $\hat{p}$ is p, and the standard error of $\hat{p}$ around p is 0, $\hat{p}$ must equal p. ■

EXERCISES

7–7. A survey is taken to determine the mean number of bath tubs and showers in the houses of a certain area of the city. Suppose that, unknown to the person taking the sample, the true value of μ is 1.8 and the standard deviation is .4. If a random sample of 36 houses is selected, what is

a) the expected value of the sample mean $\overline{X}$?
b) the expected value of the sample median *Md*?
c) the standard error of $\overline{X}$?
d) the standard error of the sample median *Md*?
e) the probability that the mean number of tubs and showers will be within .1 of 1.8?
f) the probability that the median number of tubs and showers will be within .1 of 1.8?

On the basis of your answers to (e) and (f), what conclusions can you draw about the relative advantages of these two statistics in estimating μ?

7–8. An advertiser believes that about 30% of doctors smoke El Ropo cigars. If he takes a random sample of 21 doctors

a) what is the sampling distribution of $\hat{p}$, the proportion of doctors in the sample who smoke El Ropo cigars?
b) what is the probability that over 50% of the doctors in the sample smoke El Ropo?
c) If the advertiser insists on getting a sample that supports the claim that a majority of doctors smoke El Ropo, how many random samples do you think he should expect to have to take before he gets one that shows a majority who smoke El Ropo?

7–9. By what proportion is the standard error of the mean reduced if

a) the sample size is increased from 2 to 8?
b) the sample size is increased from 30 to 270?
c) the sample size is increased from 256 to 1024?

7–10. If a large population has a standard deviation of 10 minutes, what is the probability that a random sample of 50 from that population will be within 2 minutes of the population expected value?

7.3 Confidence Intervals

Up to this point I have discussed only single-value estimators. These estimators are single numbers that estimate the exact value of a parameter.

> A statistic used to estimate the exact value of a parameter is known as a POINT ESTIMATOR. The specific value that a point estimator takes on is known as a POINT ESTIMATE.

The problem with a point estimator, even an unbiased one, is that it is almost sure to be wrong. The chance of hitting the true value of the parameter precisely is in most cases very slim.

Confidence intervals are a method of estimation that gives you a better chance of estimating the parameter. When you use a confidence interval instead of a point estimator, you lose precision but you gain confidence.

> A CONFIDENCE INTERVAL is an interval, based on a sample statistic, that contains a parameter value with a specified probability.

CONFIDENCE INTERVALS FOR MEANS

Suppose you take a large sample ($n \geq 30$) and compute the sample mean $\overline{X}$. What is the probability that $\overline{X}$ will be within 2 standard errors of the expected value μ? We want

$$P\left(\mu - \frac{2\sigma}{\sqrt{n}} < \overline{X} < \mu + \frac{2\sigma}{\sqrt{n}}\right)$$

If you make the transformation

$$z = \frac{\overline{X} - \mu}{\sigma/\sqrt{n}}$$

This probability is equivalent to (using the Central Limit Theorem)

$$P(-2 < z < 2) = .4772 + .4772 = .9544$$

By similar reasoning you can find the probability that $\overline{X}$ is within any given number of standard errors of the expected value. The probability of being within 1.96 standard errors of the expected value is

$$P\left(\mu - 1.96\frac{\sigma}{\sqrt{n}} < \overline{X} < \mu + 1.96\frac{\sigma}{\sqrt{n}}\right) = .9500$$

This expression can be shown to be equivalent to

$$P\left(\overline{X} - 1.96\frac{\sigma}{\sqrt{n}} < \mu < \overline{X} + 1.96\frac{\sigma}{\sqrt{n}}\right) = .9500 \qquad 7.4$$

Equation 7.4 is of particular interest, because confidence intervals are frequently stated with a probability, or confidence, of .95.

The interval

$$\left(\overline{X} - 1.96\frac{\sigma}{\sqrt{n}}, \overline{X} + 1.96\frac{\sigma}{\sqrt{n}}\right) \qquad \textbf{7.5}$$

is known as the .95 CONFIDENCE INTERVAL, or the 95% CONFIDENCE INTERVAL FOR THE MEAN.

If you want an interval that gives you a higher degree of confidence, you can go to Table II and find the z value associated with the appropriate probability. If, for example, you want to find a 99% confidence interval, look up the z value for .495 (a symmetrical interval will have probabilities of .495 on either side, which sum to .99). This z value is about 2.58. The 99% confidence interval for the mean will therefore be

$$\left(\overline{X} - 2.58\frac{\sigma}{\sqrt{n}}, \overline{X} + 2.58\frac{\sigma}{\sqrt{n}}\right)$$

This confidence interval is an interval which contains μ with a .99 probability, or 99% degree of confidence. The 99% confidence refers to the procedure, not to this particular interval. It means that if you adopted this procedure in setting confidence intervals, 99% of the intervals so established (in the mythical "long run") would contain the parameter μ.

EXAMPLE 7.7 In order to determine the mean amount of money spent on expense accounts by senior managers in the bilbo-processing industry, a random sample of 49 senior managers from that industry was selected and analyzed. The mean expenditure $\overline{X}$ was found to be \$8345. Assuming that the standard deviation of the size of expense accounts is \$3500, set up a 95% confidence interval for μ, the mean of all expense-account spending for senior managers in the industry.

Solution: Use Expression 7.5 to find the 95% confidence interval

$$\$8345 - (1.96)\frac{\$3500}{\sqrt{49}} < \mu < \$8345 + (1.96)\frac{\$3500}{\sqrt{49}} = \$7365 < \mu < \$9325$$

We have 95% confidence that μ is between \$8345 and \$9325. ■

The confidence intervals we have looked at up to this point were based on a known or assumed value of σ. In many cases you may not know σ or be willing to assume a value for σ. For large samples, however, it is not necessary to know σ in setting up confidence intervals for μ, if you can take the standard deviation s from your sample. For samples of at least size 30, the sample standard deviation is a sufficiently good estimator of σ to say that

$$P\left(\mu - 1.96\frac{s}{\sqrt{n}} < \overline{X} < \mu + 1.96\frac{s}{\sqrt{n}}\right) = .9500 \qquad \mathbf{7.6}$$

The interval gets smaller as the sample size n increases.

Confidence Intervals for Proportions

You can follow a very similar procedure if you want to set up a confidence interval for a population proportion, or the probability of a "success" in a two-valued population.

If you take a large sample from a universe with a proportion p of successes, the probability that your sample proportion $\hat{p}$ will be within 1.96 standard errors of p is .95. That is,

$$P(p - 1.96\sigma_{\hat{p}} < \hat{p} < p + 1.96\sigma_{\hat{p}}) = .95$$

which is equivalent to

$$P(\hat{p} - 1.96\sigma_{\hat{p}} < p < \hat{p} + 1.96\sigma_{\hat{p}}) = .95 \qquad \mathbf{7.7}$$

There is a problem in using Expression 7.7 for a confidence interval. The standard error of $\hat{p}$ (that is $\sigma_{\hat{p}}$) is given by $\sqrt{p(1-p)/n}$, but you do not know p. However, you can use the sample proportion $\hat{p}$ as an estimator of p, as long as you have a fairly large sample.

> When n is large, the 95% CONFIDENCE INTERVAL FOR A POPULATION PROPORTION is approximately
>
> $$\hat{p} - 1.96\sqrt{\frac{\hat{p}(1-\hat{p})}{n}} < p < \hat{p} + 1.96\sqrt{\frac{\hat{p}(1-\hat{p})}{n}} \qquad \mathbf{7.8}$$

EXAMPLE 7.8 In a random sample of 75 people who were treated for severe drinking problems, 25% had their problem under control a year after completing the treatment. Set up a 95% confidence interval for p, the probability that a person undergoing this treatment will have the problem under control after a year.

Solution: Use Formula 7.8 to get

$$P\left[.25 - 1.96\sqrt{\frac{(.25)(.75)}{75}} < p < .25 + 1.96\sqrt{\frac{(.25)(.75)}{75}}\right] = .95$$

or

$$P(.152 < p < .348) = .95$$

The 95% confidence interval for p is the interval (.20, .30). ■

How Big a Sample Should I Take?

The discussion on confidence intervals will enable you to handle one of the most common questions people ask in doing survey research: How big a sample do I need?

If you inspect the expressions for confidence intervals, Formulas 7.6 and 7.7, you can see how important the sample size n is in reducing the size of these intervals. Since the larger the sample is, the smaller the confidence interval will be, the natural answer to the question on sample size is "the bigger the better." But this is not a practical answer if sampling is costly or time consuming, or in some cases destructive of the product you are sampling. To determine what size sample you should take, you should answer another question: "Given the costs of sampling, how big an error am I willing to tolerate?"

In the case of estimating a mean, the expression $1.96(\sigma/\sqrt{n})$ is added to and subtracted from $\overline{X}$ to give you the limits of the confidence interval. You will be within $1.96(\sigma/\sqrt{n})$ of the true mean or expected value with a probability of .95. The value $1.96(\sigma/\sqrt{n})$ is the "fudge factor" or error, E.

Error term for the 95% confidence interval for μ

$$E = 1.96\frac{\sigma}{\sqrt{n}} \qquad \textbf{7.9}$$

In order to determine the size you want, determine the size of the error (E) you can reasonably tolerate, keeping in mind that setting E very low can get you involved in an intolerably large sample. Once you decide on what size of E you will allow, you can solve Equation 7.8 for n.

Optimal sample size when estimating μ

$$n = \frac{(1.96)^2\sigma^2}{E^2} \qquad \textbf{7.10}$$

You will not be able to use Equation 7.10 until you know or assume a value for σ^2, and in most cases in which you do not know μ, you will not know σ^2 either.

There are several possible ways to estimate σ. One is from knowledge of comparable data. For example, if you wanted to estimate the mean value of houses in an urban area, you could find the standard deviation of house values that you already knew from a comparable area in another study. This standard deviation could be used to estimate σ^2 for use in Equation 7.10.

A second way of estimating σ is by taking a pilot sample. A pilot sample is a small preliminary sample designed to give you some clues about taking your main sample. The standard deviation of the values in the pilot sample can be used to estimate the standard deviation of the population.

A third method of estimating σ is by the "quick and dirty" formula

$$\sigma = \frac{\text{Range of 99.7\% of all values}}{6}$$

The basis for this formula is that in the normal probability distribution the probability is about .997 that a value is within 3 standard deviations of the expected value. That is a total range of 6 standard deviations.

EXAMPLE 7.9 Suppose you want to determine the mean number of days of absence for the employees of a county government over the past year. You want to be accurate within .25 days with a confidence of .95. To estimate σ, you try to guess what is the range containing 99.7% of all cases. The smallest number of days of absence is probably going to be zero, and somebody tells you that the most absences he has ever heard of for anybody was 60. You decide that the probability is .997 that a person will be absent between 0 and 60 days a year, giving you a .997 range of 60 − 0 = 60. Then you estimate σ as 60/6 = 10.

This estimate of σ is based on the assumption of a normal distribution, but it is unlikely that the number of days of absence will be normally distributed. However, the use of the normal assumption is in this case conservative, inasmuch as it will lead to an overestimation of σ rather than an underestimation. Since an underestimation of σ would probably be a more serious error than an overestimation (lulling you into a false sense of security about your sample size), it may not be a bad procedure to estimate σ by a method more likely to lead to an overestimation than an underestimation.

Letting $E = .25$, you can compute the optimal sample size as

$$n = \frac{(1.96)^2(10)^2}{(.25)^2} = 6147$$

At this point you may say "I can't take a sample of 6147. That's too much!"

If this is a bigger sample size than you can afford to take, you will have to do an agonizing reappraisal. This will be a very useful exercise, because it forces you to examine more thoroughly the "tradeoffs" between how much information you want and how much time and money you are willing to spend getting the information. You may decide that you really do not need to be within .25 of the population mean—that being within .5 will be good enough. In other words, you are willing to relax your requirement for E in a decision that weighs the benefit of precision against the cost of sampling. If you let $E = .5$, then your revised sample size becomes

$$n = \frac{(1.96)^2(10)^2}{(.5)^2} = 1537$$

By doubling your allowable error, you have reduced the required sample size by a factor of 4. This still may be too big a sample for your budget; if so, you may have to do a little more reassessing. ■

Suppose you want to determine the optimal sample size for estimating a proportion. As in the case of estimating μ, you need to decide on how large an error term you can allow. The sample size is given by

$$n = \frac{1.96^2 p(1 - p)}{E^2} \qquad \textbf{7.11}$$

To apply this formula you need to have an estimate of p, but p is the very parameter that your sample is attempting to estimate. As with getting a preliminary estimate on the standard deviation, there are several approaches to getting an initial approximation for p. One is to use information on a similar variable for which you know p. A second method is to take a pilot sample, and use the sample proportion as your estimate of p. A third method is simply to assume initially that p is .5. This apparently simplistic rule has the virtue of being conservative; it will yield the highest value of n. It is also consoling to know that n is not very sensitive to departures of p values from .5, as long as p is somewhere between .3 and .7. Therefore even substantial errors in your original estimate of p will not likely have a very profound effect on n.

EXAMPLE 7.10 A research institute on public opinion is planning to take a poll of the voters in Anthony County to find the proportion who favor Ms. Addams. How large a sample should be taken in order to have a .95 probability of being within 2 percentage points of the true proportion who favor Ms. Addams?

Discussion: Equation 7.11 is based on the assumption of random sampling, but public opinion polls are seldom conducted on a purely random basis. However, a well-designed poll does protect itself against many of the pitfalls of bad sampling design, and the formulas based on random sampling, though not strictly applicable, can serve as a general guide to the magnitude of error. If you use Equation 7.11, you need to make an initial assumption about p, the proportion who favor Ms. Addams. If you believe the election is close, it would be reasonable to determine n on the assumption that p is .5, since such an assumption will insure that you do not underestimate n. Therefore,

$$n = \frac{(1.96)^2(.5)(.5)}{(.02)^2} = 2401 \blacksquare$$

EXERCISES

7–11. Four patients in a dental clinic were selected at random. The times (in minutes) that the patients had to wait for treatment were 10, 20, 40, and 50 minutes.

Assume that waiting times are normally distributed, and that the standard deviation of waiting times is 10 minutes. Set up a 95% confidence interval for the expected waiting time for all patients in the dental clinic.

7–12. A random sample of 36 staff sergeants at Camp Burnside showed that the mean amount of time they had spent at Camp Burnside was 7.4 months, and the standard deviation was 2 months. (Assume that there are over 500 staff sergeants at Camp Burnside.)

a) Set up a 95% confidence interval for the expected amount of time spent by all staff sergeants of Camp Burnside.
b) Set up a 99% confidence interval.

7–13. Out of 100 students who attempted to pass a statistics waiver examination, 10% passed. Set up a 95% confidence interval for the proportion of all students who pass the waiver examination. Assume that the 100 students are a random sample of the population you are interested in.

7–14. A random sample of 45 prescriptions revealed that 60% were for brand-name drugs. Set up a 95% confidence interval for the proportion of all prescriptions that are written for brand-name drugs.

7–15. Assuming that the standard deviation of weekly food expenditures for households in a certain area is \$5.00, how large a sample would you need to take in order to estimate the mean amount of weekly food expenditures within \$1?

7–16. How large a sample would you need in order to estimate the proportion of the population of people between the ages of 15 and 30 who take "hard" drugs? Assume that this proportion is not more than .20, and that you want to be within 2 percentage points with a probability of .95.

7.4 Sampling Designs

Almost all of the discussion and examples of sampling have thus far been of simple random sampling.

> SIMPLE RANDOM SAMPLING is a sampling procedure in which every sample of n elements in the population has an equal probability of being selected.

It is not a simple task to obtain a simple random sample. It requires getting a list of every element in the universe or population, and following a procedure that insures that each element has an equal probability of being selected for the sample. One method of insuring equal probability is to use a table of random digits, such as that of Table IX. Assign each element in the population a number from 1 to N, where N is the number of elements in the population. If there are 5120 elements in the population, and you want to pick 100 at random, you have to assign each element a different number. Then go to the Table of Random Numbers, starting anywhere, and take in order all the 4-digit sequences of numbers that are between 0001 and 5120. If the sampling is to be without replacement, which is the more usual and efficient way to sample, then you skip over any repetitions and take 100 *different* 4-digit numbers. The elements corresponding to these numbers are the sample. Where the sample is to be a large one, you can use a computer to pick your random sample for you.

You can see that picking a simple random sample can be a very time-consuming task. An even more difficult task that is often involved is that of finding and getting information about the elements that have been selected for the sample, particularly when they are geographically spread out. When the population is large or diverse, it is often difficult or impossible to get a list of all the elements in the population. For this reason simple random sampling is not a common procedure in most survey designs.

Why, then, do we devote so much space to simple random sampling, if it is not used more? There are two principal reasons.

1. Many nonrandom sampling procedures (such as quota sampling) are designed to resemble simple random sampling in many respects. Although nonrandom methods generally give the sampler some choice in what or whom to select for the sample, restrictions on the choice can be made so as to minimize judgment biases. If the nonrandom sample is well

designed, it will bear considerable resemblance to the random sample; the biases will be small and the errors will be not much more than those of simple random sampling. Studying simple random sampling will help to design a nonrandom sample that comes close to the ideal of a simple random sample.

2. Many of the more elaborate sampling designs use simple random sampling in some aspects of the design. An understanding of simple random sampling is essential to an understanding of these designs.

Systematic Sampling

You can usually achieve the necessary randomness of a simple random sample with less time and effort by using a systematic sample.

> If there are N elements in the population, listed in order from 1 to N, a SYSTEMATIC SAMPLE of n elements is one that starts with some element 1 to N/n (picked at random) and includes every N/nth element thereafter, counting from the first element chosen.

In picking a systematic sample, you need to pick only *one* random number to get started, and then pick the other elements at equal intervals. For example, to pick a systematic sample of 100 elements from a universe of 5000 elements, you randomly select a number from 1 to 50, and then take every 50th element for your sample. This procedure is quicker and easier than simple random sampling, and it is about as good as long as the elements in the universe are not arranged in cycles (especially cycles that are multiples of 50).

Stratified Random Sampling

Suppose you wanted to determine the mean number of children under 12 years of age for the 2500 employees in a government bureau, and you wanted to restrict your sample size to 100 employees. The employee records give information on employees' age, sex, and religion. You may be able to use some of this information to reduce the standard error of the mean.

Suppose, for example, that the records show that 40% of the employees are between 20 and 40 years of age, and that this age group tends to have the greatest number of children under 12. It would be desirable if exactly 40% of your sample were in this age group. But from what you know about the standard error of a proportion, you know that the actual proportion in a simple random sample of 100 who are in this age group has a standard error of $\sqrt{(.6)(.4)/100}$, or nearly 5 percentage points. A 95% confidence interval for the proportion of this age group in the sample would extend from just over 30% to nearly 50%. But you can reduce this standard error to zero by deciding you will take exactly 40% of your sample from this age group and 60% from other age groups. Such a sampling procedure is known as stratified random sampling.

> STRATIFIED RANDOM SAMPLING is a sampling procedure in which one takes simple random samples from subsets (or strata) of the population in proportion to the size of these subsets.

Stratified random sampling makes most sense when you have good reason to believe that the strata differ markedly from each other in the characteristic that you are measuring. When determining the mean number of (employees') children under 12, it is very advantageous if you can stratify the employees by age, since age groups are likely to differ substantially in the average number of children. You could stratify by sex and religion as well, or by any characteristic for which you know the proportions in the whole population. Whether it is practical to stratify depends on your knowledge of the possible strata; if employees with children under 12 are much more likely to be men than to be women, then sex is a good candidate for stratification; otherwise, there is little to be gained from stratifying by sex.

Cluster Sampling

Stratified sampling is made possible by the fact that you already know some characteristics of the population other than the one you want to find out about. Another sampling design that makes use of prior information about the population is cluster sampling. This is a method of reducing cost and time in getting information. The method consists of selecting one or more subpopulations that you believe provide a "cross-section" of the entire population.

Suppose you wanted to determine the proportion of voters in a state who favor legalization of off-track betting on horse races. You would like to base your estimate on a sample of about 500 voters, but you realize that a simple random sample of 500 voters across the entire state would be extremely difficult and expensive to obtain. Suppose, however, that we know that Ferriss County is typical of the state in a variety of respects, including those that you suspect are most closely related to views on legalization of off-track betting. For example, you know that the distribution of incomes, the ethnic mix, and the religious mix of Ferriss

County is similar to that of the state, and you believe that income, ethnic status, and religion are 3 of the variables most related to views on legalization. Because of its cross-sectional aspects, Ferriss County may be an ideal "cluster," an area in which you might take a sample of 500 voters. More likely, you will find several such clusters and take samples from each of them.

> CLUSTER SAMPLING is a sampling procedure that limits the sampling to only one or a few subsets (or clusters) of the population, in the expectation that these clusters are representative of the population as a whole.

Both stratified sampling and cluster sampling involve random sampling, either within the strata or within the clusters. But sometimes it is simply not practical to get a random sample. If you do not have a list of all the elements in the universe, or if you do not have the time it takes to select or obtain information about a random sample, you do not know the probability that an element will be included in the sample. In many cases its inclusion or noninclusion depends to some degree on the judgment of the person taking the sample. The problem with relying on a person's judgment in selecting a sample, even if the person has good judgment, is that you cannot measure the sampling error due to judgment, as you can when the sampling is done on the basis of probability.

Quota Sampling

One of the most widely used forms of judgment sampling is quota sampling. The method is commonly used in interviews for public opinion polls. Quota sampling is a method of controlling the subjective element in the selection process by strictly specifying the number of individuals in certain categories that the interviewer may select. The interviewer is required to get information about a specified number of people in certain age, sex, race, income, etc., categories. Although such a method is not random, by setting the quotas in the categories that are most closely related to the variables that one is trying to get information about, you can achieve results that are in many cases not very inferior to those from random sampling. The big advantage of quota sampling is speed; you do not have to spend many hours tracking down those people who were randomly selected but hard to find or uncooperative. In quota sampling the interviewer picks people that are readily available. The result is that for the same amount of time and money one can obtain a much larger sample using quota sampling than using random sampling. Quota sampling is especially useful when the results must be up-to-date, such as polls that attempt to measure the very current state of public opinion.

A troublesome problem in quota sampling is that interviewers are usually able to fill their quotas more quickly by going to places where people are concentrated in a small area, such as bus depots, shopping centers, etc. By going to such places, interviewers may be introducing unsuspected biases (e.g., people who avoid shopping centers or crowds may be underrepresented).

When one has the time and resources to design a careful survey, when the estimates are going to be generalized to future cases as well as the present, and when the methods and results are going to be subjected to the appraisal and criticism of scholars in the field, there is no substitute for random sampling.

An understanding of the theory of sampling is indispensable both in social science research and in managerial decision making. People unfamiliar with sampling principles are often predisposed to put little trust in a sample and great faith in a census. In many cases, however, it is impossible or impractical to take a census. And even when a census is feasible it is usually a foolish waste of time and resources to look at every individual. In fact, surprising though it may seem, in many cases a sample will give you better results than a census. Among the reasons for this is fatigue; the time, the effort, and the cumbersomeness of obtaining and collecting information on all the 87,473 citizens of East Bolingbroke permit more biases and recording errors to creep in than if we took a random sample of 5000 of these citizens.

It seems incredible to many that a sample of 10%, 1%, or even 1/100 of 1% of the universe can give very accurate results. The reasons why they can lie in the theory of probability and the theory of random sampling that was discussed in this chapter. Even though the individuals in the population vary in what seems to be a haphazard and unpredictable manner, probability theory has enabled us to measure and anticipate this variability, and to reduce the uncertainty almost to any degree we want merely by adhering to random methods and by choosing the proper sample size. The knowledge of the behavior of samples has enabled research on human behavior to draw conclusions that go far beyond the limited data that investigators have at their disposal. In managerial decision making this knowledge has enabled administrators to take action much more quickly and decisively than would have been possible if it had been necessary to wait until "all the facts were in."

EXERCISES

7–17. In manufacturing a component for a missile system, a manufacturer uses 20 parts of a certain type. If one of the parts is defective, the entire system will be defective. In testing to see whether any of these parts is defective, should the manufacturer take a random sample or a census? Discuss.

7–18. In determining the effects on human beings of taking large doses of saccharin, should one take a census or a sample of the population? Discuss.

7–19. The governor of a state has just made a speech on television urging the adoption of an increased state income tax in order to avoid a financial crisis in the state. In order to get an assessment of the initial reaction of the voters to the proposal, would you suggest taking a census, a random sample, a quota sample, or using some other method of getting this information? Discuss the reasons for your suggestion.

7–20. In order to assess the attitude of the alumni toward a policy of awarding athletic scholarships to women on an equal basis with men, it is planned to select 300 alumni for interviewing. Would a stratified sampling design be useful for this study? If so, on the basis of what variables would you suggest that the stratification be made?

7–21. If you were to select areas of a city as clusters in order to determine voter attitudes toward a proposal to integrate the schools through busing, what characteristics would you look for in selecting the cluster?

7–22. An interviewer used his judgment in selecting a sample of 36 out of 10,000 participants in a health care program. Of these 36, the mean number of visits to the clinic was 4.6 per year and the standard deviation was 2.0. Can you set up a 95% confidence interval for μ, the expected number of visits for all participants in the program? Why or why not?

Summary of Chapter

An arithmetic mean or a proportion computed from a sample is subject to sampling variability. If the sample is a random one, the expected value and standard error of the mean of the proportion can be determined. The expected value of the sample mean or proportion is simply the universe mean or proportion (regardless of the sample size). But the standard error decreases as the sample size increases. It is this fact that enables one to have more confidence in a sample mean or proportion when it comes from a larger sample.

We use the sample mean $\overline{X}$ to estimate μ and the sample proportion $\hat{p}$ to estimate p because these statistics are unbiased and efficient estimators of the respective parameters.

In an independent random sampling procedure, the sampling distribution of a mean or proportion tends toward the normal distribution as the sample size increases. For this reason one can use the Table of Normal Areas when setting up a confidence interval for μ or p.

In determining the optimal sample size it is necessary to determine how large an error one is willing to tolerate. Formulas for the optimal sample size include this tolerable error as well as an estimate of the universe standard deviation or proportion.

An introductory discussion of the theory of sampling assumes a simple random sampling procedure. There is a variety of more elaborate sampling designs that enable one to get more information from a given sample size or a given budget. However, an understanding of simple random sampling is essential to an understanding of these other designs.

REVIEW EXERCISES FOR CHAPTER 7

7–23. The expected number of days in a school year that third grade children lose from sickness is 11.4 and the standard deviation is 3.0 days.

a) What is the sampling distribution of the mean number of days lost from sickness of a randomly selected group of 36 third grade children?

b) What is the probability that the mean of the random sample of 36 will be more than 13.0 days?

7–24. In a random sample of 64 deans of graduate schools, 20% were able to give the name of the Secretary of Labor. Set up a 95% confidence interval for the proportion of all deans of graduate schools who can identify the Secretary of Labor.

7–25. If you take a random sample of 20 civil servants at Grade GS-11 in order to estimate μ, the expected value of the number of years of service of all GS-11 civil servants,

a) find a statistic V that would have an upward bias in estimating μ, (i.e., on the average it would overestimate, so that $E(v) > \mu$).

b) find a statistic W that would have a downward bias in estimating μ.

c) find an unbiased estimator of μ.

7–26. A random sample of 73 retired employees from Berea Enterprises, Inc., yielded a mean annual pension of $1763 and a standard deviation of $420. Set up a 95% confidence interval for the mean size of the annual pension for all retirees from Berea Enterprises.

7–27. A new surgical procedure has been tried on 84 patients. Of these, 59, or approximately 70%, recovered. Assuming that these 84 patients constitute a random sample of all future patients undergoing this procedure, set up a 90% confidence interval for the probability of recovery from this operation.

7–28. A state senator wants to know the proportion of voters in his district who favor a proposed bill that would prohibit the hiring of homosexuals in state jobs. To find this proportion, the senator's research staff is considering conducting a stratified sample, a cluster sample, or a combination of both. Which of the following groups do you think would more likely qualify as (1) a stratum (singular of strata), (2) a cluster, or (3) neither? Give reasons for your choices.

a) registered Republicans

b) voters over 65 years of age

c) voters whose names start with R

d) homosexuals

e) members of the American Civil Liberties Union

f) voters living in a precinct that is considered "typical"

g) voters who live in a two-block area known for its gay population

7–29. There has been a mass inoculation program in Bourbon County of a new experimental influenza vaccine. A group of scientists want to learn the proportion of vaccine recipients who contract influenza in the succeeding 6-month period. Since the rate of influenza in the whole population is anticipated to be no more than 20%, the scientists are willing to assume that among the vaccine recipients at the very most 20% will come down with the flu. In order to be within 3 percentage points of the true proportion with a .95 degree of confidence, how large a sample should the scientists take?

7–30. Which of the following statements are true?

a) If you quadruple the sample size, you will reduce your error to 1/4 of what it was.

b) If you quadruple the sample size, you will reduce your error to 1/2 of what it was.

c) If you quadruple the sample size, you will reduce the 95% of confidence interval to 1/2 of what it was. Assume in each case that the population is large or infinite.

7–31. Example 7.1 presented a problem involving a sample of 2 fathers out of a population of 6 fathers whose numbers of children were 1, 2, 3, 4, 5, and 6. Is there any difference in principle between the problem of finding the mean number of children in a sample of 2 and the mean number of spots showing on the roll of two dice? Why or why not?

SUGGESTED READINGS

COCHRAN, W. G. *Sampling Techniques,* 2nd ed. New York: John Wiley and Sons, Inc., 1963.

HAMBURG, MORRIS. *Statistical Analysis for Decision-Making,* 2nd ed. New York: Harcourt, Brace, Jovanovich, Inc., 1977. Chapter 5.

HANSEN, M. H.; W. N. HURWITZ; and W. G. MADOW. *Sample Survey Methods and Theory,* Vol. I: *Methods and Applications.* New York: John Wiley and Sons, Inc., 1953.

MORONEY, M. J. *Facts from Figures,* 3rd ed. Harmondsworth, Middlesex: Penguin Books, Ltd., 1956. Chapters 10 to 12.

EIGHT

Hypothesis Testing and Decision Making

Outline of Chapter

8.1 The Null Hypothesis

The null hypothesis is the "pure chance" hypothesis. It says that any difference between the result you would normally expect and the result you got is purely due to chance.

8.2 Statistical Tests for Means from Large Samples

The sample mean is used to make a decision about the mean or expected value of the population.

8.3 Tests for Proportions from Large Samples

The sample proportion is used to make a decision about the proportion or probability of success in the population.

8.4 Tests for Means: Small Samples with σ Unknown

The test statistic for estimating the mean of a normally distributed population when σ is estimated from the sample has a t distribution.

8.5 Decision Making and Policy Analysis

In making a policy or a decision it is often useful to use the criterion of maximizing expected value.

*8.6 Making Decisions on the Basis of Sample Information

By use of Bayes' Theorem one can incorporate prior probabilities, sample information, and utilities or payoffs to arrive at a decision.

Objectives for the Student

1. Understand what is meant by accepting or rejecting the null hypothesis.
2. Run a statistical test for a mean from a large sample.
3. Run a statistical test for a proportion from a large sample.
4. Run a t test for a mean.
5. Be aware of the conditions that warrant the use of the t distribution.
6. Draw up a payoff table in a problem of decision making under uncertainty.
7. Understand the meaning of a utility function.

*8. Use Bayes' Theorem to incorporate sample information into a decision.

The study of probability and sampling distributions should give you an idea of how deceptive sample data can be. A run of bad luck in sampling can lead you to reject a shipment of goods that was really up to standard; pure chance can lead you to conclude that a new treatment is an improvement over the old, when in fact it is no more effective; the conclusion that blondes have more fun in life than brunettes can arise from an unrepresentative sample rather than from any fundamental differences associated with hair color.

The testing of hypotheses differs from the methods of estimation that you studied in Chapter 7 in that hypothesis testing involves a decision: to accept or reject a lot of goods, to go ahead with a new method or to stick with the old, to accept a new proposition or to stay with the "conventional wisdom."

The testing of hypotheses is particularly relevant when one needs to establish, or "prove," a proposition beyond a reasonable doubt. For example, a prospective blood donor will generally be rejected if there is a reasonable doubt that the donor's blood is free of hepatitis.

Statistical decision theory, on the other hand, is an alternative approach to hypothesis testing that lends itself to choices among several courses of action without requiring conclusive "proof" that one course of action is better than another. Decision theory can be used to determine which of 2 or more competing policies or courses of action should be adopted. Making use of Bayes' Theorem, decision theory makes use of sample information to revise prior probabilities; it also is used to determine the value of information that one can expect to get from a sample. The presentation of decision theory in this chapter will be very brief and introductory; a more thorough account of the subject belongs in a more advanced course in statistics.

8.1 The Null Hypothesis

Betty Bishop was full of excitement when she came home saying that her boyfriend Pete had extrasensory perception. "I picked a card out of the deck and asked him to guess what card it was, and he correctly answered 'four of clubs' without looking at the card."

"Could be a trick," suggested her father.

"I don't see how," she said. "It was my deck of cards, Pete had not touched the deck, and anyway, Pete is not a card trickster."

"Could be he was just lucky," said her mother.

"But there was just one chance in 52 he would have guessed right," Betty pointed out," so I don't think it was just luck."

"On the other hand, it may be easier to believe in a lucky call than to believe in ESP," her mother said.

Statisticians are skeptical people. No matter how intriguing the result of a particular experiment or other experience may be, they tend to pose the unwelcome question, Was it simply due to chance? If 13 out of 13 patients who took Dr. Quack's Amazing Remedy were cured of their headaches, the statistician will suggest that this might be due to pure chance. And if someone picked the right card out of the 52, the statistician will throw cold water on the whole discussion by posing the hypothesis that it was all due to chance.

The compelling force of the hypothesis of pure chance is that it is so mundane and believable, unless, of course, the probability of the events occurring by chance is very low. The philosopher David Hume, in writing on miracles, put forward a convincing argument against belief in reported miracles, by proposing that the less improbable, or less miraculous, of two ideas carries a greater weight of belief. "No testimony is sufficient to establish a miracle, unless the testimony be of such a kind that its falsehood would be more miraculous than the fact which it endeavors to establish." If, for example, Hume believed that having ESP was more miraculous, or improbable, than guessing the right card out of 52, he would opt for chance as a more reasonable explanation than ESP. In proposing the hypothesis of pure chance the statistician is proposing that pure chance is the least miraculous, and therefore, by Hume's standards, the most believable, hypothesis. We give the hypothesis of pure chance a primary consideration among the possible explanations of a phenomenon, largely to protect us from jumping to exciting and elaborate conclusions not warranted because a more mundane and believable explanation will do. This hypothesis therefore initially warrants the focus of our attention. Only if we show that pure chance is not a reasonable explanation, can we consider the other alternatives.

> The hypothesis that we focus our attention on and test is known as the NULL HYPOTHESIS, and it is designated as H_0. It is a statement *about* the value of a universe parameter. It gets its name from the fact that, according to the hypothesis, there is *no difference* between the hypothesized value of the parameter and a sample statistic, other than what we can reasonably attribute to chance.

If I were to flip a coin and you called it "heads" correctly, would I conclude that you have psychic predictive powers? Of course not, since you have a .50 probability of calling the coin right, even without such powers. The null hypothesis is that p, the probability of your calling the coin correctly, is .5.

If you called the coin correctly twice in a row, I would still stick with the null hypothesis, since the probability of your guessing correctly is .25, and I consider this more believable than that you have psychic powers.

However, if you called the coin correctly 5 times in a row (with no previous false calls), then I may be tempted to reject the null hypothesis, since the probability of 5 straight correct calls, according to the hypothesis, is only 1/32, or .031. It is only when the probability of the sample result according to H_0 is very low (.05 or less), will a cautious person be inclined to reject the null hypothesis. This follows from a reluctance to reject an easily believable hypothesis (that you have a 50–50 chance of guessing right on each call) in favor of a more fantastic hypothesis (that you have psychic powers), unless the evidence is overwhelming.

There is another reason why we might want to stick to the nullhypothesis unless the evidence to the contrary is overwhelming. This reason lies in the seriousness of the consequences of error. Suppose, for example, that you found a mushroom and that you were unsure of whether or not it was poisonous. If we can assume that you are a mushroom lover, you would eat the mushroom if you believed that it was nonpoisonous. When you decide on a course of action, there are two possible errors to consider: you could eat a poisonous mushroom, or you could throw away a good mushroom. Everyone but the most passionate mushroom lover would probably say that the first of these two errors is the more serious error. In case of doubt, therefore, you would act as though the mushroom were poisonous and throw it away. Therefore the hypothesis that receives the benefit of the doubt is that the mushroom is poisonous. This becomes H_0, the null hypothesis; it is the hypothesis that you would not reject unless the evidence to the contrary were overwhelming. The two errors can be displayed as in Table 8.1.

Rejecting the null hypothesis H_0 when it is true is known as a Type I Error; accepting the null hypothesis when it is false is known as a Type II Error. Because the null hypothesis is the one we want to favor in case of doubt, we should set up the hypotheses in such a way as to make the Type I Error the more serious error.

We set the hypotheses up in this manner because when we design a test for a null hypothesis, we can directly control the probability of a Type I Error. It is naturally desirable that we have the more direct control over the more serious of the errors.

In testing a null hypothesis, H_0 is always tested against an alternative hypothesis.

> The ALTERNATIVE HYPOTHESIS, designated as H_1, is a statement that the value of the population parameter is different from that specified by H_0.

If you reject H_0, then you accept H_1. Since H_0 is given the benefit of the doubt, H_1 is saddled with the burden of proof. You may reject H_0

Table 8.1. *POSSIBLE ERRORS ARISING FROM DECISION ON THE NULL HYPOTHESIS*

		State of Nature	
		H_0 is true	H_0 is false
Action	Accept H_0	No error	Type II error
	Reject H_0	Type I error	No error

and accept H_1 only if you can show beyond a reasonable doubt that H_0 is false.

In order to verify a scientific hypothesis the scientist must place the burden of proof on that hypothesis. The hypothesis that is to be proved, therefore, becomes the alternative hypothesis; the null hypothesis H_0 becomes the one to be disproved. That is, the scientist must discredit the hypothesis (H_0) that his or her sample results can be attributed to chance by demonstrating that there is a greater probability they can be attributed to an alternative (H_1) agent. This puts the research scientist in the position of assuming and giving the benefit of doubt to the hypothesis that she or he is trying to disprove. It is small wonder that those who are insecure about the validity of their theories and hypotheses are reluctant to put them to statistical tests.

EXAMPLE 8.1 It is a premise of the English legal heritage that it is a more serious error to convict an innocent person than to acquit a guilty person.. If a person is brought to be tried for a crime, what is the null hypothesis and what is the alternative hypothesis?

Discussion: The more serious error should be designated as the Type I Error. This is the error of convicting an innocent person. The null hypothesis is therefore that the person is innocent, and the Type I Error is to reject that hypothesis when it is true. This is another way of saying that a person is presumed innocent until proven guilty, and that the burden of proof must rest on the prosecution, which must show "beyond a reasonable doubt" that the person is guilty. ■

Levels of Significance

In a world of certainty we could always avoid both Type I and Type II Errors. But in the real world, and especially that of administration, sample evidence is unreliable, and we are inevitably going to make errors. One way we can cope with this situation is to decide what probability of error we are going to allow ourselves. Naturally, we would like the probabilities of both the Type I and the Type II Errors to be very low—as close to zero as possible. But reducing both of these two errors will require bigger samples, and we may be limited in the amount of sampling we can do. For a given sample size, we will find ourselves in the frustrating position of being able to decrease the probability of one error only by allowing the probability of the other error to increase. To cope with this situation we can focus our attention on the more serious error, the Type I Error, and ask ourselves, "How large a probability of a Type I Error are we willing to tolerate?"

> In testing a null hypothesis, the LEVEL OF SIGNIFICANCE of the test, denoted by the Greek letter α (alpha), is the probability of committing a Type I Error.

The more benefit of the doubt we wish to give to the null hypothesis, the lower we will want to set α. It is a common practice to set α at .05. In situations where a Type I Error is particularly costly, we might set α at .01. In exploratory studies, or in studies based on small samples, people are commonly less stringent about the Type I Error, and let α go as high as .10 or even .20.

In determining whether Betty Bishop's boyfriend has ESP on the basis of his correctly guessing one card, we would have rejected the null hypothesis if the level of significance α had been set at .05. According to this level of significance, a result that has a probability of less than .05 would be a result sufficient to reject the null hypothesis. Since the probability of guessing the right card is only $1/52 = .019$, we can reject H_0 and conclude that he has ESP.

EXAMPLE 8.2 In order to test the effectiveness of a new vaccine, a public health official has inoculated 1000 randomly selected elderly individuals with the vaccine and compared the results with those for 1000 randomly selected elderly individuals who were inoculated with a placebo. If the results show conclusively that the new vaccine is effective, then a major program of mass inoculation of elderly persons would be undertaken. Because of the cost of such a program, the official considers that it would be a very serious error to go ahead with the mass inoculation unless the results of the test are conclusively in favor of the new vaccine.

What should be the null hypothesis, and what should be the alternative hypothesis? What level of significance would you suggest be used?

Discussion: The error of greater concern to the official is to say that the new vaccine is effective when it really is not. This would be the Type I Error. The null hypothesis would therefore be that the vaccine is not effective, and the alternative hypothesis, which must be conclusively established, is that the vaccine is effective. Because of the seriousness of the Type I Error, the level of significance should be set quite low—perhaps at .01. ■

EXAMPLE 8.3 In a preliminary test to see whether a vaccine is effective, the official gives the vaccine to 20 people. If there is some evidence that the vaccine is effective, then there will be a follow-up study which will involve the inoculation of 2000 people. In the initial study, what should be the null hypothesis and the alternative hypothesis? What level of significance would you suggest be used?

Discussion: As in the previous example, the burden of proof should rest on the hypothesis that the vaccine is effective, so that H_0 would be that the vaccine is not effective. On the other hand, it would be a mistake to make the burden of proof too severe. In a small sample you would need spectacular results to reject H_0 at a low level of significance. Furthermore, a Type I Error is not as serious here as in the previous example, since it would lead us to a follow-up test which we would expect would accept the null hypothesis. A level of significance of .10 or even .20 might be appropriate. ■

The whole question of where the "burden of proof" lies is very crucial to policy-oriented research. Public policy questions can be influenced profoundly by where we put the burden of proof, and what we set up as the null hypothesis. For example, in testing a drug, is it incumbent on the tester to prove that the drug is safe, or to prove that it is unsafe? The same question of burden of proof pertains to questions about the possible harmfulness of alcohol, tobacco, marijuana, air pollutants, and food preservatives, to name a few examples. The answer to the question of burden of proof will lie largely in the analysis of the costs of the two types of errors.

EXERCISES

8–1. A patient has a small tumor that might be malignant. If the tumor is malignant, it should be removed at once; if it is benign, there is no necessity to remove the tumor.

a) If you were the patient, what is the null hypothesis that you would want to be tested?

b) Describe, in terms of this example, the four possible consequences as outlined in Table 8.1.

c) What level of significance would you think is appropriate in this problem if you were the patient?

8–2. Dr. Davis, the Commissioner of Forbes County, has declared that there is no need to impose stricter air pollution control standards in the county, because there is insufficient scientific evidence that current levels of pollution are harmful to human life and health.

a) What null hypothesis and alternative hypothesis can you infer Dr. Davis is setting up when he makes this statement?

b) How else might he set up the hypotheses?

c) Which way do you believe is more appropriate?

8–3. You are driving a car on a narrow winding road behind a slow truck. You would like very much to pass the truck. Because there is very little traffic coming the other way, you believe that it is very unlikely that you will meet an oncoming vehicle when you try to pass the truck, but you cannot be absolutely certain because the winding road does not permit you to see more than 100 yards in front of you. In deciding whether or not to take a chance and pull out to pass the truck,

a) what is the null hypothesis that you would test?
b) what are the consequences of a Type I Error and of a Type II Error?
c) what level of significance would you suggest is appropriate?

8.2 Statistical Tests for Means from Large Samples

In testing a null hypothesis about a parameter from sample evidence, we look at a statistic from the sample and decide whether or not the value of the statistic is a reasonably likely value according to the null hypothesis. If it is reasonably likely, we accept the null hypothesis; if it is not, we reject. For example, if you wanted to test the hypothesis that half of all aardvarks are male, and if 11 aardvarks are male in a random sample of 20 aardvarks, you would accept the null hypothesis, since 11 out of 20 appears to be a reasonable outcome according to the hypothesis. But if 19 out of 20 were male, you would probably reject the hypothesis. Although such an outcome is possible according to the hypothesis, it is so improbable that you don't believe the hypothesis. It would be more miraculous, to paraphase Hume, to get 19 out of 20 males, than to believe the sex ratio of aardvarks was something other than 50–50.

In this section we will discuss in some detail the procedure for testing a hypothesis about a population mean or expected value.

In a campaign to reduce gasoline consumption, the Mayor of Oldsborough has led a publicity campaign urging citizens not to drive downtown to work unless one and preferably more than one passenger was in the car. It was felt that the campaign would be successful if cars going downtown in the morning had a mean of more than 2.8 persons per car. In order to determine whether the campaign was successful, a sample of 100 cars crossing the Jimmy Foxx Bridge into Oldsborough between the hours of 8:00 and 9:00 was taken. The mean of the sample was 2.9 persons per car. Assuming that the standard deviation of persons per car is 1.00, and that the sample is a random one of cars going to work in Oldsborough, can we say with a reasonable degree of certainty that the goal of more than 2.8 persons per car has been achieved?

In order to establish a hypothesis with reasonable certainty, we must place the burden of proof on that hypothesis. Therefore, the hypothesis that we want to show, that the mean occupancy rate of cars is more than 2.8 persons, must be the alternative hypothesis. The null hypothesis, which is accepted as true until proven otherwise, is that the mean occupancy rate of cars is not more than 2.8. The hypothesis can be stated, therefore, as

H_0: $\mu \leq 2.8$
H_1: $\mu > 2.8$

Assume, for the time being, that H_0 is true. Under this hypothesis, the mean $\overline{X}$ of a sample of size 100 will have a sampling distribution that

1. is normal
2. has an expected value $\mu_{\overline{X}} = 2.8$
3. has a standard error $\sigma_{\overline{X}} = \sigma/\sqrt{n} = 1/\sqrt{100} = .1$

The reasons for these characteristics of the sampling distribution of $\overline{X}$ are

1. The Central Limit Theorem says that the sampling distribution of $\overline{X}$ approaches normal as n gets large.
2. The Central Limit Theorem says that $\mu_{\overline{X}}$ is equal to the population expected value μ. Although the null hypothesis says that $\mu \leq 2.8$, we actually test the hypothesis that $\mu = 2.8$. We do this because we need to *specify* the value of the parameter in order to specify the sampling distribution of the statistic we use to test H_0. It is for this reason that in setting up a pair of competing hypotheses, we must always include an equal sign in the null hypothesis because it is the "equality" hypothesis that we specifically test. The null hypothesis is therefore stated as an *equality*.
3. The standard error of the mean is based on the assumed value of σ, which is 1.00. If we do not know or assume σ when we set up the null hypothesis, we must estimate σ from s, the sample standard deviation. When the sample is large $(n \geq 30)$, we can use the method presented in this section with $\sigma_{\overline{X}} = \frac{s}{\sqrt{n}}$.

Notice that the standard error of the mean is given by Equation 7.3 rather than by Equation 7.2. Unless otherwise stated, you can assume that the population from which the sample is drawn is either infinite or so large that the finite population correction factor $\sqrt{(N-1)/(N-n)}$ is close enough to 1 that we can ignore it. The sampling distribution of $\overline{X}$ is shown in Figure 8.1.

Suppose we are willing to commit a Type I Error with a probability of .05. Then the statistic $\overline{X}$ should give us a "rejection signal" with a probability of .05, even though H_0 is true. That means that 5% of the area, or probability, under the curve in Figure 8.1 will be designated as the rejection region, or critical region.

Where is this critical region going to be? To answer this question, you should consider what the alternative hypothesis says. If you reject H_0, then you accept H_1. In the present case, H_1 says that μ is *greater* than 2.8. It is reasonable to make the rejection region the *high* values of $\overline{X}$, those that would give a strong signal that μ is greater than 2.8, or that H_1 is true. Therefore we designate the *highest* 5% of the area under the curve in Figure 8.1 as the region of rejection, or critical region, for H_0.

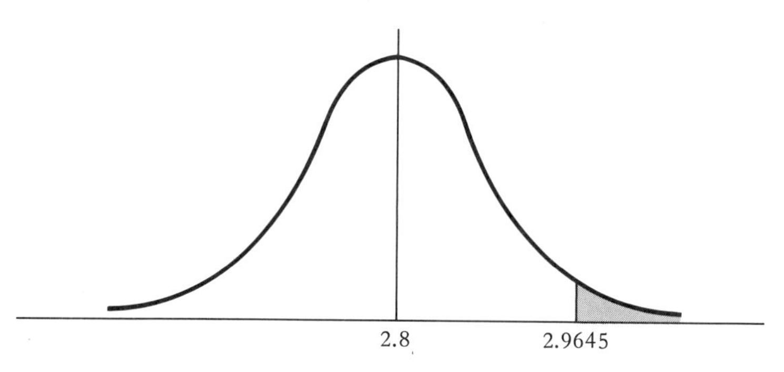

Figure 8.1. *SAMPLING DISTRIBUTION OF $\overline{X}$ UNDER THE NULL HYPOTHESIS*
H_0: $\mu \leqslant 2.8$ ($\sigma = 1.0$, $n = 100$)
The Shaded Area Is the Rejection Region of $\overline{X}$ (Where H_1 Is Accepted) at the .05 Level of Significance

> The CRITICAL VALUE of $\overline{X}$, denoted as $\overline{X}_\alpha$, is that value of $\overline{X}$ above which the critical region having an area of α lies.

In this example, we want to find $\overline{X}_\alpha$, so that

$$P(\overline{X} > \overline{X}_\alpha) = \alpha = .05$$

To find $\overline{X}_\alpha$, find z_α, where

$$P(z > z_\alpha) = \alpha = .05$$

From Table II, we find, by looking up the probability .4500 in the body of the table, that $z_{.05} = 1.645$. Applying Equation 6.2 we have

$$\overline{X}_{.05} = \sigma_{\overline{X}} z_{.05} + \mu$$

so that

$$\overline{X}_{.05} = (.1)(1.645) + 2.8 = 2.9645$$

The critical region in Figure 8.1 is the shaded area to the right of $\overline{X}_{.05} = 2.9645$. This critical value gives you a *decision rule* for testing H_0.

If $\overline{X} > 2.9645$, reject H_0 (and say that μ is greater than 2.8).
If $\overline{X} \leqslant 2.9645$, accept H_0 (and say that μ has not been shown to be greater than 2.8).

Notice that the conclusion is a much stronger one when we reject H_0 than when we accept it. Since the burden of proof is on the alternative hypothesis, a rejection of H_0 means that there is substantial evidence that H_0 is false. When we accept H_0, however, we do so not on the basis of strong evidence that it is true, but only on the basis of lack of strong evidence that it is false.

An equivalent method of running this test is to use the standard normal variable

$$z = \frac{\overline{X} - \mu}{\sigma/\sqrt{n}}$$

as the test statistic. Figure 8.2 shows the sampling distribution of z and the critical region to the right of $z_{.05} = 1.645$. The decision rule can then be stated:

If $z > 1.645$, reject H_0.
If $z \leqslant 1.645$, accept H_0.

It is only now, after we have set up the decision rule, that we should look at the sample result. The reason for maintaining this "decision rule first: sample later" rule is to prevent the sample result from influencing our decision rule. Most investigators come to their research with a bias: they want to reject the null hypothesis because they are trying to show that the alternative is true. Being able to reject may mean the difference between having the study published or not. Suppose, therefore, that an ardent young untenured social scientist comes up with a result that is significant at the .10 level but not at the .05 level. Can he or she resist the temptation of saying, "I think the .10 level of significance is appropriate to use in this situation."

Having set up a decision rule, we may now look at the data. The mean of a sample of 100 observations is 2.90. According to the decision rule stated in terms of $\overline{X}$, we should accept H_0, since 2.90 is not in the critical region; it is below the critical value of 2.9645.

Figure 8.2. *SAMPLING DISTRIBUTION AND CRITICAL REGION OF THE STANDARD NORMAL VARIABLE z*
Transformed from $\overline{X}$ in Figure 8.1

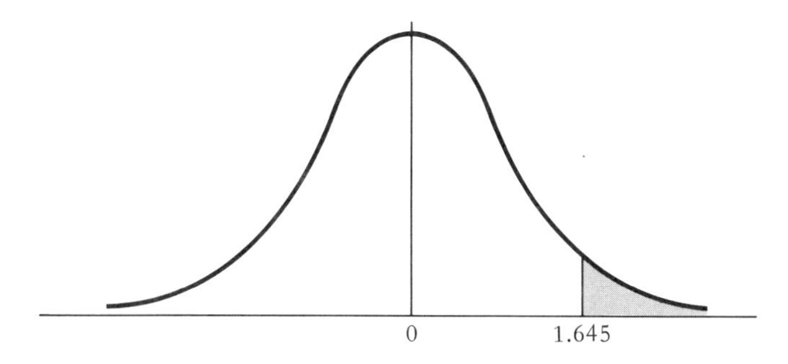

Alternatively, we could use the test statistic z. The value of z from the sample is

$$z = \frac{2.9 - 2.8}{1/\sqrt{100}} = +1.00$$

According to the decision rule stated in terms of z, we should accept H_0, since $+1.00$ is not in the critical region; it is below the critical value of 1.645.

The tests using the test statistic $\overline{X}$ and the test statistic z are equivalent, so that the decision will be the same whichever test you use. In this case we accepted the null hypothesis. In terms of this problem, this means that we do not have sufficient evidence to say that the expected value of the number of persons per car is more than 2.80; chance alone could have accounted for the 2.90 mean that we got, so that the null hypothesis stands, perhaps slightly bruised and battered by the sample result, but still unrefuted. It would have taken a higher value of $\overline{X}$ or z to knock it down at the .05 level of significance.

When the sample statistic falls in the critical region, it is commonly said that the results are statistically significant at the .05 level, or that the sample result is significantly different from the hypothesized value of μ.

> A sample result is said to be STATISTICALLY SIGNIFICANT, or SIGNIFICANTLY DIFFERENT, from a hypothesized parameter, if we can reject the null hypothesis and say we are not willing to attribute the difference to chance. If the sample statistic falls in the critical region of size α, then we say that the sample result is SIGNIFICANT AT THE α LEVEL OF SIGNIFICANCE.

EXAMPLE 8.4 A small religious community has been so isolated from the rest of society that marriages between those in the community and outsiders never occur. The result has been many generations of inbreeding, so that almost everyone in the community is a not-too-distant cousin of everyone else. A random sample of 64 members of the community took a standard aptitude test, and their mean score was 530. If it is known that the expected value of aptitude scores in the population as a whole is 500 and the standard deviation is 100, can we say that the expected value of aptitudes in this community is different from that in the rest of the country? Assume that the community is large enough that we can ignore the finite population correction factor.

Discussion: In order to establish the hypothesis that the community's expected aptitude score is different from 500, we must reject the null hypothesis that μ is 500. The alternative hypothesis we seek to establish is that μ is different from 500. The hypotheses can be stated as follows:

$$H_0: \mu = 500$$
$$H_1: \mu \neq 500$$

Under the assumption that H_0 is true, the test statistic $\overline{X}$ will be normally distributed, have an expected value $\mu_{\bar{x}}$ of 500 and a standard error $\sigma_{\overline{X}} = 100/\sqrt{64} = 12.5$.

The critical region will have a total area of .05, which will be split into two areas, the upper "tail" of the sampling distribution of $\overline{X}$ and the lower "tail" of the sampling distribution of $\overline{X}$. Each tail will have an area of .025. The reason for the two tails is the nature of the alternative hypothesis H_1. If $\overline{X}$ is far away from $\mu = 500$ in *either direction,* we will be inclined toward the alternative hypothesis that μ is *not equal* to 500. The sampling distribution of $\overline{X}$ and the critical region are shown in Figure 8.3.

We want to find two critical values of $\overline{X}$, each of which cuts off .025 in the tail. To do this go to Table II and find the 2 z values $z_{.025}$ and $-z_{.025}$ that cut off .025 of the area in their respective tails. Since

$$P(-z_{.025} < z < z_{.025}) = .95$$

it follows from the symmetry of the normal distribution that

$$P(0 < z < z_{.025}) = .475 \qquad \text{and} \qquad P(-z_{.025} < z < 0) = .475$$

Look up the probability .475 in the table, find $z_{.025}$ to be 1.96 and $-z_{.025}$ to be -1.96. The critical values of $\overline{X}$ are, from Equation 6.2

$$\sigma_{\overline{X}}z + \mu \qquad \text{and} \qquad \sigma_{\overline{X}}(-z) + \mu$$

which equal, respectively,

Figure 8.3. *SAMPLING DISTRIBUTION OF $\overline{X}$ AND CRITICAL REGION FOR EXAMPLE 8.4*

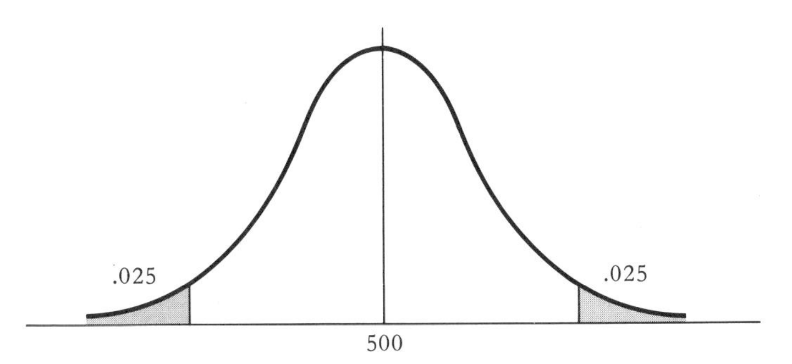

$$(12.5)(1.96) + 500 = 524.5 \qquad \text{and} \qquad (12.5)(-1.96) + 500 = 475.5$$

The decision rule is:

If $\overline{X}$ is between 475.5 and 524.5, accept H_0.
Otherwise, reject.

Stated in terms of z, the decision rule is:

If $-1.96 < z < 1.96$, accept H_0.
Otherwise, reject.

The sample mean is 530.0. According to our decision rule, we reject.

If we calculate z, we find it to be (from Equation 6.1)

$$z = \frac{530 - 500}{12.5} = +2.4$$

This is greater than 1.96, and is, therefore, in the critical region.

The conclusion, whichever test statistic you use, is that the expected value of the aptitude scores in the religious community is different from that of the population as a whole. Having disposed of the null hypothesis, those doing the study are now in a position to explore other hypotheses about why the community differs in this respect, and what hereditary factors, environmental factors, etc. may contribute to this difference. ■

EXERCISES

8–4. The sulfur dioxide count in a particular area is supposed to be no more than 0.12 parts per million. A random sample of 36 readings of SO_2 counts at 8:00 A.M. on weekdays yielded a mean 0.10 parts per million. Assume that the standard deviation of readings is 0.6 parts per million. Can you say, at the .05 level of significance, that the mean level of readings is below the maximum standard of 0.12?

8–5. A maintenance crew replaces fluorescent lights in an office building on a regularly scheduled basis. The assumption underlying the scheduling is that lights last on the average 1250 hours and the standard deviation is 90 hours.

In a controlled experiment, 225 lights were selected at random and the times were recorded until they burned out. The mean time was 1230 hours. Does this result give evidence, at the .05 level of significance, that the assumption of an expected value of 1250 hours was incorrect?

8–6. Which of the following pairs of hypotheses would call for two-tail tests, and which for one-tail tests?

a) H_0: People on Diet #313 do not gain weight on the average.
H_1: People on Diet #313 do gain weight on the average.

b) H_0: People on Diet #314 do not change weight on the average.
H_1: People on Diet #314 do change in weight on the average.

c) H_0: The expected time to drive from Shanklin to Dover is 45 minutes.
H_1: The expected time to drive from Shanklin to Dover is not 45 minutes.

d) H_0: These potatoes are not big enough.
H_1: These potatoes are big enough.

8–7. A random sample of 49 young men from the Hale and Harty Club had a mean time of 7 minutes and 52.5 seconds in running the H and H mile. If you assume that the standard deviation of running times is 30 seconds, is mean running time of the 49 significantly lower than 8 minutes? Let $\alpha = .05$.

8–8. A survey is to be taken to determine the mean amount of time spent on coffee breaks during the working day. If the result of the survey leads Mr. Folger to believe that the mean amount of time is more than 20 minutes a day, he plans to write a nasty memo to all the employees about it.

Assume that the sample size is to be 36 employees, and the standard deviation of time spent on breaks is known to be about 6 minutes. Set up a decision rule, based on the sample result, so that Mr. Folger will run a risk of no more than .01 of sending out the nasty memo, if the true mean is no more than 20 minutes.

8–9. The Lo-Cal Cola Company advertises that its drinks contain no more than 30 calories per 12-ounce bottle. Investigators from a federal agency took a random sample of 40 bottles and found that the mean caloric content per bottle was 32.2.

Is there sufficient evidence to prosecute the company for a fraudulent claim? Assume that the standard deviation of caloric content per bottle is 5 calories. Set up your decision rule so that prosecution of the company when it is meeting standards is the Type I Error, and you want to keep α at .01.

8.3 Tests for Proportions from Large Samples

Statistical tests for proportions are used when we want to determine whether the proportion of the elements in a population having a certain characteristic is greater than (or less than) some specified value. The logic of the approach is identical with that outlined in the previous section for means. As in the case for means, the hypothesis you test (the null hypothesis) is the one you are trying to refute. You initially assume that the null hypothesis is true, find the sampling distribution of a test statistic under the assumption that H_0 is true, select a critical region which will lead you to reject H_0, and then take the sample.

EXAMPLE 8.5 A psychologist conducted a motivational research study among highway repairmen to determine whether wages were a more important incentive to perform well in their jobs than other factors. The psychologist's research hypothesis was that over 60% of highway repairmen were motivated primarily by factors other than wages. In order to demonstrate this contention, the psychologist took a random sample of 96 repairmen and found that 64 of them (66.66%) were measured as being motivated primarily by factors other than wages. Does this result establish the psychologist's original claim?

Discussion: In order to establish the original hypothesis, the psychologist must reject the null hypothesis that no more than 60% are primarily motivated by nonwage factors. The null hypothesis to be tested is:

$$H_0: \quad p \leq .60$$
$$H_1: \quad p > .60$$

Where p is the proportion in the universe who are motivated primarily by factors other than wages.

The statistic $\hat{p}$, the proportion in a sample of 96 who indicate primary motivation from nonwage factors, is used as the test statistic. In a sample of size 96,

$$E(p) = p = .60$$
$$\sigma_{\hat{p}} = \sqrt{\frac{p(1-p)}{n}} = \sqrt{\frac{(.60)(.40)}{96}} = .05$$

and the sampling distribution of $\hat{p}$ will be a normal probability distribution, as shown in Figure 8.4. The critical region will be in the upper tail, since the alternative hypothesis H_1 says that $p > .60$. If we test H_0 at the .05 level of significance, we can find the critical value of $\hat{p}$ by referring to Table II. This table tells us that 5% of the area of the normal

Figure 8.4. *SAMPLING DISTRIBUTION OF $\hat{p}$ WHERE $n = 96$, $p = .60$*
The Shaded Area Is the Critical Region of the Test for H_0: $p = .60$ Against H_1: $p > .60$ at the .05 Level of Significance

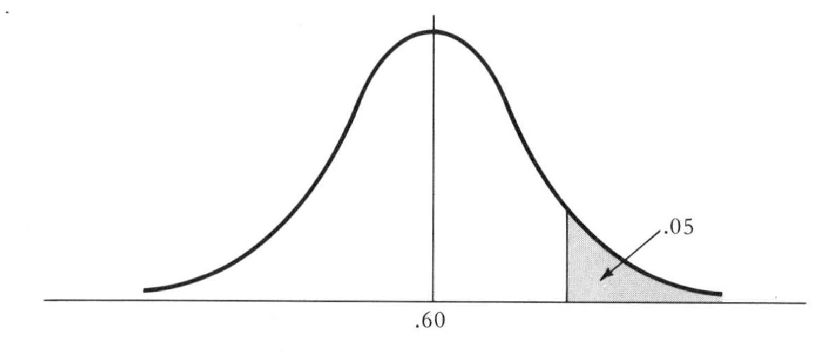

distribution is to the right of $z = 1.645$. The critical value of $\hat{p}$ will be that which is 1.645 standard errors of $\hat{p}$ above its expected value, or

$$p + 1.645\,\sigma_{\hat{p}} = 0.60 + 1.645(.05) = 0.68225$$

The decision rule is:

If $\hat{p} > 0.68225$, reject H_0.
If $\hat{p} \leqslant 0.68225$, accept H_0.

Since the observed value of $\hat{p}$ is .666, we accept H_0. The psychologist's claim is not established.

If we use the standard normal variable z as the test statistic, the decision rule becomes:

If $z > 1.645$, reject H_0.
If $z \leqslant 1.645$, accept H_0.

The statistic z is computed to be

$$z = \frac{\hat{p} - p}{\sigma_p} = \frac{.666 - .60}{.05} = 1.333$$

which is in the acceptance region. ■

A common error made by students in testing for a proportion is to use the observed value of $\hat{p}$, rather than the hypothesized value of p, in determining the standard error of $\hat{p}$. Remember that the standard error of $\hat{p}$ is $\sqrt{p(1 - p)/n}$, where p is the value of p that we *assume under the null hypothesis.* If you set up your decision rule before observing the sample statistic, you will avoid this error.

The logic of the testing procedure is identical with that of the test for means. The logic may become a little clearer if you compare it with the dialectic form of argumentation that Plato employed in his dialogues. The argument is of the form: if you want to show that A is true, assume that its contrary A' is true. Then show that A' implies some consequence that we know is absurd or contrary to our experience. Since the assumption of A' leads to an absurdity, then A' must be false and hence, A must be true. The argument of the statistical test is of the form: if you want to show that H_1 is true, assume that its contrary H_0 is true. Then show that H_0 implies a consequence that makes the thing we observed *very unlikely.* Since the assumption of H_0 leads to something that is hard to believe, then we reject H_0 and accept H_1.

The difference between the Platonic dialectic and the statistical test is that the former deals in certainties while the latter deals in probabilities. The former deductively *proves* that A' is false, and hence A is true. The latter inductively *casts doubt on* H_0 and leads you to H_1 as the more reasonable conclusion. The former is error free; the latter accepts the risk of error, but tries to minimize it. The Platonic form of argument is an ideal that is appropriate in a world of propositions that we believe with certainty; but the hypothesis-testing approach is applicable to the real world, in which our information is limited.

EXERCISES

8–10. At Leviathan University, 50% of the sophomores are women. Of the 36 students who signed up for Professor Macho's course in sophomore Psychology, only 12, or 33.3% were women. A colleague suggested to Professor Macho that there was something about his course that turned women off, since the proportion of women who enrolled was so much lower than 50%. Professor Macho counters that the low enrollment of women can be explained by chance. Is there evidence, at the .01 level of significance, that Professor Macho's course tends to attract fewer women than men?

8–11. A random sample of 48 juvenile prisoners put in a special rehabilitation program showed that 29 of them, or slightly over 60%, remained free of arrest for over a year. However, the proponents of the program had claimed that it would result in at least 70% of the prisoners going for a year without being arrested. Do the sample results refute the claim at the .01 level of significance?

8–12. A coin is tossed 5 times. Every time it comes up heads. Test the hypothesis, at the .05 level of significance, that the coin is fair.

8–13. Of the people that are hired, 20% are members of a minority group. In a random sample of 64 people promoted within 6 months of hiring, only 10, or 15.6% were minority-group members. Do these results show, at the .05 level of significance, that fewer than 20% of those promoted in the organization are minority group members?

8.4 Tests for Means: Small Samples with σ Unknown

In the discussion thus far, I have assumed a sample size large enough to use the Central Limit Theorem. In testing for means from smaller samples, it is permissible to use the normal distribution if

1. the probability distribution of the random variable X is normal or approximately normal, and
2. σ is known or assumed.

The first condition is not often perfectly fulfilled, but it is fairly closely approximated in a wide variety of situations. For example, if the probability distribution is symmetrical and unimodal (i.e., it has one peak in the middle) the sampling distribution of $\overline{X}$ will usually be very close to normal even for a sample of size 8 or 10. And a sample size of 15 will probably be sufficient to use the normal approximation for the sampling distribution of $\overline{X}$, even if the underlying probability distribution of X is far from normal.

The second condition, however, is seldom fulfilled. If you are running a test for μ, it is not likely that you are going to be all that certain about σ. When you do not know or assume a value of σ, you may use the sample to estimate it. The statistic we use is the sample standard deviation s. If we assume that the underlying probability distribution of X is normal, then we can use the statistic

$$t = \frac{\overline{X} - \mu}{s/\sqrt{n}} \qquad \textbf{8.1}$$

as the test statistic for μ. The formula for t looks very similar to that for z, except that instead of the parameter σ in the denominator we have the estimating statistic s. Since s is itself subject to sampling error as well as $\overline{X}$, the statistic t is a ratio of two variables. Both the $\overline{X}$ in the numerator and the s in the denominator can vary from sample to sample. The probability distribution of the statistic t has the form of a symmetrical bell-shaped curve, with an expected value of 0. As the sample size n increases, the standard deviation of the t distribution decreases and approaches 1. When n gets over 30, the distribution of t is almost identical with that of the standard normal variable z, and we can then use z to approximate t.

The t distribution is really a family of distributions, one for each value of n from 2 up to infinity. (Remember that the statistic s is undefined for $n = 1$, so that the smallest sample size for which we can use the t distribution is 2.) Table III gives the critical values of t for two-tail tests at the .10, .05, .02, and .01 levels of significance. The first column is labeled "degrees of freedom." The number of degrees of freedom in a t test for a mean is $n - 1$. We can illustrate the use of the table by an example.

EXAMPLE 8.6 A random sample of 9 students who wrote their first computer programs showed that the mean number of submittals of their programs before getting significant output without error messages was 6.8 and the standard deviation was 3.0. It had been previously predicted that the average number of submittals for all students writing their first program would be 9.0. Does this result show, at the .05 level of significance, that μ, the expected number of submittals prior to error-free output, is really different from 9.0? Assume that the number of submittals per student is a normally distributed random variable.

Solution: The null hypothesis is that the number of submittals is a normally distributed random variable with $\mu = 9$. The competing hypotheses can be stated as

H_0: $\mu = 9.0$
H_1: $\mu \neq 9.0$

The standard deviation σ is not known or assumed. But we do have a sample that provides an unbiased estimator of σ, which in this case happens to be 3.0. Since we assume that the distribution is normal, the t statistic is appropriate as a test statistic.

Before calculating t, formulate a decision rule by finding the critical values of t from Table III. There are $9 - 1 = 8$ degrees of freedom. Go across the row marked 8 degrees of freedom to the column headed by .05, and read the t value as 2.306. This tells us that the two-tail critical region at the .05 level of significance consists of all the t values more than than 2.306 and less than -2.306. The decision rule is therefore:

If $-2.306 \leq t \leq 2.306$, accept H_0.
Otherwise, reject H_0.

Next, calculate the value of t from the sample. This is

$$t = \frac{\overline{X} - \mu}{s/\sqrt{n}} = \frac{6.8 - 9.0}{3/\sqrt{9}} = 2.20$$

According to the decision rule, we accept H_0. There is as yet insufficient evidence to persuade us to abandon the null hypothesis that $\mu = 9$.

Notice that had we *assumed* that the population standard deviation was 3, then we would have calculated a z statistic of 2.20. We could have rejected H_0, because the z statistic is significant at the .05 level if it is greater than 1.96 or less than -1.96. The necessity to use t instead of z cost us the right to reject the null hypothesis. ■

Example 8.6 was solved by use of a two-tail test. A two-tail test was appropriate, because the question was: "Does this result show that μ is *different* from 9?" The wording led us to write H_1 as $\mu \neq 9$.

Suppose the question had been: "Does this result show that μ is *less than 9?*" In this case the alternative hypothesis H_1 would be: $\mu < 9$. A one-tail test would then be appropriate. Let us run a one-tail test.

For the one-tail test we put the entire critical region in the lower tail of the t distribution, as shown in Figure 8.5. The reason the critical region is in the lower tail is that the alternative hypothesis, $\mu < 9.0$, is of a "less than" variety. It is very low values of t that would lead us to reject H_0 in favor of a "less than" alternative.

We want to find the (negative) critical value of t that cuts off an

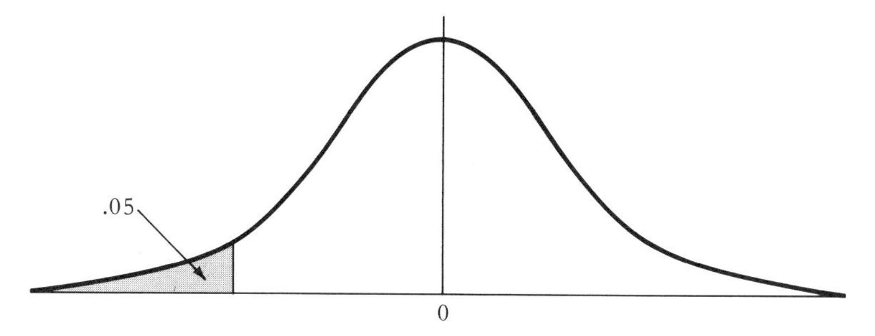

Figure 8.5. *t DISTRIBUTION AND CRITICAL REGION IN A ONE-TAIL TEST Degrees of Freedom = 8 and α = .05*

area of .05 in the lower tail. Table III gives us critical values for two-tail tests. To use Table III for a one-tail test, we enter the column listed for *double* the level of significance of the test. The t values that cut off .05 in each tail will cut off a total of .10 in two tails. Therefore we look in row $9 - 1 = 8$ degrees of freedom under the column headed .10 to find the critical value of t. Since the critical region is in the lower tail, we add a negative sign to the 1.860 reading to get a critical value of -1.860. The decision rule for the one-tail test is:

If $t \geqslant -1.860$, accept H_0.
If $t < -1.860$, reject H_0.

As before, the value of t from the sample was calculated to be -2.20. This is a sufficiently negative value of t to allow rejection of the null hypothesis. We have sufficient evidence to say that the expected value of μ is less than 9.00.

It may seem odd that the two-tail test fails to reject the null hypothesis, whereas the one-tail test rejects the null hypothesis at the same level of significance. The reason for this is that in the case of the one-tail test the alternative hypothesis is more specific. It makes the assumption that any difference you are looking for is a negative one. The two-tail test is merely looking for a difference, whichever direction the difference is in. Its critical region is scattered in two directions. It assumes less, and is in a sense a more cautious test. In general, in preliminary research when it is not known for sure what kind of differences (positive or negative) might turn up, a two-tail test may be more appropriate. If you start with a one-tail test you may be assuming some preconceived ideas that are not justified. Only at a later stage of research, when you are more sure of which direction to look for differences, can you focus more specifically on the research question with a one-tail test.

> The t distribution should be used
>
> 1. when the probability distribution of the random variable is reasonably close to normal;
> 2. when the sample size n is too small to use the Central Limit Theorem; and
> 3. when you do not know or assume a value for σ but must estimate σ from the sample standard deviation s.

The Number of Degrees of Freedom

Students often have trouble understanding what the number of degrees of freedom means. Perhaps the best thing to say at this point is not to worry about understanding the concept, but simply accept it as a parameter of the distribution. Hereafter in this book you will frequently encounter the parameter "degrees of freedom." In every case the number of degrees of freedom will be closely related to the number n in the sample, but in every case it will be a little bit less than n. In general, the more degrees of freedom you have, the more information you have in estimating a parameter or testing a hypothesis.

You can gain some insight into the value of sample information by inspecting the critical values of t in Table III. If we have a sample of size 2 (so that d.f. = 1), the statistic s is going to be a very unreliable estimator of σ, and the statistic t will have a very high standard error, which reflects this uncertainty in s. At the .05 level of significance (two-tail), you need a t value of 12.706 before you can reject a null hypothesis. If you are able to increase your sample size from 2 to 3 (so that d.f. increases from 1 to 2) the critical value of t drops to 4.303; and further increases in the sample size decrease t to 3.182 and to 2.776. The gain (in terms of reduction of the critical value of t) gets less for each unit increase in sample size, until, with a sample size greater than 20, the critical value of t is close to 1.96, the critical value of the normal variable at the .05 level.

EXAMPLE 8.7 To illustrate the problem of a very small sample size, consider a case where $n = 2$. Suppose that you wanted to show that the mean height of adult women was less than 72 inches. Two randomly selected women were 60 and 64 inches tall. The mean $\overline{X}$ of the sample is 62 inches and the standard deviation is $\sqrt{8} = 2.83$ inches. The calculated value of t is

$$\frac{62 - 72}{\sqrt{8}/\sqrt{2}} = -5.000$$

In order to use t as a test statistic, we must assume that the probability distribution of heights is normal, and with a sample size this small, we cannot tolerate much deviation from the assumption of normality. We are justified in this case by the knowledge that heights are normally distributed.

For one degree of freedom the critical value of t is -6.314 at the .05 level of significance in a one-tail test. The calculated value of -5.000 fails to reject, even though the women in the sample may have seemed short enough to give strong evidence of the falsity of the hypothesis. The example illustrates the near impossibility of rejecting a null hypothesis when you have only one degree of freedom. ■

EXERCISES

8–14. A random sample of 9 acres showed a mean yield of 168 bushels per acre and a standard deviation of 12 bushels. If you assume that the yield per acre is a normally distributed random variable, is there evidence, at the .05 level of significance, that the mean yield per acre of this crop is below 175 bushels?

8–15. The manager of Aaron Burr Hospital has made many of his policy decisions on the assumption, based on a study made in 1927, that the mean duration of a patient's stay in the hospital is 10 days. A part-time orderly, who was earning money while taking a statistics course in an MPA program, decided to test this assumption. He randomly observed 16 patients and found that the mean length of time that they stayed was 8.2 days and the standard deviation of the time they stayed was 5 days.

a) If the orderly assumes that the probability distribution of the length of stays is fairly symmetrical about its mode, can he conclude, at the .05 level of significance, that the manager's assumption is incorrect?

b) What course of action might you suggest the orderly take next?

8–16. The oxygen readings, in parts per million, of a random sample of 9 specimens of water from Lake Wellington, showed an $\overline{X}$ of 3.8 and a standard deviation of 0.8. Test the hypothesis, at the .01 level of significance, that the mean oxygen count, is 4.0.

8–17. The incomes of retired workers in St. Paulsburg are assumed to be at least $5000 a year by the chairman of the Senior Citizens Council. A member of the SCC who challenged this assumption set out to "disprove" it by taking a random sample of 3 retired workers. The 3 had incomes of $2000, $3000, and $1000 a year.

a) Does this result suggest that the assumption that incomes average at least $5000 a year is erroneous?

b) Does this result show, at the .05 level of significance, that the assumption is erroneous? (Assume incomes are normally distributed.)

8.5 Decision Making and Policy Analysis

Public managers may often be reading the results of surveys or studies in the social sciences, where statistical hypothesis testing is particularly useful when it is desirable to establish a proposition beyond a reasonable doubt. It is therefore incumbent on public managers to understand the logic of hypothesis testing and the meaning of statistical significance.

However, in setting policies, a manager may often have to make a decision that does not fit well into the hypothesis-testing model. She or he may not want to gather evidence to "prove" or "disprove" a hypothesis at a level of significance of .05 or .01, but simply to select what appears to be the better of two or more possible courses of action or policies.

The increasing emphasis on policy analysis in recent years has generated a controversy between those who believe in a strictly rational approach to policymaking and those whose views are sometimes described as favoring an "incremental" or "muddling through" approach. The former group stresses the need to set forth all possible alternative courses of action and all possible outcomes under each of these; to evaluate each of these outcomes in some quantitative way; and, by use of a logical or mathematical criterion, to arrive at an optimal policy or course of action. The latter group emphasizes that policymaking is a political process, that it is often impractical or even impossible to list all the alternative actions and outcomes or set quantitative values on these, and that objective criteria for deciding on courses of action are impossible to apply when conflicting interest groups have widely differing points of view on what the aims of a policy should be.

The approach of statistical decision theory is a rational approach. By presenting this approach I do not maintain that making decisions on matters of public policy does not involve political or nonrational elements. The fact that there are more than purely rational elements in the policy-making process does not justify ignorance about rational approaches to decision making. Even in cases where political or nonrational elements are very much in evidence, there may be possible areas of agreement and common interest among apparently conflicting points of view. An understanding of the rational approach to decision making on the part of the various parties can provide a basis for finding areas of agreement. The presence of political elements in a policy decision does not imply the total absence of rational elements. And in decision processes in which there is a high degree of consensus among the parties, there is every reason to use rational approaches that are relevant to the question under consideration.

THE PAYOFF TABLE

In order to drive to work at her new job in Bison Bend, Jenny Bishop has bought a car that cost her $3000. Her insurance agent recommends that she buy $250 deductible collision coverage at a cost of $100 a year. Jenny and Max, in pondering the question of whether it is a good idea to buy this coverage, draw up a payoff table.

> A PAYOFF TABLE is a two-dimensional table having the following characteristics:
>
> 1. The rows represent a set of mutually exclusive events, or "states of nature," denoted by $e_1, e_2, \ldots, e_i$. The decision maker has no control over which of these events will occur.
> 2. The columns of the table represent alternative courses of action. These are denoted by $a_1, a_2, \ldots, a_j$.
> 3. The cells of the table contain the number of dollars or other units the decision maker would gain for each $e_i - a_j$ pair. These gains are called payoffs, and may be positive or negative (representing losses).

Table 8.2 is a simplified payoff table for the Bishops in their insurance-buying situation.

Jenny and Max consider three possible events: no damage to the car; partial damage, which they estimate would be $500 worth; and total destruction, which is equivalent to losing $3000. The analysis oversimplifies, inasmuch as it treats all levels of partial damage as amounting to $500, but the analysis *in principle* will be the same as though they meticulously looked at a whole series of possible levels of damage.

The entries in the first column of Table 8.2 represent the $100

Table 8.2. *PAYOFF TABLE*
For Decision Whether to Buy $250
Deductible Collision Insurance

		ACTIONS	
		a_1 (Insure)	a_2 (Do Not Insure)
	e_1: No damage	−$100	$0
EVENTS	e_2: Partial damage	−350	−500
	e_3: Total damage	−350	−3000

premium plus the $250 deductible that Jenny would have to pay in case of damage. The entries in the second column are the out-of-pocket expenses necessary either to repair or replace the car.

Maximization of Expected Payoff

Table 8.2 displays the payoffs, but it does not tell the Bishops which course of action is better. A criterion for making a decision that is very useful is the maximization of expected payoff. This criterion has the feature that, over the long run, if it were used in identical or similar situations, it would lead to the greatest overall total payoff.

Suppose that, based on actuarial records, the probability of partial damage, e_2, is determined to be .05, and the probability of total destruction, e_3, is set at .01. Using these probabilities, Jenny can compute expected payoffs for each course of action, using Equation 5.8:

$$E(X) = \Sigma xP(X = x)$$

where the x values are the actual payoffs.

Table 8.3 shows the calculation of the expected payoffs of actions a_1 and a_2. Since the expected payoff of action a_2 (not insuring) is −$55—a smaller loss than the −$115 of a_1—Jenny decides that the better action is not to buy the coverage.

Utility Functions

In almost any insurance-buying situation, the criterion of maximizing expected payoff would lead to a decision not to insure. The reason for this is simply that insurance companies are interested in selling poli-

Table 8.3. *CALCULATION OF EXPECTED PAYOFFS For Decision Whether to Buy Collision Insurance*

State of Nature		Action			
		a_1 (Insure)		a_2 (Do Not Insure)	
(1) Event	(2) $P(e_i)$	(3) Conditional Payoff	(4) Expected Payoff	(5) Conditional Payoff	(6) Expected Payoff
Ke_1	.94	−$100	−$94.00	$0	$0
e_2	.05	−$350	−17.50	−$500	−$25
e_3	.01	−$350	−3.50	−$3000	−$30
			−$115.00		−$55

Expected payoff of a_1 = −$115.00
Expected payoff of a_2 = −$55.00

cies that have a positive expected payoff to the insurance company, and, hence, a negative expected payoff to the buyer. But does this mean that buying insurance is always a bad idea? Not necessarily. The reason one still might want to insure against a possible catastrophic loss by paying a moderate premium can be seen in terms of what the moderate cost means to the individual in relation to what a catastrophic loss would mean.

Ask yourself the following question: Would I prefer having one million dollars, or a 50–50 chance of having three million dollars? If you are like 98% of the students to whom I have posed that question, you would prefer taking the million dollars. Yet according to the criterion of maximizing expected payoff, you should take the gamble, since the expected payoff is

$$E(X) = .5(\$0) + .5(\$3 \text{ million}) = \$1.5 \text{ million}$$

Why, then, would people prefer the one million dollars? The typical answer might run something like this: One million dollars means almost as much to me, in terms of the satisfaction or enjoyment I can get, as three million dollars.

When dollars vary like this in terms of the benefit or enjoyment one gets out of them, depending on how many dollars an individual already has, a person will not want to use the criterion of maximizing the expected monetary payoff. In such a situation, a more reasonable criterion is that of maximization of the expected utility of money.

> Let x be the amount of money an individual has. Let $u(x)$ be the utility of x dollars. The criterion of MAXIMIZATION OF EXPECTED UTILITY is one that maximizes the expected value of $u(x)$.

The reason why an individual might prefer one million dollars to a 50–50 chance on three million dollars is that the utility of three million dollars is not sufficiently greater than the utility of one million dollars to warrant taking the chance. Let me illustrate this point by assigning a set of utilities for a particular individual.

Let $u(\$0)$ be 0.
Let $u(\$1 \text{ million})$ be 80.
Let $u(\$3 \text{ million})$ be 100.

The expected utility of taking a 50–50 chance on three million dollars or nothing is

$$E(u) = .5(0) + .5(100) = 50$$

This is less than 80, the utility of one million dollars. By this criterion, therefore, the million dollars can be shown to be preferable to the gamble.

The concept of utility can be operationally defined in terms of the probabilities that one would require to take gambles such as this.

The concept is of particular importance to managers in the public or nonprofit sector. Although expected monetary payoffs may very likely be the criterion appropriate to a business executive, public managers are frequently looking at "payoffs" that cannot be measured in dollars. In some cases the manager may be interested in the utility of certain dollar costs or payoffs; in other cases the outcomes can be expressed in utilities that bear little relationship to money at all.

EXERCISES

8–18. A company offers to maintain a copying machine in good working order for $250 a year. If the machine is not maintained under the contract, it is estimated that it will have a major breakdown with a probability of .2 and that the cost of making repairs would be $750. What is the optimal decision, according to the criterion of maximization of expected payoff?

8–19. Suppose on a visit to Graymare Race Track, a friendly neighbor gives you the "hot tip" that Guinivere will win the second race. You notice that the odds quoted on Guinivere's winning are 20 to 1. If you are deciding whether or not to bet $2 on Guinivere to win, your payoff table will look like this:

		a_1 Bet on Guinivere	a_2 Do Not Bet
e_1	Guinivere will not win.	−$2	$0
e_2	Guinivere will win.	+$40	$0

a) Supposing that the true probability of Guinivere's winning is .04, which action maximizes the expected payoff?

b) Supposing that you believe the "hot tip" to the extent that you would increase the probability of Guinivere's winning to .10, calculate the expected payoff of each action.

8–20. If A sues B he will win $150,000 net of costs if he wins his suit. He will lose $40,000 in attorney's fees if he loses the suit. If the probability that he wins the suit is .4,

a) should he sue, if he uses the criterion of maximization of expected payoff?

b) should he sue, if he uses the criterion of maximization of expected utility? (Assume the following.)

$$u(-\$40{,}000) = 0$$
$$u(\$0) = 60$$
$$u(\$150{,}000) = 100$$

***8–21.** Mrs. Welborne has $100,000. There is an investment opportunity in which Mrs. Welborne can turn that $100,000 into $500,000 with a probability of .4, but with a probability of .6 she will lose all of her $100,000. The odds and payoff are just good enough to tempt her, but she is equally prone to keep her $100,000. What is Mrs. Welborne's utility for $100,000 if we assign

$$u(\$0) = 0 \qquad \text{and} \qquad u(\$500{,}000) = 100$$

*8.6 Making Decisions on the Basis of Sample Information

It is sometimes naively proposed that decisions should be based strictly on empirical information without any preconceived or prior notions of values or probability. This proposal can be justified only if one has an overwhelming amount of empirical evidence; but if one is dealing with a small sample, the decision should take into consideration prior probabilities. Consider, for example, a baseball player who gets 3 hits in his first 3 times at bat in the major leagues. It would not be reasonable to take this sample alone as the indicator of his batting ability. To do so would say that he would get a hit 100% of the time. It makes more sense to use a "prior probability" distribution that says it is very improbable that in the long run a hitter will hit safely more than 30% of the time.

It is possible to make a policy-making decision on the basis of prior probabilities, with little or no "hard" statistical evidence. It is also possible to gather evidence to revise these probabilities and reconsider the decision.

Let us suppose that a new "safe" oral contraceptive pill is being proposed for marketing. If it is 100% effective, you believe that it should be released to be sold; but if it is only 99% effective, you believe that it should not be marketed. Unfortunately, testing for effectiveness takes time, and you are considering the possibility of putting the pill on the market without further testing. Suppose your utilities for the several actions and events are as expressed in Table 8.4.

Suppose that preliminary laboratory tests up to this time lead you to assign the following probabilities before extensive field research gives you further evidence:

$$P_0 = (S_1) = .50$$
$$P_0 = (S_2) = .50$$

The symbol P_0 means "prior probability." These probabilities are assigned *prior* to sample evidence. On the basis of your prior probabilities, the expected utility of action a_1 is given by

Table 8.4. *UTILITY MATRIX*
For Marketing Oral Contraceptive

		ACTION	
		a_1 (Release for Marketing)	a_2 (Hold Off the Market)
EVENT	S_1 (100% effective)	+100	0
	S_2 (99% effective)	−200	0

$$E[u(a_1)] = .5(100) + .5(-200) = -50$$

The expected utility of a_2 is

$$E[u(a_2)] = .5(0) + .5(0) = 0$$

Action a_2 has the higher expected utility. If expected utility is your criterion for decision making, you should not put the contraceptive on the market.

But suppose that it is possible to get some additional information based on field experiments. Suppose the results of 100 trials showed that the pill was effective in every case. Such information will naturally encourage you to believe that the pill is 100% effective, but of course the result is not conclusive, since if the pill is only 99% effective, purely by chance it could result in 100 out of 100 successes. But what is the probability that it is 100% effective in light of this sample evidence?

Originally you assigned prior probabilities of .5 to each of the events S_1 and S_2. You now want to use the sample information to revise these prior probabilities.

This kind of problem can be approached by the use of Bayes' Theorem, which was discussed in Section 4.4. Recall that Bayes' Theorem, for two possible prior events A_1 and A_2 is given by

$$P(A_1|B) = \frac{P(A_1) \cdot P(B|A_1)}{P(A_1) \cdot P(B|A_1) + P(A_2) \cdot P(B|A_2)} \qquad \textbf{4.19}$$

In Equation 4.19, A_1 and A_2 are the two possible events, or "states of nature," and B is an observed outcome of an experiment or a sample result. In the present example, we can use Bayes' Theorem to find $P_1(S_1|x)$, the "posterior probability" of the event S_1 in light of the sample evidence x. Using Bayes' Theorem, we have

$$P_1(S_1|x) = \frac{P_0(S_1) \cdot P(x|S_1)}{P_0(S_1) \cdot P(x|S_1) + P_0(S_2) \cdot P(x|S_2)} \qquad \textbf{8.2}$$

We have $P_0(S_1) = .5$ and $P_0(S_2) = .5$. In order to find $P(x|S_1)$ and $P(x|S_2)$, you need to know something about the method by which the experimental evidence "100 successes in 100 trials" was obtained. If you can assume that the 100 trials were independent, then you can use the binomial distribution, to find, using Equation 5.7,

$$P(x|S_1) = \frac{100!}{100!\ 0!}(1)^{100}(0)^0 = 1 \qquad \textbf{8.3}$$

$$P(x|S_2) = \frac{100!}{100!\ 0!}(.99)^{100}(.01)^0 = .3660 \qquad \textbf{8.4}$$

Substitute Equations 8.3 and 8.4 into Equation 8.2 to get

$$P_1(S_1|x) = \frac{(.5)(1)}{(.5)(1) + (.5)(.3660)} = .732 \qquad \textbf{8.5}$$

In light of the sample information, the probability that the pill is 100% effective, or $P_1(S_1|x)$, is .732, and the probability that it is only 99% effective is $1 - P_1(S_1|x) = .268$.

Using the posterior probability distribution of the effectiveness of the pill, you can assess the expected utility of each course of action.

Expected utilities of actions a_1 and a_2 in light of sample evidence x

$$E[u(a_1)] = P_1(S_1|x) \cdot u(a_1, S_1) + P_1(S_2|x) \cdot u(a_1, S_2) \qquad \textbf{8.6}$$

$$E[u(a_2)] = P_1(S_2|x) \cdot u(a_2, S_1) + P_1(S_2|x) \cdot u(a_2, S_2) \qquad \textbf{8.7}$$

Use Equations 8.6 and 8.7 to compute

$$E[u(a_1)] = .732(100) + .268(-200) = +19.6$$
$$E[u(a_2)] = .732(0) + .268(0) = 0$$

The sample information changes the expected utilities of the two actions so that a_1 (release for marketing) has now become the optimal action.

The Value of Information

One of the major contributions of statistical decision theory is that it provides a method for estimating how valuable a piece of sample information will be even before the decision to obtain the information is made. A major problem that an executive or decision maker will have is how much information to get before taking action.

In the example of the oral contraceptive, we first determined the expected utility payoff of each action prior to getting information, and we selected the optimal course of action accordingly. We then used sample information to compute expected utility payoffs in light of this information. In general, the more information we obtain, the greater the utility payoff we expect to gain. But there is a limit to this gain in payoff..Once we have overwhelming or complete information, we will know what course of action to take and what the utility payoff will be.

The difference between what we would expect the payoff to be if we knew which event *e* would take place and the expected payoff of the best action we would take under uncertainty is the value of that knowledge of the event *e*.

> The EXPECTED VALUE OF PERFECT INFORMATION, or the COST OF UNCERTAINTY, is the difference between the expected payoff one could have obtained under certainty and the expected payoff one can get by taking the optimal course of action under uncertainty.

The expected value of perfect information is an upper limit on what you should be willing to pay for information about the future, no matter how good or reliable such information might be. This concept is important, not because you are likely to find a perfect predictor in the real world, but because it tells you how costly is your present degree of uncertainty. Even the best decision makers, in environments of uncertainty, can be expected to incur such cost, and this uncertainty should be taken into account when you are evaluating the decision maker's performance.

The concept of the value of information is important to any executive who wants to guard against two common but opposite tendencies:

1. a tendency to be indecisive, to go to great lengths and costs to get information before making a decision, and to defer a decision until large quantities of information are in, and
2. a tendency to act too quickly, to make snap judgments on the basis of too little information, when a modest investment of time and money could have significantly reduced the expected losses of a decision.

The first type of person sets too high a value on additional information; the second type sets too low a value.

The material presented in this section can only help give you a feeling for assessing the value of information. A more advanced course in statistics that includes decision theory will give you the tools to determine precisely the value of the information you may obtain.

EXERCISES

***8–22.** The City of Oz is considering whether to build a children's recreational area at a cost of $100,000. The total benefit of the recreational area, discounted over a future period of time, is $15.00 for each child who uses it.

There are 50,000 children who are potential users of the playground. The probability distribution for the proportion of these children who use the playground is:

p	$P(p)$
.10	.5
.20	.5

a) Set up a payoff table, showing the net benefits of building the recreational area for each of the 2 possible states of nature.
b) According to the criterion of expected payoff maximization, should the city build the recreational area?
c) A sample of 10 children reveals that 4 would use the playground facilities. Assess the probability distribution of p, the proportion of children using the playground, posterior to this sample result.
d) Using the probability distribution you calculated in (c), determine the expected payoff of each action (to build or not to build), and the expected losses.

***8–23.** Given the following prior probability distribution of state of nature e and the conditional probabilities of a sample result X. Consider that X is an economic indicator that occurs more commonly when the unemployment rate is going to be high than when it is going to be low.

State of Nature	$P_0(e)$	$P(X/e)$
e_1: The unemployment rate will be over 10%	.4	.7
e_2: The unemployment rate will not be over 10%	.6	.4

Find the posterior distribution of e, given the sample result X.

***8–24.** Suppose, in anticipation of a high unemployment rate, a central bank is considering an expansionary monetary policy.

Let e_1 be: The unemployment rate will be over 10%.

Let e_2 be: The unemployment rate will not be over 10%.

The payoff table (in utilities) is

State of Nature	a_1 *(expansionary policy)*	a_2 *(nonexpansionary policy)*
e_1	10	0
e_2	-10	0

a) Use the prior probabilities in Exercise 8–23 to compute the expected utility payoff of each action. Which action should the central bank take?

b) Use the posterior probabilities you calculated in Exercise 8–23 after observing the economic indicator X to calculate the expected utility of each action. In view of X, which action should the central bank take?

***8–25.** Let e be the rate of inflation next year. An econometric study gave the following probability distribution to this rate:

e	$P_0(e)$
$e_1 = 5\%$	.10
$e_2 = 6\%$	.20
$e_3 = 7\%$	.30
$e_4 = 8\%$	.20
$e_5 = 9\%$	.10
$e_6 = 10\%$	.10

After this study, a new economic indicator y was observed. This indicator has the following conditional probabilities, given the various possible states of nature:

$$P(y|e_1) = .10 \qquad P(y|e_4) = .40$$
$$P(y|e_2) = .20 \qquad P(y|e_5) = .50$$
$$P(y|e_3) = .30 \qquad P(y|e_6) = .60$$

Find the posterior probability distribution of e, given y.

Summary of Chapter

When we take a sample, we will generally be using the sample result to make a decision, a decision that will be based on our belief about some population parameter. Because sample results are unreliable, we protect ourselves against jumping to unwarranted conclusions by setting up a null hypothesis and refusing to reject that hypothesis unless the sample result is so highly improbable according to the hypothesis that we are reasonably confident in rejecting it.

Methods were presented to test hypotheses about a population mean and a population proportion. Each of these tests requires that we make assumptions and carefully check to see whether these assumptions are justified.

Statistical decision theory is an approach to making decisions on the basis of maximization of expected payoff or expected utility. Decision theory makes use of both prior probabilities and sample information in assessing expected payoffs.

The approach of hypothesis testing is appropriate when we want to verify or "prove" a hypothesis beyond a reasonable doubt. Decision theory lends itself to situations in which we want to decide between a set of alternative courses of action when the possible outcomes and their probabilities are known or can be estimated.

REVIEW EXERCISES FOR CHAPTER 8

8–26. A survey of 64 randomly selected houses showed that 25 (about 39%) had at least one occupant per room. Test the hypothesis, at the .05 level of significance, that at least 50% of the houses in this statistical universe have at least one occupant per room.

8–27. A random sample of several thousand junior high school boys and girls showed that the girls had higher mean verbal aptitude score than the boys. This difference was significant at the .01 level. Because the difference was so highly significant, someone concluded that there is a very large difference between boys and girls in their mean verbal aptitude. Criticize this conclusion.

8–28. There are 31,360 pupils in a school district, and 1,120 teachers. Can we say, with 95% confidence, that the pupil–teacher ratio in the district is not more than 33 to 1?

8–29. There are 200 secretaries working for a Department of Planning and Research. A random sample of 100 secretaries showed that they worked on the average 41.2 hours per week, with a standard deviation of 4.0 hours. Do the results show, at the .01 level of significance, that the expected value of the working time for all 200 secretaries is more than 40 hours?

8–30. A random sample of 16 university professors showed that their mean spending on magazine subscriptions per year was $127.20 and the standard deviation of the 16 was $20. If the spending by professors on magazines is a normally distributed random variable, can we say, at the .05 level of significance, that professors spend more than $120 a year on the average on magazines?

8–31. Regulations call for a mean gas mileage of at least 30 miles per gallon on a particular model of car. A test of 5 cars yielded readings of 28, 30, 36, 34, and 32 miles per gallon. Assuming that these readings are a random sample of a normal population, can you say, at the .05 level of significance, that this model meets the standard?

8–32. Suggest levels of significance for the following pairs of hypotheses. Discuss the reasons for your suggestions and the conditions you assume in making your assumption.

a) H_0: A suspected disease is contagious.
   H_1: The suspected disease is not contagious.

b) H_0: A driver was not exceeding the speed limit.
   H_1: The driver was exceeding the speed limit.

c) H_0: A lot of goods is up to specifications.
   H_1: The lot is not up to specifications.

d) H_0: Cigarette smoking does not produce hypertension.
   H_1: Cigarette smoking produces hypertension.

8–33. a) If the value of a test statistic falls in the critical region, what is your decision?

b) If the value of a test statistic falls in the acceptance region, what is your decision?

c) Which of these two decisions implies that the sample results are significant?

8–34. Suppose you want to test the hypothesis that the expected value of the calorie content of 12 ounces of "Low Calorie" beer was 100, against the alternative that it was less than 100.

a) What would be the Type I Error in this problem?

b) What would be the Type II Error in this problem?

8–35. A sample of 144 pickerel caught in a lake in the Adirondack Mountains yielded a mean length of 18.8 inches with a standard deviation of 3.6 inches.

a) If you assume that the sample is random, how strong is your evidence that the mean length of pickerel in the lake is more than 18 inches?

b) If the 144 pickerel were all caught in the first week of August 1977, how strong is your evidence that the mean length of all pickerel in the lake is more than 18 inches?

8–36. A sample of 4 pike caught in the same lake in the Adirondacks had a mean length of 20.4 inches, and their standard deviation was 3.6 inches. (Assume that the length of pike is a normally distributed random variable, of which these 4 are a random sample.)

a) Do these results show, at the .05 level of significance, that the mean length of all pike in the lake is greater than 18 inches?

b) Do these results show, at the .05 level of significance, that the mean length of all pike in the lake is greater than 16 inches?

c) Do these results show, at the .05 level of significance, that the mean length of all pike in the lake is greater than 14 inches?

8–37. The FDA is testing to determine whether a new drug is harmless. The FDA's decision will determine whether or not the drug is put on the market.

a) What hypothesis do you believe should be the null hypothesis?

b) As a result of your answer to (a), state specifically the meaning of a Type I Error and a Type II Error.

8–38. A research project is being carried out to determine whether chewing tobacco contributes to lung cancer.

a) What hypothesis do you believe should be the null hypothesis?

b) As a result of your answer to (a), state specifically the meaning of a Type I Error and a Type II Error.

c) Compare your answer in part (a) to your answer to part (a) in Exercise 8–37. Is there a difference in principle between the two situations?

d) Might a representative of the Tobacco Institute give a different answer to part (a) than a representative of the American Cancer Society? Discuss.

8–39. In a random sample of 2 people, both were left-handed. Does this result indicate that, at the .05 level of significance, more than 20% of people are left-handed?

8–40. A political candidate is considering whether to challenge her opponent to a televised debate. She believes she has only a .4 probability of "winning" the debate, but if she "wins" the debate, she expects to gain 25,000 votes. She expects to lose only 10,000 votes if she "loses" the debate.

a) Set up a payoff table for her decision.
b) Find the expected payoff for each course of action.
c) According to the criterion of maximizing expected payoff, what is the optimal course of action?

8–41. In the situation of Exercise 8–20, suppose the probability distribution of settlement of the suit for various possible amounts, and the utility functions for these amounts, are as follows:

e (net payoff)	$P(e)$	$u(e)$
−\$40,000	.5	0
+\$10,000	.2	70
+\$30,000	.1	80
+\$60,000	.1	90
+\$150,000	.1	100

a) What is the expected payoff if A sues? On the basis of the criterion of maximizing expected payoff, should he bring suit?
b) If the utility of \$0 is 60, should he sue if he wants to maximize expected utility?

***8–42.** a) If you were faced with the risk of losing \$10,000 with a probability of .01, how much would you be willing to pay to get out of this risky situation?

b) If $u(-\$10{,}000) = 0$ and $u(\$0) = 100$, what is the utility of the amount you are willing to pay?

***8–43.** A shipment of fuses has arrived from the Reputable Supply Company. From past experience we know that 60% of the lots from Reputable have only 2% defectives in them, and 40% of the lots contain 5% defectives. Our agency has the choice of accepting or rejecting the shipment. The following table shows the payoffs associated with each state of nature.

	a_1 (Accept)	a_2 (Reject)
e_1: $p = .02$	\$300	\$0
e_2: $p = .05$	−\$200	\$0

a) On the basis of prior probabilities, which action has the higher expected payoff?
b) Suppose a sample of size 2 is taken, and both are defective. In view of the sample information, which action has the higher expected payoff?

SUGGESTED READINGS

HAMBURG, MORRIS. *Statistical Analysis for Decision-Making,* 2nd ed. New York: Harcourt Brace Jovanovich, Inc., 1977. Chapter 7.

HARNETT, DONALD L. *Introduction to Statistical Methods,* 2nd ed. Reading, Mass.: Addison-Wesley Publishing Co., 1975. Chapter 8.

LOETHER, HERMAN J., and DONALD G. McTAVISH. *Descriptive and Inferential Statistics: An Introduction.* Boston: Allyn and Bacon, Inc., 1976. Chapter 15.

MENDENHALL, WILLIAM, and LYMAN OTT. *Understanding Statistics,* 2nd ed. North Scituate, Mass.: Duxbury Press, 1976. Chapter 7.

NINE

Tests for Two or More Samples

Outline of Chapter

9.1 Tests for Differences Between Two Means

We often want to determine whether two groups differ on the average with respect to some measured characteristic.

9.2 Tests for Differences Between Two Proportions

When we want to determine whether two populations differ in the proportions possessing a certain characteristic, we test the null hypothesis that the proportions are equal.

9.3 Tests Using the Chi-Square Statistic

The chi-square statistic is used to test for differences among two or more sample proportions, and for tests for independence between two nominal-level random variables.

9.4 Which Statistical Test Should You Use?

A review of the assumptions of the tests you have studied can help you determine which tests to use. An examination of the nature of the problem can help you determine the level of significance and whether a test should be one-tailed or two-tailed.

Objectives for the Student

1. Perform tests for differences between two means.
2. Understand the assumptions underlying each test for differences between two means.
3. Perform tests for differences between two proportions.
4. Understand the assumptions underlying tests for differences between two proportions.
5. Perform chi-square tests for independence. Understand the conditions under which it is appropriate to run a chi-square test.
6. Identify the appropriate test to run in different situations.

Chapter 8 introduced the concept of hypothesis testing, placing special emphasis on the logic of the procedure. The procedure was illustrated by some of the simpler types of tests, such as those for means and for proportions. The overwhelming majority of interesting statistical questions, however, involve tests for differences between two or more samples. Investigations are usually concerned with such questions as: Is this group better (on the average) than that group? Do people who are treated by one method differ from people who are treated by another method? Do individuals (whether people or things) having certain characteristics differ in a specified measure from individuals having other characteristics? Do people's attitudes differ this year from their attitudes of last year?

Chapter 3 was concerned with measuring the effect of one or more variables on another. The present chapter does the same thing, with one fundamental difference. Whereas Chapter 3 was concerned with describing data per se, this chapter will consider the data only as a random sample of a larger population. It is the population, and the parameters that describe the population, that command our interest. If two statistics from different samples differ from each other, we will no longer be content simply to describe or measure this difference. We will try to determine whether the sample differences are merely due to chance, or due to fundamental differences between the populations from which they are drawn. In order to make this determination, we will employ and extend the methods of testing hypotheses that were presented in Chapter 8.

9.1 Tests for Differences Between Two Means

Suppose you draw a random sample of size n_1 from one large population with expected value μ_1 and standard deviation of σ_1, and a sample of size n_2 from another large population with expected value μ_2 and standard deviation σ_2. It can be shown that, if the probability distributions of the two populations are normal, or if the sample sizes n_1 and n_2 are both large, then the sampling distribution of the statistic $(\overline{X}_1 - \overline{X}_2)$, the difference between the two sample means, is normally distributed, the expected value of the difference between the means is given by

$$\mu_{\overline{X}_1-\overline{X}_2} = \mu_1 - \mu_2 \qquad \textbf{9.1}$$

and the standard deviation of the statistic $\overline{X}_1 - \overline{X}_2$ is given by

$$\sigma_{\overline{X}_1-\overline{X}_2} = \sqrt{\frac{\sigma_1^2}{n_1} + \frac{\sigma_2^2}{n_2}} \qquad \textbf{9.2}$$

which is known as the STANDARD ERROR OF THE DIFFERENCE BETWEEN MEANS.

This result enables us to use the standard normal variable z, defined here as

$$z = \frac{(\overline{X}_1 - \overline{X}_2) - (\mu_1 - \mu_2)}{\sigma_{\overline{X}_1-\overline{X}_2}}$$

to test for the significance of the difference between means.

EXAMPLE 9.1 In order to determine the effect of certain air pollutants on the severity of bronchopulmonary diseases, a study was conducted to compare the severity of cases of bronchial infections of patients confined to hospitals in two areas which differed markedly in the quality of the air in their immediate environment. Of 32 patients in Bellaire Hospital, the mean number of days before the infection "cleared up" was 12.8; of 32 patients in Carbonville Hospital, the mean number of days until the infection "cleared up" was 15.5. If we assume that the standard deviation of the time until infections "clear up" is 8 days, can we say, at the .05 level of significance, that the expected time for the infections to clear up differs in the two hospitals?

Discussion: Notice first of all that the question is asked only about differences between the hospitals and not about differences in air quality. The reason for this is that the two hospitals may differ in many respects other than air quality, so that a conclusion that a difference in recovery rates can be attributed to air quality would probably not be warranted. Only if the experiment had been so thoroughly controlled that all other variables such as types of patients admitted, differences in methods of care, different standards of measurement, etc., could be ruled out as explanatory variables, would we be justified in attributing a difference in recovery times to air quality.

The question asks for a two-tail test: it asks for a difference, regardless of whether the difference favors one hospital or the other. The two-tail approach leaves open the possibility that highly polluted air could actually decrease the recovery time as well as increase it.

The set of hypotheses is

$$\begin{aligned} H_0&: \mu_1 = \mu_2 \quad \text{or} \quad \mu_1 - \mu_2 = 0 \\ H_1&: \mu_1 \neq \mu_2 \quad \text{or} \quad \mu_1 - \mu_2 \neq 0 \end{aligned}$$

where μ_1 is the expected number of days to clear up the infection in Hospital #1 (Bellaire), and where μ_2 is the expected number of days to clear up the infection in Hospital #2 (Carbonville).

Under the assumption of H_0, the test statistic $\overline{X}_1 - \overline{X}_2$ is normally distributed (because each sample size is at least 30), with expected value $\mu_1 - \mu_2 = 0$, and the standard error of the difference,

$$\sigma_{\overline{X}_1-\overline{X}_2} = \sqrt{\frac{\sigma_1^2}{n_1} + \frac{\sigma_2^2}{n_2}} = \sqrt{\frac{8^2}{32} + \frac{8^2}{32}} = 2.00$$

For a two-tail test at the .05 level of significance, we want the critical values to cut off .025 of the area in each tail of the normal curve. From Table II you can find that the required values of z are 1.96 and -1.96. The critical region of the statistic $\overline{X}_1 - \overline{X}_2$ is therefore that region that is more than 1.96 standard errors from its expected value of 0. The critical values are $1.96 \times 2 = 3.92$ and $-1.96 \times 2 = -3.92$. Therefore the decision rule (in terms of the statistic $\overline{X}_1 - \overline{X}_2$ is:

If $-3.92 \leq \overline{X}_1 - \overline{X}_2 \leq 3.92$, accept H_0.
If $\overline{X}_1 - \overline{X}_2 < -3.92$, or if $\overline{X}_1 - \overline{X}_2 > 3.92$, reject H_0.

The sample result is

$$\bar{X}_1 - \bar{X}_2 = 12.8 - 15.5 = -2.7$$

This is in the acceptance region. We cannot reject H_0.

We could have used the standard normal variable z as the test statistic. For a two-tail test at the .05 level of significance, the decision rule would be:

If $-1.96 \leq z \leq 1.96$, accept H_0.
If $z < -1.96$, or if $z > 1.96$, reject H_0.

To convert the statistic $\bar{X}_1 - \bar{X}_2$ to z, subtract $\mu_{\bar{X}_1-\bar{X}_2}$ and divide by $\sigma_{\bar{X}_1-\bar{X}_2}$ to get

$$z = \frac{(\bar{X}_1 - \bar{X}_2) - (\mu_1 - \mu_2)}{\sigma_{\bar{X}_1-\bar{X}_2}} = \frac{\bar{X}_1 - \bar{X}_2 - 0}{\sqrt{\frac{\sigma_1^2}{n_1} + \frac{\sigma_2^2}{n_2}}}$$

In the present example, z is found by

$$z = \frac{-2.7 - 0}{\sqrt{\frac{8^2}{32} + \frac{8^2}{32}}} = -1.35$$

As before, the test statistic is in the acceptance region. Notice that the standard error of the difference $\sigma_{\bar{X}_1-\bar{X}_2}$, depends on assumed values of σ_1 and σ_2. Under the null hypothesis that the two expected values are the same, any assumptions that we make about the standard deviations will very likely assume the two are equal, as we did in this example. ■

When the samples are large (each $n \geq 30$), you can use the sample standard deviations s_1 and s_2 as approximations to σ_1 and σ_2 in Equation 9.2. This gives us an estimated standard error of the difference between two means, which can be written as follows.

Estimated standard error of difference between means of large samples

$$s_{\bar{X}_1-\bar{X}_2} = \sqrt{\frac{s_1^2}{n_1} + \frac{s_2^2}{n_2}} \qquad (9.3)$$

EXAMPLE 9.1 (continued) Suppose we felt uncomfortable with the assumption that the standard deviation of times for infections to clear up was 8 days in each of the hospitals. With samples of size 32 from each population, however, we do not have to make such an assumption, since we can estimate the standard error of the difference using Equation 9.3. Suppose we calculated s_1 to be 7.0 and s_2 to be 6.0. Then

$$s_{\overline{X}_1-\overline{X}_2} = \sqrt{\frac{(7.0)^2}{32} + \frac{(6.0)^2}{32}} = 1.63$$

If we use the test statistic z, we can compute z as

$$z = \frac{(\overline{X}_1 - \overline{X}_2) - (\mu_1 - \mu_2)}{s_{\overline{X}_1-\overline{X}_2}} = \frac{(12.8 - 15.5) - 0}{1.63} = 1.66$$

By the decision rule using the test statistic z, we accept the null hypothesis. Although the decision is the same as before, notice that the use of the estimated standard error from the sample might have changed the decision from what we decided before using assumed standard deviations. The discrepancy could arise either from an erroneous assumption about σ or by sample standard deviations that, by bad luck, were not representative of the population from which they were drawn. ■

Tests for Differences Between Two Means for Small Samples with Standard Deviations Unknown

The normal probability distribution can be used to test differences between means only when the standard deviations are known or assumed, or if the samples are large enough that we can accept the sample standard deviations as reliable estimates of the population standard deviation.

When we do not know or assume the standard deviations, and the two samples are small, we cannot use the normal variable as a test statistic, even if the underlying probability distribution of the random variable we are measuring is normal. In this case, we can compute a test statistic that has the t distribution with $n_1 + n_2 - 2$ degrees of freedom. The null hypothesis is that both samples come from the same normal population. The set of hypotheses can be stated as:

H_0: $\mu_1 - \mu_2 = 0$.
H_1: $\mu_1 - \mu_2 \neq 0$ (for a two-tail test).

The test statistic t is defined as

$$t = \frac{\overline{X}_1 - \overline{X}_2}{s_{\overline{X}_1-\overline{X}_2}}$$

The standard error of the difference between means is given by:

Standard error of the difference between means of small samples with unknown population variances

$$s_{\bar{X}_1-\bar{X}_2} = \sqrt{\frac{s^2}{n_1} + \frac{s^2}{n_2}} \qquad \textbf{9.4}$$

The statistic s^2 is an unbiased estimator of σ^2, the variance of the hypothesized normal population. It is computed as the weighted mean of the two sample variances.

$$s^2 = \frac{(n_1 - 1)s_1^2 + (n_2 - 1)s_2^2}{n_1 + n_2 - 2} \qquad \textbf{9.5}$$

EXAMPLE 9.2 A consumers' research organization ran a test to compare the gasoline mileages of 2 models of compact cars. Six randomly selected Corsairs were driven by 6 randomly selected test drivers over a test course. Their mean mileage was 32.8 miles per gallon and their standard deviation was 1.2 miles per gallon. Eight randomly selected Buccaneers, driven over the same course by 8 randomly selected drivers, had a mean mileage of 33.9 miles per gallon and a standard deviation of 0.9 miles per gallon. Do these results show, at the .05 level of significance, that Buccaneers have a higher gasoline mileage than Corsairs?

Discussion: Before you can answer this question, you must assume that higher gasoline mileage *over this test run* indicates higher gasoline mileage in general. Alternatively, you can define a car's gasoline mileage as the mileage that it would get over this test run driven by a test driver. Having defined gasoline mileage in this way, you can proceed to run the statistical test.

The question calls for a one-tail test. The hypotheses are:

H_0: $\mu_1 - \mu_2 = 0.$
H_1: $\mu_1 - \mu_2 < 0.$

where μ_1 = expected gas mileage of all Corsairs
μ_2 = expected gas mileage of all Buccaneers

Check carefully to see that the setting up of these hypotheses is in accordance with the question that is asked. This is a good exercise, inasmuch as the setting up of the hypotheses is the most basic, and often the trickiest, procedure in running a hypothesis test.

The critical value of t at the .05 level of significance for $6 + 8 - 2 = 12$ degrees of freedom is -1.782. (Remember that in a one-tail test you double the probability in order to locate the level of significance column in Table III.)

The decision rule is therefore:

If $t \geqslant -1.782$, accept H_0.
If $t < -1.782$, reject H_0.

To calculate t you need to find s^2. From 9.5 you have

$$s^2 = \frac{(6-1)(1.2)^2 + (8-1)(0.9)^2}{6+8-2} = 1.0725$$

You can now use Equation 9.4 to calculate $s_{\bar{X}_1-\bar{X}_2}$.

$$s_{\bar{X}_1-\bar{X}_2} = \sqrt{\frac{1.0725}{6} + \frac{1.0725}{8}} = .5593$$

Then

$$t = \frac{32.8 - 33.9}{.5593} = -1.967$$

According to the decision rule, you should reject the null hypothesis. There is evidence at the .05 level of significance that Buccaneers have higher gasoline mileage than Corsairs.

Had the question been, "Do the results show that Buccaneers have a *different* expected gasoline mileage from Corsairs," you should have run a two-tail test. The critical values of t would have been +2.179 +2.179 and −2.179, and the test result would not have been significant. It was the fact that the research question was focused on differences in one direction (superiority of Buccaneers over Corsairs) that made it possible to run a one-tail test and get significant results. ■

The Importance of the Size of the Samples

Equations 9.2, 9.3, and 9.4 all reveal how critical the size of the smaller of the two samples is in controlling the standard error of the difference. For example, if we assume $\sigma^2 = 4$ in Equation 9.2, let us compare the standard errors from two proposed sampling designs.

Design 1: $n_1 = 8$ and $n_2 = 8$. Then

$$\sigma_{\bar{X}_1-\bar{X}_2} = \sqrt{\frac{4}{8} + \frac{4}{8}} = 1$$

Design 2: $n_1 = 2$ and $n_2 = 100$. Then

$$\sigma_{\bar{X}_1-\bar{X}_2} = \sqrt{\frac{4}{2} + \frac{4}{100}} = 1.43$$

Although Design 1 entails a total of only 16 observations, while Design 2 entails 102 observations, Design 1 achieves a lower standard error of the difference because it samples equally from the two groups. It is the smaller sample that contributes most to this standard error, so that in making a comparison between two groups it is important to keep the smaller of your samples as large as possible. The most efficient design, in terms of the size of the standard error in relation to the total number sampled, is to have equal-size samples from each group.

EXAMPLE 9.3 A survey is to be undertaken to determine the differences in attitudes of public school teachers and public school pupils. A variety of questions is to be asked, and many of the responses will be of numerical form, so that means and standard deviations can be calculated. There are 1000 teachers and 22,000 pupils in the school system. It has been proposed to take a 1% sample of both teachers and pupils, so that the samples would consist of 10 teachers and 220 pupils. Criticize the approach taken here.

Discussion: Since the survey is concerned with differences between the 2 groups, you will have a better chance of getting significant results if you keep the standard error of the difference between means as small as possible. Assume the responses of both teachers and pupils to a question have a standard deviation of σ. The standard error, according to the proposed design, is

$$\sigma_{\bar{X}_1-\bar{X}_2} = \sqrt{\frac{\sigma^2}{10} + \frac{\sigma^2}{220}} = \sigma\sqrt{\frac{1}{10} + \frac{1}{220}} = .323\sigma$$

If you could increase the sample size of the teachers to 50 and reduce the number of pupils in the sample to 100, you would be not only reducing the total number that you need to sample from 230 to 150, but you would decrease this standard error to

$$\sigma_{\bar{X}_1-\bar{X}_2} = \sqrt{\frac{\sigma^2}{50} + \frac{\sigma^2}{100}} = \sigma\sqrt{\frac{1}{50} + \frac{1}{100}} = .173\sigma$$

An "ideal" sample design would be one that sampled equally from teachers and pupils. This might not be a realistic ideal, since because of their greater number, it may be easier to get a larger random sample of pupils than of teachers. But the example does illustrate the considerable gain to be obtained by increasing the number of teachers in the sample if you otherwise would have a very small number. The gain from adding 40 teachers to the sample far outweighed the loss from taking away 70 pupils. ■

Paired Observations

The previous discussions were all concerned with two samples that were *independent* of each other. The individuals for each sample were randomly selected, and there was no connection between the procedure for selecting individuals for the first sample and the procedure for selecting individuals for the second sample.

But when we run a test of paired observations, the selection of an individual for the first sample *determines* the selection of an individual for the second sample, whether the individual in the second sample is the same individual, or an individual related to or naturally paired with, the corresponding individual in the first sample. For example, to determine whether a weight-reducing plan is effective, there are two ways we might do the research. One way is to take independent samples. This method would involve selecting a set of people who had not followed the plan and compared their weights (or weights adjusted for height) with those of another set of people who had followed the plan. There will be many problems of comparability between groups if we try to do the research in this manner. A second, and probably much more efficient, way to undertake the research is to compare the *same* individuals before and after following the plan, and observe the weight *changes.* Again, there are going to be problems in controlling for extraneous variables, but at least you will no longer have the problem of individuals not being comparable. In this second method, the two samples are dependent. If an individual is selected for the first sample, we know that individual will also be in the second sample.

The method of testing for differences using paired observations is essentially a one-sample method. Instead of testing two different sample means, you look at the difference for each pair, and treat these differences as a single sample.

EXAMPLE 9.4 A highway has been undergoing repaving, one stretch of road at a time, over a period of 8 months. In order to determine the effect of the repaving on accident rates, 5 3-mile segments were randomly selected for observations over a 2-week period both before and after repaving. The results of observations are given in Table 9.1.

Because of the smallness of the sample, it was decided to set α at .10; the high level being justified by the fact that a Type I Error would lead to a follow-up study, not a final, possibly disastrous, action.

The sample consists of 5 pairs of observations of differences δ, and not 10 independent observations, as would have been the case if these were independent samples. The null hypothesis is that the expected value of the differences in accident rates is zero. The hypotheses can be stated as:

$$H_0: \quad \mu_\delta = 0$$
$$H_1: \quad \mu_\delta \neq 0$$

where μ_δ is the expected value of δ, the difference in accidents. If we assume that the differences are normally distributed, we can run a sin-

Table 9.1. *NUMBER OF ACCIDENTS ON FIVE SEGMENTS OF HIGHWAY, BEFORE AND AFTER PAVING*

Segment	Accidents Before	Accidents After	δ	$\delta - \bar{\delta}$	$(\delta - \bar{\delta})^2$
A	6	7	+1	−4	16
B	4	9	+5	0	0
C	10	21	+11	+6	36
D	6	11	+5	0	0
E	4	7	+3	−2	4
			+25	0	56

$$\Sigma\delta = 25 \qquad \Sigma(\delta - \bar{\delta}) = 56$$

$$\bar{\delta} = \frac{\Sigma\delta}{n} = \frac{25}{5} = 5.00$$

$$s_\delta = \sqrt{\frac{\Sigma(\delta - \bar{\delta})^2}{n - 1}} = \sqrt{\frac{56}{4}} = 3.74$$

$$s_{\bar{\delta}} = \frac{s_\delta}{\sqrt{n}} = \frac{3.74}{\sqrt{5}} = 1.67$$

gle-sample t test, using the s_δ calculated in Table 9.1 as our estimated standard error. The critical values of t (for $5 - 1 = 4$ degrees of freedom; $\alpha = .10$) are $\pm$ 2.132. From Equation 8.1, we calculate

$$t = \frac{\bar{\delta} - 0}{s_\delta/\sqrt{n}} = \frac{5.00}{1.67} = 2.99$$

By the decision rule, we reject the null hypothesis. We have evidence that the paving made a difference. ■

The above example casts an interesting side light on the merit of using a two-tail test. If the question has been: "Does the paving reduce the accident rate?" the answer would clearly have been "no." The two-tail test, however, leaves us open to observe a change in the accident rate in *either* direction. In this example, the test was able to spot a significant *increase* in the accident rate following the repaving. The result might have been unexpected, but for that very reason the research might have been valuable.

EXERCISES

9–1. In order to determine whether the use of the computer by daytime students enrolled in a course in Public Policy differed from the mean usage of the computer by evening students enrolled in the same course, random samples of 50 daytime students and 50 evening students were taken. The mean computer time used by daytime students was 22.4 minutes; for evening students this mean was 20.8 minutes. Assume that the standard deviation of usage is 5.0 minutes for both daytime and evening students. At the .05 level of significance, can you say that the two groups of students differ in their mean usage of the computer?

9–2. An official in the Transportation Department is conducting an experiment to determine which of two routes between the two campuses of Northern University takes less time to drive. He had a test driver drive each route 8 times at randomly selected times. The mean time for Route A was 36.4 minutes and the variance s^2 was 50. The mean time for Route B was 44.5 minutes and the variance was 78. At the .05 level of significance, can you say that the two routes differ in expected driving time? Assume that driving time is a normally distributed random variable.

9–3. A random sample of 50 people over 45 years old who had been laid off from their jobs in November, 1977, received Unemployment Compensation checks for a mean of 10.6 weeks and a standard deviation of 2 weeks. A random sample of 50 people under 25 years old who had been laid off received Unemployment Compensation checks for a mean of 9.4 weeks and a standard deviation of 2 weeks. At the .01 level of significance, can you say that there is a difference in the expected number of weeks that people over 45 and people under 25 will receive their Unemployment Compensation checks?

9–4. A random sample of 5 pupils in the fourth grade received scores of 6, 8, 10, 12, and 14 on a reading achievement test. A random sample of 5 pupils in the sixth grade received scores of 8, 10, 12, 14, and 16 on the same reading achievement test. If you assume that the two samples are independent, do you have evidence, at the .05 level of significance, the sixth grade pupils have higher mean reading achievement scores than fourth grade pupils? Assume that test scores are normally distributed.

9–5. A random sample of 5 pupils in the fourth grade was given reading achievement tests. Two years later, when they were in sixth grade, the same 5 pupils were tested again. The scores on both tests were

Pupil	*4th Grade*	*6th Grade*
A	6	8
B	8	10
C	10	12
D	12	14
E	14	16

a) Is there evidence, at the .05 level of significance, that sixth grade pupils will test higher on the average than fourth grade pupils? Assume a normal distribution of test scores.

b) Compare your answer to your answer to Exercise 9–4. How are the two situations different?

9–6. A random sample of 40 feminists had a mean weight of 130 pounds and a standard deviation of 20 pounds. A random sample of 40 antifeminists had a mean weight of 160 pounds and a standard deviation of 25 pounds. Is there evidence, at the .01 level of significance, that antifeminists outweigh feminists?

9–7. Does your answer to Exercise 9–6 have important implications about the effect of one's political stance on feminism on one's weight? Discuss.

9.2 Tests for Differences Between Two Proportions

Many research topics involve the question of whether two populations differ in the proportion having a particular characteristic. Do people who receive Treatment A have a higher recovery rate than people who receive Treatment B? Do older employees have a different probability of achieving some goal than younger employees? Which of two different methods of rehabilitating prisoners has a higher probability of keeping them from getting into trouble again with the law?

The approach to such questions is the same in principle and only slightly different in detail from the approach taken in Section 9.1.

In a study comparing the effects of a "permissive" and a "severe" policy in treating young people with previous arrests, a sample of 48 young people who had received a "permissive" treatment was randomly selected for a follow-up study, as was a sample of 48 who had been given a "severe" treatment. Of the 48 who had received the "permissive" treatment, 14 got into trouble within a year; of those receiving the "severe" treatment, 24 got into trouble within a year. Do these results show, at the .05 level of significance, that the two treatments give different probabilities of getting into trouble later?

The null hypothesis is that the probabilities of getting into trouble given each of the treatments are equal, so that

$$H_0: \quad p_1 = p_2, \text{ or } p_1 - p_2 = 0.$$
$$H_1: \quad p_1 \neq p_2, \text{ or } p_1 - p_2 \neq 0.$$

Let $\hat{p}_1$ and $\hat{p}_2$ be the observed sample proportions. It can be shown that the sampling distribution of $\hat{p}_1 - \hat{p}_2$, the difference between two sample proportions when both samples are large, is approximately normal. The expected value of $\hat{p}_1 - \hat{p}_2$ is given by

$$\mu_{\hat{p}_1-\hat{p}_2} = p_1 - p_2 \qquad \mathbf{9.6}$$

The standard error of the difference between proportions is given by

$$\sigma_{\hat{p}_1-\hat{p}_2} = \sqrt{\frac{p_1(1-p_1)}{n_1} + \frac{p_2(1-p_2)}{n_2}} \qquad \mathbf{9.7}$$

You cannot use Equations 9.6 and 9.7 unless you know or assume values for the parameters p_1 and p_2. If you are testing the null hypothesis that $p_1 = p_2$, Equation 9.6 becomes

$$\mu_{\hat{p}_1-\hat{p}_2} = 0 \qquad \textbf{9.8}$$

The null hypothesis says that p_1 and p_2 take on a common value which we can call p. The standard error of the difference becomes

$$\sigma_{\hat{p}_1-\hat{p}_2} = \sqrt{\frac{p(1-p)}{n_1} + \frac{p(1-p)}{n_2}} = \sqrt{p(1-p)\left(\frac{1}{n_1} + \frac{1}{n_2}\right)} \qquad \textbf{9.9}$$

The parameter p can be estimated as the weighted mean of the two sample proportions. This estimate, known as $\hat{p}$, is a "pooled estimator" of the common proportion and is given by

$$\hat{p} = \frac{n_1\hat{p}_1 + n_2\hat{p}_2}{n_1 + n_2} \qquad \textbf{9.10}$$

We can use this pooled estimator $\hat{p}$ to estimate $\sigma_{\hat{p}_1-\hat{p}_2}$. The estimator of $\sigma_{p_1-p_2}$ is given by

$$s_{\hat{p}_1-\hat{p}_2} = \sqrt{\frac{\hat{p}(1-\hat{p})}{n_1} + \frac{\hat{p}(1-\hat{p})}{n_2}} = \sqrt{\hat{p}(1-\hat{p})\left(\frac{1}{n_1} + \frac{1}{n_2}\right)} \qquad \textbf{9.11}$$

To test the null hypothesis that $p_1 = p_2$ when you have two proportions $\hat{p}_1$ and $\hat{p}_2$ from large independent random samples of sizes n_1 and n_2, from two populations, use either the test statistic $\hat{p}_1 - \hat{p}_2$ or the test statistic z.

The probability distribution of $\hat{p}_1 - \hat{p}_2$ will be normal, with $\mu_{p_1-p_2} = 0$ and $\sigma_{p_1-p_2}$ estimated by

$$s_{\hat{p}_1-\hat{p}_2} = \sqrt{\hat{p}(1-\hat{p})\left(\frac{1}{n_1} + \frac{1}{n_2}\right)}$$

where

$$\hat{p} = \frac{n_1\hat{p}_1 + n_2\hat{p}_2}{n_1 + n_2}$$

The test statistic

$$z = \frac{\hat{p}_1 - \hat{p}_2 - 0}{s_{\hat{p}_1-\hat{p}_2}}$$

has the standard normal distribution.

9.12

In the present example, $n_1 = 48$, $n_2 = 48$, $\hat{p}_1 = 14/48 = .29$, and $\hat{p}_2 = 24/48 = .5$. At the .05 level of significance in a two-tail test, the critical values of the test statistic z are -1.96 and $+1.96$, so that the decision rule, stated in terms of z, is

If $-1.96 \leqslant z \leqslant 1.96$, accept H_0.
If $z < -1.96$ or if $z > 1.96$, reject H_0.

In order to calcuate z, find $s\hat{p}_1 - \hat{p}_2$, which depends on $\hat{p}$. Using Equation 9.10 we find

$$\hat{p} = \frac{(48)(14/48) + 48(24/48)}{48 + 48} = .40$$

and from Equation 9.11,

$$s\hat{p}_1 - \hat{p}_2 = \sqrt{(.4)(.6)\left(\frac{1}{48} + \frac{1}{48}\right)} = .1$$

Then, from Equation 9.12 the value of z is

$$\frac{(14/48) - (24/48)}{.1} = -2.08$$

By the decision rule, we reject H_0. We conclude that the two methods do differ in their results.

EXAMPLE 9.5 Last year, in a random sample of 100 employees in a large organization, 28 rated the relationship between the administration and the secretarial staff as "abominable." This year, in a similar survey of 100 employees selected independently of last year's survey, only 20 rated the relationship between the administration and the secretarial staff as "abominable." Is there evidence, at the .05 level of significance, that a smaller proportion of all employees rate the relationship between administration and secretarial staff as "abominable" this year than last year?

Solution: Let p_1 be the proportion of all employees last year who thought the relationship was "abominable." Let p_2 be the proportion of all employees this year who think the relationship is "abominable." Then the hypotheses are:

H_0: $p_1 = p_2$ or $p_1 - p_2 = 0$.
H_1: $p_1 > p_2$ or $p_1 - p_2 > 0$ (one-tail, since the question asked whether p_2 was smaller).

The critical value of z (.05 level of significance for a one-tail test) is $+1.645$. The decision rule, in terms of z, is

If $z \leqslant +1.645$, accept H_0.
If $z > +1.645$, reject H_0.

Calculate p (from Equation 9.10) to be

$$p = \frac{(28/100)(100) + (20/100)(100)}{100 + 100} = .24$$

Then, from Equation 9.11,

$$s_{\hat{p}_1 - \hat{p}_2} = \sqrt{(.24)(.76)\left(\frac{1}{100} + \frac{1}{100}\right)} = .003648 = .0604$$

and, from Equation 9.12,

$$z = \frac{.28 - .20}{.0604} = +1.32$$

We accept H_0, and conclude that there is insufficient evidence to say that there is a decline in the proportion who give the relationship between administration and secretaries an "abominable" rating. ■

EXERCISES

9–8. Of 32 randomly selected professors of economics, 4 favored a new tax proposal which was intended to stimulate private investment. Of 32 randomly selected economists who were employed by corporations, 6 favored the proposal. Test the hypothesis, at the .10 level of significance, that there is no difference in the proportion of professors and corporation economists who favor the proposal.

9–9. A random sample of 100 women showed that 84 favored a bill that would remove all sex discrepancies in hours and working conditions in factory jobs. A random sample of 100 men showed that 76 favored the bill. Do women differ from men in their attitude toward the bill? Test at the .05 level of significance.

9–10. When Colonel Bellingham was in charge of the 3304th Air Group, the group flew 104 missions, and 86 of these were deemed to be "successful" missions. When Colonel Everett was in charge of the group, the group flew 80 missions, of which 48 were rated as "successful." Do these results show, at the .01 level of significance, that

a) Colonel Bellingham had a higher proportion of successful missions than Colonel Everett?

b) that the difference between Colonel Bellingham's proportion of successes and Colonel Everett's was not simply due to chance?

9–11. Under a Republican mayor the City of Sainte Cloud had 104 days of rain out of 730 days. Under the Democratic mayor who succeeded her, the City of Sainte Cloud had 142 days of rain over a 730 day period. You have been asked to run a statistical test based on these figures to determine whether there is a significant difference in the proportion of rainy days in Sainte Cloud between Republican and Democratic administrations.

a) What assumptions do you have to make before running a statistical test?

b) What level of significance would you suggest be used?
c) Run the test, based on the level of significance you selected in (a).
d) What, specifically, are your conclusions from the test in terms of the relationship between political party and rainfall in Sainte Cloud?

9.3 Tests Using the Chi-Square Statistic

Suppose 12 out of 16 social studies teachers are coaches.
Suppose 6 out of 24 teachers in the humanities are coaches.
Suppose 2 out of 20 math and science teachers are coaches.

Do these figures show that there is a systematic tendency for there to be a higher proportion of coaches in some teaching fields than in others, or could the observed differences be simply due to chance? Assume the data are randomly selected from the population of social studies teachers, humanities teachers, and math and science teachers.

The hypotheses can be written as:

H_0: $p_1 = p_2 = p_3 = p$.
H_1: The 3 proportions are not all equal.

The statistic used to test the null hypothesis is the χ^2 (chi-square) statistic that was discussed in Chapter 3. Recall that

$$\chi^2 = \sum \frac{(f - f_e)^2}{f_e} \qquad \textbf{3.2}$$

where f = observed frequency in each cell
f_e = expected frequency in each cell

We need to set up a table relating the nominal variables "coaching status" and "teaching area." This is done in Table 9.2.

The null hypothesis states that the proportions of coaches in each area are all the same proportion p. The estimate of p is the total number of coaches in the sample (20) divided by the total number of teachers in the sample. Therefore, under the null hypothesis we expect $20/60 = 1/3$ of the teachers in each area to be coaches. The expected frequency (f_e) of each teaching area is calculated as

$$\frac{\text{row total} \times \text{column total}}{\text{grand total}}$$

The calculations of f_e are entered in parentheses in each cell of Table 9.2.

If you inspect Equation 3.2, you can see that χ^2 will be zero if the frequencies f in each cell are identical with the expected frequencies f_e,

Table 9.2. *FREQUENCY TABLE*
Coaching Status Related to Teaching Area
for 60 Teachers in
Dolly Madison High School

		Teaching Area						
		Natural Sciences		Social Studies		Humanities		
Coaching Status	Coach	2	(6.7)	12	(5.3)	6	(8.0)	20
	Not a coach	18	(13.3)	4	(10.7)	18	(16.0)	40
		20		16		24		60

and that the more the actual frequencies depart from the expected frequencies, the greater χ^2 will be. Under the assumption of H_0, we can normally expect some departure of the actual from the expected frequencies due to random variation, but if the departure is too great, as indicated by too high a value of χ^2, we will reject the null hypothesis.

Table IV shows the critical values of χ^2 for 1 to 30 degrees of freedom. The number of degrees of freedom is given by $(r - 1)(c - 1)$, where r is the number of rows, and c is the number of columns. If χ^2 is computed to be greater than the critical value at a given level of significance, then we have evidence sufficient to reject H_0 at that level of significance. In the present example, the number of degrees of freedom is $(2 - 1)(3 - 1) = 2$, and the critical value of χ^2 at the .05 level of significance is 5.991. The decision rule is, therefore:

If $\chi^2 < 5.991$, accept H_0.
If $\chi^2 \geqslant 5.991$, reject H_0.

We calculate χ^2 from the sample (using Equation 3.2) to be

$$\chi^2 = \frac{(2 - 6.7)^2}{6.7} + \frac{(12 - 5.3)^2}{5.3} + \frac{(6 - 8)^2}{8} + \frac{(18 - 13.3)^2}{13.3} + \frac{(4 - 10.7)^2}{10.7} + \frac{(18 - 16)^2}{16} = 18.373$$

Since $18.373 > 5.991$, the value of χ^2 at the .05 level of significance, we reject H_0. We have sufficient evidence to conclude that the population proportions are unequal.

The χ^2 distribution is a continuous probability distribution with one parameter, the number of degrees of freedom.

The theoretical distribution of χ^2 goes from 0 to infinity. However, the computed statistic $\Sigma[(f - f_e)^2/f_e]$ is a discrete random variable, and is only an approximation of the theoretical probability distribution of Equation 9.2. The approximation is very close when all the theoretical frequencies are large, but if any of the expected frequencies is less than 5, the approximation is generally considered to be too rough to use for most statistical tests. For this reason, when cells have expected frequencies below 5, χ^2 tests are not run unless the cells can be combined with other cells to give expected frequencies of at least 5 in each cell.

An exception to the restriction of a minimum expected frequency of 5 is provided in the case of a 2×2 table. To correct for the upward bias in the statistic when some of the f_e are small, we estimate χ^2 by using

Yates correction for continuity

$$\chi^2 = \sum \frac{(f - f_e - 1/2)^2}{f_e} \qquad 9.13$$

where $f - f_e$ is an absolute (or nonnegative) value.

TESTS FOR INDEPENDENCE

The χ^2 test can be used to test the hypothesis that two nominal variables are independent. The test is identical in procedure to the test for differences among proportions, but there is one difference in principle. In the test for independence, we treat two variables as random variables, whereas in tests for differences among proportions only one variable is random. In a test for differences among proportions, we control the number we observe in each category, and take random samples within each category. The column totals are fixed. In a test for independence, we take a single random sample of the whole; the numbers in both the row and the columns are random.

EXAMPLE 9.6 In an attempt to determine whether religion was related to one's views on divorce, a random sample of 100 people was taken, and information was obtained on their religion and their views on a proposed no-fault divorce bill. The results of the survey are shown in Table 9.3.

There are 8 cells in the table. The expected frequencies in each cell are calculated as (row total × column total/grand total). Unfortunately, the expected frequency in the lower right cell is 3.00. We cannot run a χ^2 test using the data in this form without violating the rule that all cells must have expected frequencies of at least 5. We can get around the problem by combining two religious categories, as shown in Table 9.4. But Table 9.4 conveys less information than Table 9.3. The combining of categories was necessary to run a valid statistical test. The loss of information is the price you may have to pay in order to run a valid χ^2 test on the basis of a small sample.

Table 9.3. *CONTINGENCY TABLE*
Random Sample of 100 Persons Classified by Religion and Views on No-Fault Divorce Bill

		Views on Divorce Bill				
		In Favor		Opposed		
Religion	Protestant	24	(28)	16	(12)	40
	Roman Cath.	22	(21)	8	(9)	30
	Jewish	16	(14)	4	(6)	20
	Other	8	(7)	2	(3)	10
		70		30		100

Suppose you want to use a .01 level of significance in testing the hypothesis that religion and views on no-fault are independent. For $(3 - 1)(2 - 1) = 2$ degrees of freedom, $\chi^2_{.01} = 9.210$. The decision rule is therefore:

If $\chi^2 < 9.210$, accept H_0.
If $\chi^2 \geqslant 9.210$, reject H_0.

We calculate the test statistic as

$$\chi^2 = \frac{(24-28)^2}{28} + \frac{(16-12)^2}{12} + \frac{(22-21)^2}{21} + \frac{(8-9)^2}{9} + \frac{(24-21)^2}{21} + \frac{(6-9)^2}{9} = 3.492$$

By the decision rule, we accept H_0. We cannot reject the hypothesis of independence. We therefore have insufficient evidence to say that religion and views on no-fault divorce are related. ∎

Table 9.4. *CONTINGENCY TABLE*
The Data of Table 9.3 with Two Religious Categories Combined

		Views on Divorce Bill				
		In Favor of No-Fault		Opposed to No-Fault		
Religion	Protestant	24	(28)	16	(12)	40
	Roman Catholic	22	(21)	8	(9)	30
	Other	24	(21)	6	(9)	30
		70		30		100

EXERCISES

9–12. Things had gotten so bad in Mr. Carp's department that 100 employees (out of a total of 300) signed a petition complaining about Mr. Carp's handling of departmental matters. A breakdown of the employees by "petition signing" and by "pay grade" is as follows:

	Grade GS9 or Lower	Grade Above GS9
Signed petition	55	45
Did not sign petition	95	105

Test the hypothesis at the .05 level of significance that people in different pay grades differ with respect to whether one signs a petition of protest against the boss. (To run this test, you must assume that these individuals constitute a random sample of the universe in which you are interested.)

9–13. An aptitude test was given to 40 men, to 30 women, and to 30 dogs. Of the 40 men, 25 passed the test; of the 30 women, 12 passed; of the 30 dogs, 23 passed. Assuming that the sample were randomly selected, can you say, at the .01 level of significance, that men, women, and dogs differ in their ability to pass the aptitude test?

9–14. A random sample of Navy doctors was classified by rank and religion in the following table:

	Protestant	Roman Catholic	Jewish	Other
Flag Rank	8	16	0	1
Captain or Commander	9	15	3	3
Lt. Commander or Lt.	13	19	7	16

Is there evidence that religion and rank are related? Test at the .05 level.

9–15. People arriving at an accident clinic are either subsequently admitted to the hospital, treated and discharged, or discharged without treatment. A random sample of 100 people coming to the clinic for treatment was classified by time of arrival and ultimate action taken:

	Admitted to Hospital	Treated and Discharged	Discharged Without Treatment
12:00–8:00 A.M.	9	13	8
8:00 A.M.–4:00 P.M.	7	24	9
4:00 P.M.–12:00 midnight	4	13	13

Test the hypothesis at the .01 level of significance that "time of arrival" and "action taken" are independent random variables.

***9–16.** Mr. Snipe has been attempting for 6 months to show that his supervisor, Ms. Anthony, is prejudicial in her treatment of employees. He has tried unsuccessfully to show that promotion rates under Ms. Anthony were significantly different with respect to age, race, sex, religion, height, and handedness. Yesterday Mr. Snipe presented the following figures in support of his claim that Ms. Anthony was discriminating on the basis of first letter of last name.

	Name Starts with A to M	Name Starts with N to Z
Recommended by Ms. A for promotion	24	6
Not recommended by Ms. A for promotion	21	29

a) What level of significance do you think is appropriate to determine whether Ms. Anthony is discriminating?

b) Run a test for independence based on the level of significance you selected.

9–17. A random sample of 25 automobiles was classified by cost of automobile (A) and cost of repairs (R) in the first year in the following table:

	R = under \$50	R = at least \$50 but under \$250	R = at least \$250
A = under \$4,000	22	16	2
A = at least \$4,000 but under \$6,000	16	20	14
A = at least \$6,000	2	4	29

Test the hypothesis at the .01 level of significance that "cost of automobile" and "cost of repairs" in the first year are independent.

9–18. Given the following contingency table:

	Male	Female
Laid off	30	20
Not laid off	120	80

a) Test the hypothesis at the .05 level of significance that sex and layoff status are independent variables.

b) Comment on the value of the χ^2 statistic that you calculated. What hypothesis would you propose to explain this value?

9.4 Which Statistical Test Should You Use?

After being exposed to a discussion of a variety of statistical tests, many students are perplexed by the question of which type of test to run in which situation. This is not an easy question to answer, since real situations do not always fall neatly into a textbook category. Most of the exercises in this book will not tell you explicitly to run a t test for differences, or a test for proportion. Generally, they will describe a situation and require you to determine the appropriate test. This section will discuss the situations that permit or call for the use of certain test statistics, the use of a one-tail or two-tail test, and the determination of the level of significance.

One or Two Tails?

The number of tails depends on the alternative hypothesis, which in turn usually depends on the nature of the question. If the question simply asks whether there is a significant *difference* (whether this is between a sample mean and a hypothesized universe mean, between two sample means, between two sample proportions, etc.), then a two-tail test is appropriate. If the question asks for a difference *in a specified direction* (whether the mean is sufficiently large, whether one parameter is greater than another, etc.), then it calls for a one-tail test. If in doubt, it may be better to run a two-tail test. This two-tail test keeps your eyes open to differences in either direction, even in a direction that may violate your preconceived notions.

Level of Significance?

In general, the more costly the Type I Error is in relation to the Type II Error, then the lower α should be. If rejection of the null hypothesis will lead to a momentous decision, such as the investment of a large amount of money, or publication of a result as "scientifically proven," then a low value of α (perhaps .01 or possibly lower) is called for. In preliminary research, rejection of H_0 will usually lead to follow-up studies, and a Type I Error is not very serious. In such cases, a level of significance of .05 or .10 may be appropriate. Higher values of α are especially appropriate when the sample size is small, a situation typical of preliminary or exploratory research. It is very difficult to get enough information to reject null hypotheses with small samples, in any case, and to inflict a small level of significance will almost guarantee that the null hypothesis will not be rejected. Imposing too stringent (small) a level of significance on small samples can result in the discarding of good ideas before they had a chance to prove themselves (through larger samples). On the other hand, publication of results as "significant," when α is fairly high, borders on the dishonest, and is not much better than not running tests at all. At almost any newsstand you can find articles in the "sensationalist" publications by and about fortune tellers and clairvoyant prognosticators who do not worry about committing Type I Errors.

WHICH TEST STATISTIC TO USE?

You may use the normal variable z only if either

a. the probability distribution of the underlying random variable is normal, *or*
b. the sample size is large enough to use the Central Limit Theorem or the normal approximation to the binomial.

The use of the normal distribution also requires that the standard error of the test statistic is either

a. based on known or assumed population standard deviations or proportions, *or*
b. based on estimates from *large* samples (meaning that the smallest sample size should be at least 30).

You may use the t distribution only if the underlying probability distribution of the random variable is normal, or approximately normal.

You *must* use the t statistic rather than a normal statistic such as z if the sample size is small and you estimate the standard error of the test statistic from the sample. Figure 9.1 summarizes these rules in the form of a decision tree, where the two question marks to the right of two of the paths represent cases you have not yet studied. When the probability distribution of the random variables is unknown and you do not have a sample large enough to warrant use of the Central Limit Theorem, it is necessary to resort to what are known as nonparametric statistical tests. These will be discussed in Chapter 11. By making fewer assumptions about the population than you need to make when using the z or t statistic, nonparametric tests are very widely applicable. They suffer from the disadvantage of making less use of the information in the sample, and hence may fail to reject a null hypothesis in a situation where a z or t test would do so.

EXERCISES

9–19. Which of the following statements are true? Give a reason for your answer.

a) You should use a t test only when the standard error of the statistic must be estimated from the sample.

b) If the probability distribution of the random variable X is known or

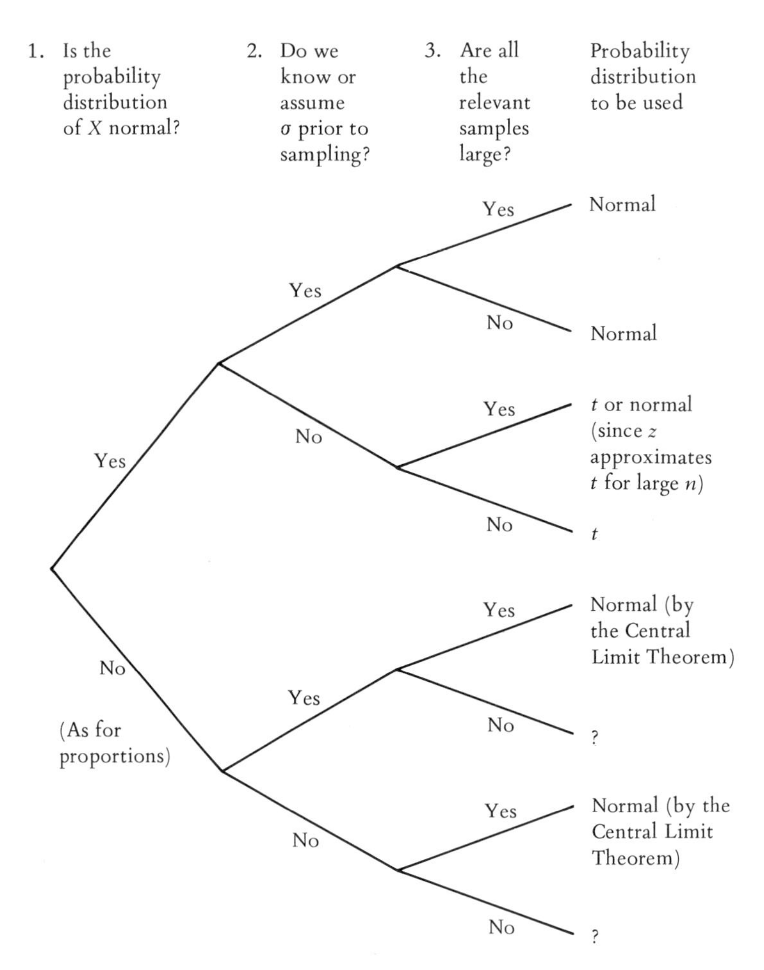

Figure 9.1. *DECISION TREE*
Uses of Statistical Tests for Means, Proportions, and Differences

assumed to be normal, then you can always use the normal distribution in running a statistical test.

c) Whenever you have a small sample you should use the t distribution in running a statistical test.

d) Whenever you have a small sample and you must estimate the standard error of the statistic from the sample, you should use the t distribution for a statistical test.

e) Whenever you have a small sample and you must estimate the standard error of the statistic from the sample, and the probability distribution of the random variable is normal, then you should use the t distribution for a statistical test.

f) You should not use the t distribution when testing for proportions or differences between proportions.

9–20. What is the probability distribution of the test statistic under the following conditions:

a) Test for a proportion; $n = 60$.
b) Test for the difference between means; $n_1 = 100$, $n_2 = 60$; σ_1 and σ_2 assumed; population normal.
c) Test for difference between proportions: $n_1 = 60$; $n_2 = 30$.
d) Test for a mean; $n = 70$; probability distribution of the random variable unknown; σ unknown.
e) Test for differences between means; $n_1 = 10$, $n_2 = 20$; population normal; no assumption about σ.
f) Test for differences; seven sets of paired observations; differences assumed to be normally distributed; population variance unknown.
g) Test for difference between means; $n_1 = 30$, $n_2 = 10$; population standard deviations unknown; probability distribution assumed to be normal.
h) Test for a proportion; $n = 5$.
i) Test for difference between 2 means; $n_1 = 5$, $n_2 = 4$; population standard deviations unknown; no assumption can be made about the probability distribution of the random variable.

Summary of Chapter

In testing for the differences between two means from large samples you can make use of the Central Limit Theorem. The test statistic will be z, the standard normal variable. If the samples are small, you must check to determine whether the underlying probability distribution is approximately normal; if it is, you can use the t statistic if the population standard deviation is estimated from the sample standard deviation s. You can test for differences between two proportions using the standard normal variable z if the proportions come from large samples. The chi-square test generalizes the test for differences between proportions to three or more cases; it is also appropriate when testing for independence of two variables.

In selecting a test, you should let the form of the alternative hypothesis be your guide as to whether to run a one or a two-tail test. The level of significance will depend largely on the seriousness of the Type I Error. Tests for means and proportions from small samples when you cannot assume an underlying normal probability distribution will be covered in a later chapter.

REVIEW EXERCISES FOR CHAPTER 9

9–21. Before a major television speech by the President, a poll of 1000 voters showed that 520 favored the President's proposal to send the Marines to the Island of Atlantis. After the speech a poll of 1000 voters, taken independently of the test poll, showed that 480 favored the proposal. Do these results show, at the .05 level of significance, that there was a change in voter sentiment toward the plan between the two polls? Assume that both polls were random samples of voters.

9–22. Describe a different method of determining changes in voter sentiment toward the President's proposal from before his speech to after his speech, using the method of paired observations. In what ways might this method be considered superior to, and inferior to, the method of taking two independent random samples described in Exercise 9–21?

9–23. A random sample of 50 recipients of MPA degrees showed that their mean IQ was 132 and the standard deviation of their IQ's was 10 points. A random sample of Ph.D.'s showed a mean IQ of 128 and a standard deviation of 10 points.

a) At the .05 level of significance, is there evidence of a difference in the mean IQ's of MPA's and Ph.D.'s?

b) Did you use a one-tail or two-tail test? Discuss the reason for the number of tails you selected.

9–24. Five doctors worked a mean of 64 hours a week, with a standard deviation of 12 hours. Seven lawyers worked a mean of 48 hours a week, with a standard deviation of 12 hours. Test the hypothesis, at the .05 level of significance, that there is no difference between doctors and lawyers in the number of hours they work per week. Assume that the samples are random and that the population is normally distributed.

9–25. A random sample of 50 lawyers revealed that their mean height was 70 inches. Further research indicated that the sample consisted of 48 male lawyers, that their mean height was 69.8 inches, and that the standard deviation of their heights was 3 inches. The two female lawyers in the sample had a mean height of 72.4 inches with a standard deviation of 3 inches

a) Do the heights of male and female lawyers differ at the .10 level of significance, according to this sample?

b) Discuss the merits of using a one-tail and a two-tail test in this exercise.

c) Do you stand a better chance of getting significant results if you take a sample of 40 males and 10 females? Discuss.

9–26. State parks have two types of campsites: a "deluxe" campsite that rents for \$5 a night, and a "basic" campsite that rents for \$2 a night. A random sample of 70 campers at Eastern State Park showed that 42 preferred to stay at a "deluxe" campsite, while a similar random sample of 52 campers at Western State Park showed that 24 preferred the "deluxe" campsite. Test the hypothesis, at the .01 level of significance, that campers at the two sites do not differ in their preference.

9–27. Before the football game between the Mastodons and the Heffalumps, a newspaper published the "vital statistics" of the two teams. These showed that the mean weight of the 48 members of the Mastodon squad was 234.2 pounds with a standard deviation of 13.7 pounds. Of the 48 Heffalumps,their mean weight was 235.7 pounds with a standard deviation of 14.9 pounds. Do these results show that the mean weights of the members of the two teams differ? Use a level of significance that you think is appropriate.

9–28. A random sample of 5 children who brushed with fluoride and 7 who did not revealed the following statistics on number of cavities:

Brushed with fluoride: 0, 2, 3, 4, 1
Did not brush with fluoride: 3, 8, 2, 0, 6, 2, 4

Can you say, at the .01 level of significance, that there is a difference in mean number of cavities between children who brush with fluoride and those who do not? Assume that the samples are random.

9–29. The diastolic blood pressures of 5 randomly selected women and their husbands were recorded as follows:

Wives		*Husbands*	
Mrs. A	80	Mr. A	88
Mrs. B	66	Mr. B	68
Mrs. C	98	Mr. C	112
Mrs. D	80	Mr. D	94
Mrs. E	76	Mr. E	78

a) Test the hypothesis, at the .05 level of significance, that there is no difference in diastolic blood pressure between husbands and wives. Assume the differences are normally distributed.

b) Can differences between blood pressures of husbands and wives be attributed solely to differences in sex? Why or why not?

9–30. A random sample of 68 school districts in urban areas showed 20 of them submitting budget requests for increases of at least 10% over the previous year's budget allocation. A random sample of 55 school districts in rural areas showed 10 of them submitting budget requests for increases of at least 10% over the previous year's budget allocation.

Is there evidence, at the .05 level of significance, that urban school districts are more likely to ask for increases of at least 10% than rural school districts?

9–31. Set up the null hypothesis and the alternative hypothesis and decide on a level of significance for each of the following situations:

a) A school of engineering has purchased 2500 electronic calculators from a supplier. A random sample of 200 is tested for defectives. If more than 3% of the whole lot are defective, it is considered an unacceptable lot and should be returned to the supplier. However, the school does not want to run a high risk of rejecting the lot if indeed the lot should be accepted.

b) It is believed that students and professors make frequent use (at least 10 times per week) of the Journal of International Politics (JIP) in the library of a School of International Affairs. A student suspects that the number of times the Journal is read per week is much lower than the commonly believed number. He records the number of times the Journal is read over 20 randomly selected hours. The study is for a term paper in one course, and is considered more of an exercise than a study for wide circulation. As a result of the limitations on time and resources, his data is admittedly very limited.

c) In the ABC Agency last year, 45% of the men were promoted and 35% of the women were promoted. An investigating commissioner assumes that if the Agency had conducted a truly nondiscriminatory policy, the expected percentage of men to be promoted would be equal to the expected percentage of women to be promoted. If the commissioner decides that the differences in percentage of promotions is due to discrimination rather than chance, the Commission will make a ruling that could seriously affect the reputation of the ABC Agency.

d) In a typing test, Sue typed 82.4 words per minute and George typed 83.5 words per minute. The office manager has decided to hire either George or Sue as a secretary, and the decision is to be based solely on the typing speeds achieved in the test.

9–32. At the College of Canine Knowledge, 30 thoroughbreds and 20 mongrels were enrolled in Course K-9. Of the 30 thoroughbreds, 16 passed the midterm examination. Of the 20 mongrels, 16 passed the midterm examination. Is there evidence, at the .05 level of significance, that thoroughbreds and mongrels have unequal probabilities of passing midterm examinations? Assume the data are random samples of all thoroughbreds and mongrels.

9–33. Betty Brown, a Management Intern at the Bureau of Economic Development, has made up a "Brain Teaser" puzzle which she has asked the various members of the Bureau to try to solve. The employees were classified by job category and their performance. The results were:

	Successful	Failed or Quit
Professional	8	12
Managerial	4	16
Secretarial	20	20

Assuming that the employees of this agency are a random sample of professional, managerial, and secretarial employees in the government, test the hypothesis, at the .05 level of significance, that job category and performance on this test are independent.

9–34. In order to compare the views of different groups of individuals on corporal punishment in schools, random samples are chosen and asked whether they did or did not favor such punishment. The frequencies of responses are as follows:

	Favorable	*Unfavorable*
Teachers	34	16
Principals	6	2
Parents	17	23
Taxpayers (general public)	18	22

Is there evidence, at the .05 level of significance, that these groups differ in their views on corporal punishment?

9–35. A university's budget permitted a 5% salary increase to the faculty members. A question that was heatedly debated was: Should the 5% increase be awarded selectively at different rates to different faculty on the basis of "merit," or should all faculty be given exactly 5% as a cost-of-living increase to help faculty catch up with a predicted 9½% inflation rate? A random sample of several groups within the university revealed the following results:

	Favored "Merit" Increase	*Favored "Cost-of-Living" Increase*
University Administrators	31	9
Professors	28	14
Associate Professors	16	22
Assistant Professors	14	32
Other faculty	23	28

a) Is there evidence, at the .01 level of significance, that the 5 groups differ in their views as to how the increase should be allotted?

b) Is there evidence, at the .01 level of significance, that the 4 faculty groups (exclusive of administrators) differ on their views on how the increase should be allocated?

SUGGESTED READINGS

HAMBURG, MORRIS. *Statistical Analysis for Decision-Making,* 2nd ed. New York: Harcourt, Brace, Jovanovich, 1977. Chapters 8, 13–17.

JOHNSON, ROBERT. *Elementary Statistics,* 2nd ed. North Scituate, Mass.: Duxbury Press, 1977. Chapter 11.

LEABO, DICK. *Basic Statistics,* 5th ed. Homewood, Ill.: Richard D. Irwin, Inc., 1976. Pp. 275–283.

LOETHER, HERMAN J., and DONALD McTAVISH. *Descriptive and Inferential Statistics: An Introduction.* Boston, Mass.: Allyn and Bacon, Inc:, 1976. Chapter 15.

OTT, LYMAN. *An Introduction to Statistical Methods and Data Analysis.* North Scituate, Mass.: Duxbury Press, 1977. Pp. 112–118.

RAIFFA, HOWARD, and ROBERT SCHLAIFER. *Applied Statistical Decision Theory.* Cambridge, Mass.: Division of Research, Graduate School of Business Administration, Harvard University, 1961.

TEN

Regression and Correlation Analysis

Outline of Chapter

10.1 Linear Regression Analysis

Linear regression analysis is a method of analyzing the linear effect of an independent variable on a dependent variable.

10.2 Linear Correlation Analysis

Linear correlation analysis looks at the interdependence of two random variables whose relationship to each other is linear.

*10.3 Nonlinear and Multiple Regression Analysis

The techniques of linear regression analysis can be extended to nonlinear relationships and to the effects of more than one independent variable on a dependent variable.

Objectives for the Student

1. Distinguish between a linear regression equation computed from a sample and the linear equation expressing the relationship between two variables in the population.
2. Test the hypothesis that the population regression coefficient B is 0.
3. Interpret the meaning of ρ^2, the population coefficient of determination.
4. Distinguish between regression and correlation analysis.
5. Test the hypothesis that the population correlation coefficient ρ is 0.

*6. Interpret the meaning of the partial regression coefficients in a multiple regression equation.

7. Be able to judge from a scatter diagram whether a linear, parabolic, or exponential regression equation would be the best fit to a set of data.

Chapter 9 discussed tests for the relationship between nominal variables, such as comparisons of religious groups, professional groups, ethnic groups, the sexes, etc. In this chapter, you will look at the relationships between two or more interval-level or ratio-level variables. These are variables that can readily be graphically depicted on a quantitative scale. We can examine the relationship between years of experience and quantity of output; between height and weight; between age of house and value of house; among air pollution levels, temperature, and number of bronchial infections; etc.

The two methods of analysis that this chapter will discuss are regression analysis and correlation analysis. The two methods bear a considerable resemblance to each other, but they differ in their fundamental approach. Regression analysis takes one or more variables as given, and examines the effect of these on a dependent random variable. Correlation analysis treats all variables as random, and examines the interrelationships (or co-relationships) between or among them.

You studied the concepts of regression and correlation in Chapter 3. The measures discussed in that chapter were purely descriptive: They described the data you observe. This chapter, however, treats the data as a random sample of an underlying population or universe, and is concerned with the question of what inferences we can make about the population from which the sample was drawn. Recognizing that sampling error is an ever-present problem as long as the data do not constitute the whole population in which we are interested, this chapter takes a more cautious approach than does Chapter 3, by insisting that we test hypotheses about universe parameters. And the smaller the sample, in general the more important it is that we exercise this caution.

10.1 Linear Regression Analysis

Betty Bishop has started a TV-radio repair business, and she works anywhere from 5 to 15 hours a week at her business. Over the course of 8 weeks she has kept records of the number of hours she has worked and her earnings for each week. These are shown in Table 10.1.

Betty sees that in general the more hours she puts in the more money she will earn. She believes that the relationship can, in general, be described as a linear one. Every additional hour that she puts in will tend to increase her earnings by a constant amount. However, the linear relationship of Betty's earnings describes only a general tendency; it does not tell *exactly* what Betty will earn on the basis of number of hours worked.

Table 10.1. *WEEKLY HOURS WORKED AND EARNINGS*
Betty Bishop's Business

No. of Hours Worked X	Earnings Y
8	$28
5	25
12	60
13	56
9	40
10	30
14	42
9	39

The relationship can be illustrated by the

> **Linear prediction equation of *Y* from *X***
>
> $\hat{Y} = a + bX$ **10.1**

which is the linear regression Equation 3.7.

To get a better picture of the effect of hours on earnings, Betty uses her records to plot the scatter diagram shown in Figure 10.1. This diagram tells Betty two things:

1. It appears that there does seem to be a linear relationship, so that a straight line could be drawn to "fit" the data.
2. It is obvious that the relationship is not deterministic, and that no matter what straight line is drawn, most of the points will be either above or below the line.

The second conclusion does not invalidate the line—it simply means that the line does not tell us everything, and that in each week there will undoubtedly be peculiar factors that will make that week a little better or worse than what she would have expected purely from the number of hours she worked.

Betty wants to know how much she would tend to earn in the long run if she works 5 hours, 10 hours, or any other specified number of hours in a week. In order to estimate earnings from hours, she would like to draw a straight line (illustrating the linear relationship) that "fits" the data points as well as possible. But how is she going to draw this straight line? Even with a straightedge, Betty has the problem of determining just where to draw the line. She needs some criteria on which to determine

Figure 10.1. *SCATTER DIAGRAM Data from Table 10.1*

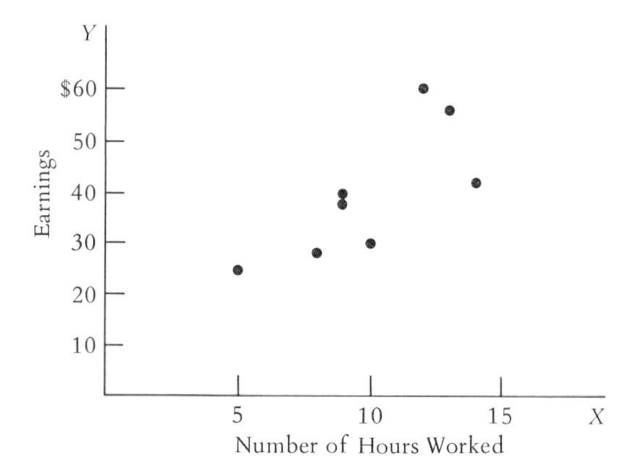

what constitutes a good fit. After considerable thought, Betty decided that the line should meet the following two criteria:

1. It should be "unbiased," in that it does not tend to overestimate the points any more than it underestimates the points. In other words, the total overestimations and underestimations should cancel each other out.
2. It should fit the points as closely as possible. The total deviations from the line (whether positive or negative) should be as small as possible.

It can be shown that these two criteria are generally contradictory; in most cases you cannot get a line that achieves both of them. But there is a method that does meet criterion 1 and comes very close to meeting criterion 2.

> The METHOD OF LEAST SQUARES is a method of fitting a line of the form $\hat{Y} = a + bX$ to a set of data. It is designed so that the sum of the squared deviations of the Y values of the data from the line $\hat{Y}$, denoted as
>
> $$\Sigma(Y - \hat{Y})^2$$
>
> is a minimum.

The method of computation of the least squares equation is presented in Section 3.4. You can use Equations 3.9 and 3.10 to find the a and b coefficients for Betty Bishop. The necessary calculations are done in Table 10.2.

Table 10.2 gives the calculation of the regression equation, $\hat{Y} = 7.30 + 3.27X$. Using this regression equation, Betty can calculate $\hat{Y}$ for each of the X values in Column 1. The results of these calculations are shown in Column 5. These are the earnings each week she would predict that she would earn on the basis of X, the number of hours worked. Of course she does not earn exactly the predicted value, as you can see by observing the differences between actual earnings (Y) and predicted earnings $\hat{Y}$. The differences $Y - \hat{Y}$ are shown in Column 6. These differences (with plus or minus sign included) add up to zero, indicating that the line $\hat{Y}$ is unbiased—its overestimations are exactly offset by its underestimations.

The values a and b are statistics, since they are values computed from sample information. The statistic a, which equals +7.30, is the value of $\hat{Y}$ when $X = 0$. You should not infer from this statistic, however, that if Betty works zero hours in a week her predicted earnings would be \$7.30. The reason for not making this interpretation is that none of the data points on which the equation was computed extend to X values below 5. Beyond the range of the data, it is generally best to interpret $\hat{Y}$ values only in a mathematical sense.

Table 10.2. *CALCULATION OF REGRESSION EQUATION*
Data from Table 10.1

(1) X	(2) Y	(3) XY	(4) X^2	(5) $\hat{Y}$	(6) $Y - \hat{Y}$	(7) $(Y - \hat{Y})^2$	(8) $X - \overline{X}$	(9) $(X - \overline{X})^2$	(10) $Y - \overline{Y}$	(11) $(Y - \overline{Y})^2$	(12) Y^2
8	28	224	64	33.46	−5.46	29.8116	−2	4	−12	144	784
5	25	125	25	23.65	+1.35	1.8225	−5	25	−15	225	625
12	60	720	144	46.54	+13.46	181.1716	+2	4	+20	400	3600
13	56	728	169	49.81	+6.19	38.3161	+3	9	+16	256	3136
9	40	360	81	36.73	+3.27	10.6929	−1	1	0	0	1600
10	30	300	100	40.00	−10.00	100.0000	0	0	−10	100	900
14	42	588	196	53.08	−11.08	122.7664	+4	16	2	4	1764
9	39	351	81	36.73	+2.27	5.1529	−1	1	−1	1	1521
80	320	3396	860		0.00	489.7340		60	0	1130	13930

$\overline{X} = 10 \qquad \overline{Y} = 40 \qquad n = 8$

$$b = \frac{n\Sigma XY - \Sigma X \Sigma Y}{n\Sigma X^2 - (\Sigma X)^2} = \frac{8(3396) - (80)(320)}{8(860) - (80)^2} = +3.27 \qquad a = \frac{\Sigma Y - b\Sigma X}{n} = \frac{58.4}{8} = +7.30$$

$$\hat{Y} = a + bX = 7.30 + 3.27X$$

The statistic b, which equals +3.27, is the one that Betty is going to be particularly interested in.

> The b statistic is known as the REGRESSION COEFFICIENT. It measures the increase in $\hat{Y}$ for every unit increase in X.

That is, it measures how much she would expect earnings to increase, on the average, for each additional hour she works. Therefore, a week in which she worked 15 hours would be expected to yield her \$3.27 more than a week in which she worked 14 hours, and $10 \times \$3.27 = \$32.70$ more than a week in which she worked only 5 hours.

TESTS OF SIGNIFICANCE OF THE REGRESSION COEFFICIENT

The statistics a and b are purely descriptive. They describe the data from which they were computed. Betty realizes that the data may not be truly representative of her long-run hours and earnings potential, and therefore wonders how much she can trust that +3.27 regression coefficient. If she had taken a different sample of 10 weeks, might she have obtained a regression coefficient that was very different from +3.27—perhaps even one that was negative? Her interest therefore goes beyond the sample. She wants to know the "true" nature of the relationship between her hours and her earnings.

The regression coefficient expressing this "true" relationship between hours and earnings is known as the parameter B. Parameter B is the increase in the expected value of Y, given X, or $E(Y|X)$ for each unit increase in X, where the relationship between $E(Y|X)$ and X is given by the linear equation

$$E(Y|X) = A + Bx \qquad \textbf{10.2}$$

The value of B can never be known with certainty, unless you are able to take a census of the entire universe. But you can use the sample regression coefficient b to estimate B. If you can assume that the conditional probability distribution of Y for any given value x is normal with a constant variance $\sigma^2_{Y.X}$, then you can set up a $(1 - \alpha)$ confidence interval for B, using the expression

$$b \pm t_\alpha s_b \qquad \textbf{10.3}$$

where t_α is found in Table III with $n - 2$ degrees of freedom and s_b is the estimated standard error of the regression coefficient.

$$s_b = \frac{s_{y.x}}{\sqrt{\Sigma(X - \overline{X})^2}} \qquad \textbf{10.4}$$

> $s_{y.x}$, known as the STANDARD ERROR OF ESTIMATE, is given by
>
> $$s_{y.x} = \sqrt{\frac{\Sigma(Y - \hat{Y})^2}{n - 2}}$$

10.5

Suppose Betty makes the assumptions necessary to set up the confidence interval. In order to calculate the interval, she first calculates $s_{y.x}$ from Equation 10.5 by summing the squared deviations $\Sigma(Y - \hat{Y})^2$, as shown in Column 7 of Table 10.2.

$$s_{y.x} = \sqrt{\frac{489.72}{8 - 2}} = \sqrt{81.62} = 9.035$$

Column 9 of Table 10.2 gives $\Sigma(X - \overline{X})^2 = 60$. This result enables her to find s_b from Equation 10.4.

$$s_b = \frac{9.035}{\sqrt{60}} = 1.166$$

Betty can now set up a 95% confidence interval ($\alpha = .95$) from Expression 10.3. For $8 - 2 = 6$ degrees of freedom, $t_{.05} = 2.447$. Therefore the confidence interval is

$$3.27 \pm 2.447(1.166) = 3.27 \pm 2.85 = .42, 6.12$$

The 95% confidence interval says that B could be as low as .42 or as high as 6.12. This seems like a large interval, but such a large interval is to be expected with a sample size of only 8. Although the interval fails to give a very precise estimate of B, it is useful to know that the entire interval lies on the positive side of zero, and that one may, therefore, have a high degree of confidence that B is not negative.

If Betty's main concern had been whether or not B was positive, she could have tested the null hypothesis:

H_0: $B = 0$, against the one sided alternative,
H_1: $B > 0$.

At the .05 level of significance for a one-tail test, the critical value of t for 6 degrees of freedom would be $+1.943$. She would reject H_0 if $t > 1.943$.

She can calculate t from the formula

$$t = \frac{b - B}{s_b} \qquad \textbf{10.6}$$

Since the hypothesized value of B is 0, Betty calculates t from Equation 10.6 as

$$t = \frac{3.27 - 0}{1.166} = 2.804$$

She can reject H_0. The regression coefficient b is significantly greater than zero.

Equations 10.4 and 10.5 are definitional formulas, but when you have to compute the statistics from lengthy data it is more efficient to use

$$s_b = \frac{s_{y.x}}{\sqrt{\Sigma(x - \bar{x})^2}} \quad \textbf{10.7}$$

and

$$s_{y.x} = \sqrt{\frac{\Sigma Y^2 - a\Sigma y - b\Sigma XY}{n - 2}} \quad \textbf{10.8}$$

THE POPULATION COEFFICIENT OF DETERMINATION

Equation 3.11 in Chapter 3 gives you the coefficient of determination, r^2, in the sample. Such a measure describes only the sample, and does not directly tell you what the strength of the relationship is between the variables X and Y in the population. Let us assume, however, as Betty Bishop did in estimating a population regression coefficient, that the relationship between X and Y can be expressed in the linear form $E(Y|X) = A + Bx$, and that the standard deviation of Y around $E(Y|X)$ is a constant $\sigma_{Y.X}$. Then we can define

THE POPULATION COEFFICIENT OF DETERMINATION

$$\rho^2 = 1 - \frac{\sigma^2_{Y.X}}{\sigma^2_Y} \quad \textbf{10.9}$$

The coefficient of determination ρ^2 tells you what proportion of the total variance of a dependent variable Y can be "explained" by an independent variable X. To illustrate this concept, suppose that Y is the amount of money families spend on clothing and X is the family income. Let us suppose that $E(Y|X)$ is a linear function of X. This means that for every dollar of increase in a family's income, there would be a constant amount B of change in a family's expected spending on clothing. Furthermore, let us suppose that for every income level, the spending of families on clothing varies by a constant amount, and this variability is measured by the variance around $E(Y|X)$, which is $\sigma^2_{Y.X}$. If spending on clothing depended on income and nothing else, then $\sigma^2_{Y.X}$ would be zero. In such a case, income would have been a perfect predictor of spending. All families with an income of X would spend exactly $E(Y|X)$. The other extreme case is that of spending having nothing to do with income. Then $E(Y|X)$

would be simply $E(Y)$, and $\sigma^2_{Y.X}$ would be σ^2_Y. Between these two extreme cases is the more likely situation in which income X explains *some* of spending Y. In this case $\sigma^2_{Y.X}$ will be greater than 0 but less than σ^2_Y. The greater the extent to which X explains Y, the closer $\sigma^2_{Y.X}$ will be to 0, and the less the extent to which X explains Y, the closer $\sigma^2_{Y.X}$ will be to σ^2_Y.

Suppose the total standard deviation of families' spending on clothing is $\sigma_Y = \$200$. Then $\sigma^2_Y = 40{,}000$. Suppose the standard deviation of families' spending on clothing, *for any given income level,* is \$100. Then $\sigma_{Y.X} = \$100$ and $\sigma^2_{Y.X} = 10{,}000$. From Equation 10.9 we can compute

$$\rho^2 = 1 - \frac{10{,}000}{40{,}000} = .75$$

We can say that 75% of families' spending on clothing is "explained" by income.

Estimation of ρ^2

The discussion on family income and spending was based on universe values. The relationship between X and Y is a "consumption function," such as those you may read about in texts on economic theory. But most economics texts discuss only theoretical functions; they do not tell you how to measure or estimate the relationships from sample data.

The estimator of ρ^2 is a statistic that bears a strong resemblance to the sample coefficient of determination which is given by

$$r^2 = 1 - \frac{\Sigma(Y - \hat{Y})^2}{\Sigma(Y - \overline{Y})^2} \qquad \textbf{3.11}$$

But r^2 is a biased estimator of ρ^2. Its expected value is greater than ρ^2. To correct for this bias, you replace the sums of squares in Equation 3.11 by $s^2_{y.x}$ and s^2_y. This gives

$$\hat{\rho}^2 = 1 - \frac{s^2_{y.x}}{s^2_y} \qquad \textbf{10.10}$$

which is the ESTIMATOR OF THE POPULATION COEFFICIENT OF DETERMINATION

where $\hat{\rho}^2$ is an unbiased estimator of ρ^2. The statistics $s^2_{y.x}$ and s^2_y are unbiased estimators, respectively, of $\sigma^2_{Y.X}$ and σ^2_Y.

You now have the equations to estimate the proportion of Betty Bishop's earnings that are explained by the number of hours that she works. Looking first at the sample data itself, you can calculate

$$r^2 = 1 - \frac{\Sigma(Y - \hat{Y})^2}{\Sigma(Y - \overline{Y})^2}$$

Column 11 of Table 10.2 gives the total $\Sigma(Y - \overline{Y})^2 = 1130$. Therefore

$$r^2 = 1 - \frac{489.73}{1130} = .567$$

In the sample, 56.7% of the variance in earnings is associated with number of hours worked. For the population, we estimate ρ^2 as

$$\hat{\rho}^2 = 1 - \frac{s_{y.x}^2}{s_y^2}$$

for which $s_{y.x}^2$ can be obtained from Equation 10.5.

$$\frac{\Sigma(Y - \hat{Y})^2}{n - 2} = \frac{489.73}{6} = 81.622$$

and s_y^2 is found from Equation 2.7.

$$\frac{\Sigma(Y - \overline{Y})^2}{n - 1} = \frac{1130}{7} = 161.429$$

Therefore,

$$\hat{\rho}^2 = 1 - \frac{81.622}{161.429} = .494$$

These results illustrate the upward bias of r^2 in estimating ρ^2. In the sample itself, the number of hours worked explains 56.7% of the variance in earnings. But we cannot use that figure to estimate the proportion of earnings that is explainable by hours worked for the whole population. The corrected estimate of this proportion is $\hat{\rho}^2 = .494$. We estimate that slightly under half of the total variance in earnings (in the universe of all weeks in which Betty works) is explainable by the number of hours worked.

Interpretations

Several parameters and statistics have been discussed in this section. In order to clarify the distinctions between these, the interpretations of the more important ones are summarized below.

The parameter B is the population regression coefficient. It has meaning only if you assume that the relationship between the variables X and Y is linear, of the form $E(Y|X) = A + Bx$. It tells you how much the Y value of the regression line $\hat{Y} = a + bX$ changes for each unit change in X. It is an unbiased estimator of B. It can be used in the setting up of a confidence interval for B and in the testing of a hypothesis about the value of B.

The parameters $\sigma_{Y.X}$ and its square $\sigma_{Y.X}^2$ are measures of the variability of Y around $E(Y|X)$. In the statistical tests presented in this chapter, the assumption is that $\sigma_{Y.X}$ and $\sigma_{Y.X}^2$ are constant for all values of X.

The statistic $s_{y.x}$, "the standard error of estimate," and its square

$s^2_{y.x}$ are measures of the scatter of the Y values of the data points around $\hat{Y}$, the least squares estimate of the Y value. They are used to estimate $\sigma_{Y.X}$ and $\sigma^2_{Y.X}$.

The parameter ρ^2, the population coefficient of determination, tells you proportionally how much less $\sigma^2_{Y.X}$ is than σ^2_Y, or how much the variable X has succeeded in explaining away the uncertainty about Y.

The statistic r^2, the sample coefficient of determination, tells you proportionally how much less $s^2_{y.x}$ is than s^2_y, or how much less is the scatter of the points around the regression line $\hat{Y}$ than around $\overline{Y}$, the mean of the Y.

The statistic $\hat{\rho}^2$ is a kind of "corrected" value of r^2, corrected to be an unbiased estimator of ρ^2.

It is important to emphasize that all of the statistics should be interpreted in a manner consistent with that in which the data were collected. For instance, most of the examples we cover will be "cross-sectional" in nature; the data will be collected from a sample of different individuals at approximately the same time. Interpretations should reflect this fact. When we predict that "spending will increase by so much for every dollar increase in income," we do not envisage *one* family's changes in spending as its income increases. Rather we are predicting the spending of *another* family with $1.00 more income. The interpretation reflects the cross-sectional manner in which the data were calculated.

Independent and Dependent Variables

Regression analysis looks at the relationship between variables in one direction only, at the effect of an independent variable, here called X, on a dependent variable, here called Y. The distinction between independent and dependent variables does not imply a cause-and-effect relationship. It merely reflects the way that the statistician looks at the variables. It is perfectly possible to take the data of Table 10.2 and find the regression equation of X *on* Y:

$$\hat{X} = c + dY$$

where

$$c = \frac{\Sigma X - d\Sigma Y}{n} \quad \text{and} \quad d = \frac{n\Sigma XY - \Sigma X\Sigma Y}{n\Sigma Y^2 - (\Sigma Y)^2}$$

The only thing you have to watch out for in doing this is the running of statistical tests. These tests, if discussed in this chapter, would require that the variable X be a normally distributed random variable around $E(X|Y)$ with variance $\sigma^2_{X.Y}$. If the X values had been controlled by the experimenter, you had better not run statistical tests that treat X as a dependent random variable.

EXERCISES

10–1. In an effort to determine whether undergraduate grades in college were a good predictor of graduate grades at the Graduate School of Public Activities, the Director of Admissions took a random sample of 10 students who had completed their first year of graduate school. The results were:

Student	*Undergrad. QPA*	*Graduate QPA*
1	3.20	3.25
2	2.92	3.75
3	3.53	3.25
4	3.00	3.25
5	2.85	3.00
6	3.30	3.50
7	2.73	3.00
8	3.20	4.00
9	2.87	3.00
10	2.40	3.00

a) Find the least squares regression equation $\hat{Y} = a + bX$.
b) Interpret the meaning of the a and b values you computed in (a).
c) Find the standard error of estimate $s_{y.x}$.
d) Find the sample coefficient of determination r^2.
e) Assuming that the relationship between X and Y can be described as $E(Y|X) = A + Bx$, and $\sigma^2_{Y.X}$ remains constant for all X, test the hypothesis that $B = 0$ at the .05 level of significance.

10–2. Give an unbiased estimate of ρ^2, the coefficient of determination, in Exercise 10–1. Using your answers to Exercise 10–1, explain, in as nontechnical terms as you can, to what extent the information about a student's undergraduate QPA is valuable in determining her or his graduate QPA in his or her first year in graduate school.

10–3. Given the following data on 5 randomly selected individuals:

Individual	*Height (inches)*	*Height (centimeters)*
A	64.0	162.56
B	68.0	172.72
C	63.0	160.02
D	60.0	152.40
E	65.0	165.10

a) Find the least squares equation that predicts height in centimeters from height in inches.
b) Find $s_{y.x}$.
c) Calculate r^2.

d) Set up a 95% confidence interval for the population regression coefficient B.
e) From knowledge that 1 inch = 2.54 centimeters, could you have answered parts (a) through (d) without observing any data?

10–4. Given: $s_y^2 = 40$, $s_{y.x}^2 = 25$, and $n = 4$:
a) Find r^2.
b) Estimate ρ^2, assuming a linear relationship with constant variance.

10–5. Given: $s_y^2 = 40$, $s_{y.x}^2 = 25$, and $n = 50$:
a) Find r^2.
b) Estimate ρ^2, assuming a linear relationship with constant variance.

10–6. Contrast your answers to Exercise 10–4 and 10–5. What effect does a larger n have on the estimated value of ρ^2?

10–7. One Christmas Eve, Santa Claus took with him in his sleigh one of his helpers who had completed a course in probability and statistics. This helper timed Santa in his visits to 5 randomly selected houses, with the following results:

Number of Children in House	*Time Santa Stayed*
2	13 minutes
4	18 minutes
1	12 minutes
0	9 minutes
3	18 minutes

a) Find the regression equation predicting the time Santa stays at a house on the basis of number of children in the house.
b) Making the necessary assumptions for the use of the t statistic, set up a 95% confidence interval for B.
c) How many minutes would you predict Santa would stay at a house with 2 children?

10.2 Linear Correlation Analysis

Section 10.1 treated Y as a dependent variable and X as an independent variable. In this section, we look at the *interdependence* of two variables. Correlation analysis is symmetrical in that the variables are treated identically; in all the formulas on correlation we could interchange X and Y and come up with identically the same results. Therefore, the correlation between X and Y is the same thing as the correlation between Y and X. In contrast, the regression of Y on X in *not* the same thing as the regression of X on Y.

The Pearson coefficient of linear correlation, r, has been defined as

$$r = \frac{\Sigma(X - \overline{X})(Y - \overline{Y})}{\sqrt{\Sigma(X - \overline{X})^2 \Sigma(Y - \overline{Y})^2}} \tag{3.5}$$

and a more efficient computational formula for r was given as

$$r = \frac{n\Sigma XY - \Sigma X \Sigma Y}{\sqrt{n\Sigma X^2 - (\Sigma X)^2}\sqrt{n\Sigma Y^2 - (\Sigma Y)^2}} \tag{3.6}$$

Notice the symmetry of Equations 3.5 and 3.6. If you interchanged all the X's with all the Y's in either formula, you would leave the formula unchanged.

In this section, we will make use of the statistic r in testing for correlation between X and Y in the universe from which the sample was drawn. We will also introduce the concept of the covariance between two random variables.

> The POPULATION COVARIANCE OF X AND Y, written $\text{cov}(X, Y)$ is defined as
>
> $$E(X - \mu_X)(Y - \mu_Y)$$

If X and Y tend to vary together, then $(X - \mu_X)$ will generally be positive when $(Y - \mu_Y)$ is positive, so that the cross-products $(X - \mu_X)(Y - \mu_Y)$ will tend to be positive. Also, when $(X - \mu_X)$ is negative, $(Y - \mu_Y)$ will also tend to be negative, and again the cross-products $(X - \mu_X)(Y - \mu_Y)$ will be positive. Therefore, when X and Y vary together the expected value of the cross-products $(X - \mu_X)(Y - \mu_Y)$ will be positive, so that the covariance of X and Y will be positive. On the other hand, if X and Y tend to vary inversely, then the cross-products $(X - \mu_X)(Y - \mu_Y)$ will tend to be negative, since they will be the products of opposite signs.

> The POPULATION COEFFICIENT OF CORRELATION, designated by the parameter ρ (rho) or ρ_{XY}, is defined to be:
>
> $$\rho = \frac{E(X - \mu_X)(Y - \mu_Y)}{\sqrt{E(X - \mu_X)^2 E(Y - \mu_Y)^2}} = \frac{\text{cov}(X, Y)}{\sqrt{\sigma_X^2 \sigma_Y^2}} \tag{10.11}$$

If the random variables X and Y are independent, then $\text{cov}(X,Y)$ will be 0, and ρ will be 0. In such a case, in a random sample from a population of (X,Y) pairs, the statistic r will have an expected value of 0. If we wanted to run a test to see whether ρ is 0, it would be convenient to know the sampling distribution of the statistic r, so that we could use r

as a test statistic. However, it is not r that we usually use as a test statistic, but a statistic derived from r and the sample size n.

> If both X and Y are normally distributed random variables with ρ equal to 0, then the statistic
>
> $$t = \frac{r\sqrt{n-2}}{\sqrt{1-r^2}} \qquad \textbf{10.12}$$
>
> has a t distribution with $n - 2$ degrees of freedom.

This result enables you to run significance tests for $\rho = 0$ for a bivariate normal population.

EXAMPLE 10.1 A psychologist was interested in whether academic achievement was related to conformity to social norms. To study this question, she obtained data on academic averages (X) of cadets in a military academy and the scores of these same cadets in "conduct" (Y). The results were:

Cadet	X	Y
A	3.2	76
B	2.3	52
C	4.0	44
D	1.8	100
E	3.7	78

In order to run a t test to test the hypothesis that $\rho = 0$, it is necessary to make the assumptions that both X and Y are normally distributed random variables. If you cannot make this assumption, there are nonparametric methods, discussed in Chapter 11, that can come to the rescue. But if you can make this assumption, then the t test is superior in that it brings more information to bear that can lead to rejection of the null hypothesis.

The null hypothesis is H_0: $\rho = 0$
The alternative hypothesis is H_1: $\rho \neq 0$

At $5 - 2 = 3$ degrees of freedom, the critical values of t at the .05 level of significance are -3.182 and $+3.182$.

In order to calculate the value of the t statistic, find r by using Equation 3.5. (The computations are shown in Table 10.3.) Then use Equation 10.12 to calculate t.

$$t = \frac{-.507\sqrt{5-2}}{\sqrt{1-(.507)^2}} = -1.02$$

Table 10.3. *CALCULATION OF r*
(X = academic average of cadets,
Y = conduct score of cadets)

CADET	(1) X	(2) Y	(3) $(X - \overline{X})$	(4) $(Y - \overline{Y})$	(5) $(X - \overline{X})(Y - \overline{Y})$	(6) $(X - \overline{X})^2$	$(Y - \overline{Y})^2$
A	3.2	76	0.2	6	+1.2	0.04	36
B	2.3	52	−0.7	−18	+12.6	0.49	324
C	4.0	44	1.0	−26	−26.0	1.00	676
D	1.8	100	−1.2	30	−36.0	1.44	900
E	3.7	78	0.7	8	+5.6	0.56	64
	15.0	350	0	0	−42.6	3.53	2000

$$\overline{X} = \frac{15.0}{5} = 3.0 \qquad \overline{Y} = \frac{350}{5} = 70$$

$$r = \frac{\Sigma(X - \overline{X})(Y - \overline{Y})}{\sqrt{\Sigma(X - \overline{X})^2(Y - \overline{Y})^2}} = \frac{-42.6}{\sqrt{(3.53)(2000)}} = -.507$$

Since t is between the critical values −3.182 and +3.182, H_0 is accepted. It cannot be concluded that there is correlation between academic average and conduct score.

In this example, the value of r was negative, a surprising result if the psychologist approached the study with the belief that higher scores on conduct were generally associated with higher academic grades. The statistical test protected her, however, from jumping too quickly to a startling conclusion; the test showed that the negative value of r was not significant. Another observation that should lead to skepticism about jumping to a conclusion about ρ is that only two out of the five cadets, C and D, contributed to this negative result. On the other hand, even though the result was not statistically significant, it might be suggestive. A curious researcher might be led to investigate the matter more fully, and take a larger sample to see whether this negative relationship persists and shows up as significant. ■

INTERPRETATION OF ρ AND r

The parameter ρ is a measure of the linear relationship of the two random variables X and Y in the population. It can range from −1 (perfect negative correlation) to +1 (perfect positive correlation). The statistic r is a measure of the relationship between the observed values of x and y in the sample. Like ρ, it can vary from −1 to +1.

But except for the values −1, 0, and 1, the values of ρ and r do not lend themselves to precise interpretation. It can be shown, however, that the *square* of ρ is the population coefficient of determination. Section 10.1 defined ρ^2 as the proportion of the variance of random variable Y that can be "explained" by the independent variable X. When both X and Y are random variables, ρ^2 can be interpreted as the proportion of the variance of one random variable (either X or Y) that can be explained by the other. In this case, the statistic r^2, the sample coefficient of determination, can be interpreted as the proportion of the variance of the observed values of one variable that can be explained by the observed values of the other variable.

It is important to distinguish between the *magnitude* of the correlation coefficient r and the *significance* of r. If the sample size is very small, even an r value as high as .6 or .7 will not be significant at the .05 level; whereas in a large sample, an r value of only .20 might be statistically significant.

The level of significance by itself does not tell us how strong the correlation is; it tells us only whether we can say that its difference from 0 can be attributed to chance. When the sample is very large, even pairs of variables that are only slightly related will show up as significant. For instance, we may find in a sample of 10,000 people that the Pearson correlation coefficient r between height and income is +.20. In a test for $\rho = 0$, the t statistic would be

$$t = \frac{.2\sqrt{10{,}000 - 2}}{\sqrt{1 - (.2)^2}} = 20.4$$

For such a large sample, the critical value of t at the .01 level of significance can be found from Table II to be +2.505. The calculated value of t is therefore highly significant. But the high level of significance does not mean that the shorties have no hope and should apply for disability compensation. Although the results show beyond a reasonable doubt that ρ is positive, that is, that there is a relationship between height and income, they also show that the relationship is weak. The square of r is only .04, meaning that only 4% of the variability in income can be attributed to height.

EXERCISES

10–8. In order to determine whether teachers who awarded higher grades tended to be rated more highly by students than teachers who gave lower grades, a faculty committee conducted a survey which compared the mean scores of teacher evaluation by students with the mean grades given by the teachers. The mean student evaluation rating (on a scale from 1 to 7) and mean grades given by the teachers (on a scale from 0 to 4) were:

Teachers	*Mean Evaluation Score*	*Mean Grade*
A	3.5	2.6
B	4.6	3.2
C	2.7	3.7
D	6.2	2.2
E	5.0	2.5
F	2.0	3.8

a) Calculate the Pearson correlation coefficient r. Is there a relationship? Does it appear to be a strong one?

b) Assuming that the student and faculty evaluations are normally distributed random variables, test the significance of r. Do you have evidence at the .05 level that ρ is not 0?

10–9. The manager of Aaron Burr Hospital has ruefully bemoaned the fact that on days when more patients are admitted to the hospital, there tend to be more staff persons who are sick. The assistant Manager did some research on the correlation of X, the number of patients admitted in a day, and Y, the number of staff employees who were sick. The results were: $n = 102$ and $r = +.30$.

a) What reasons could you suggest to explain the positive correlation coefficient r?

b) Assuming that X and Y are normally distributed random variables, test the hypothesis, at the .05 level of significance, that $\rho = 0$. Comment on the statistical significance and the importance of X in explaining Y.

10–10. A public health official compared the rates of influenza and number of inoculations per 1000 people in 10 counties. The results were:

County	*Inoculations per 1000*	*Reported Cases of Flu per 1000*
A	125	82.5
B	130	16.6
C	145	85.3
D	171	80.4
E	115	115.2
F	164	36.0
G	155	22.4
H	158	20.0
I	176	18.2
J	111	122.5

a) Test the hypothesis, at the .05 level of significance, that the inoculation rate and flu rate are independent random variables. Assume the two variables are normally distributed. [Suggestion: Make a transformation of the data by subtracting a constant from each value. As a starter, you might subtract 100 from each number of inoculations.]

b) Would a regression analysis be more appropriate than a correlation analysis? Discuss the situations in which each might be appropriate.

10–11. A random sample was taken of 24 teenagers who dropped out of school and had been arrested at least once. A correlation analysis of "most advanced grade" attained in school (X) and "number of arrests" (Y) yielded the following statistics:

$$n = 24 \qquad \Sigma Y = 90 \qquad \Sigma X^2 = 2064$$
$$\Sigma X = 216 \qquad \Sigma XY = 696 \qquad \Sigma Y^2 = 510$$

a) Find s_y^2.
b) Find $s_{y.x}^2$.
c) Find r^2 and interpret its meaning.
d) Estimate ρ^2.
e) Assuming that the sample was a random sample of a population consisting of 2 normally distributed random variables, test for $\rho = 0$.

10–12. Let X be the time in minutes that a person takes to complete a task. Let Y be the time the person has to complete the next task. A random sample of 6 people revealed the following:

X	Y
32	28
24	36
16	44
18	42
54	6
36	24

a) Find the coefficient of linear correlation r.
b) Find the standard deviation of X and the standard deviation of Y.
c) Find $s_{y.x}$, the standard error of Y given X.
d) Find $s_{x.y}$, the standard error of X given Y.
e) Assuming X and Y are random variables, test at the .01 level of significance the hypothesis that $\rho = 0$.

***10–13.** If the sample coefficient of linear correlation is -1 or 1, will it always be significant at the .01 level? Assume the random variables are normally distributed.

10–14. A random sample of 15 married couples who both earned income showed the following results:

X (*Income of Husband*)	Y (*Income of Wife*)
5000	6000
22000	12000
16000	4000
18000	24000
23000	7000
12000	16000
7000	2000
40000	22000
9300	3900
37000	5000

X (Income of Husband)	Y (Income of Wife)
15000	11000
24000	36000
13000	16000
22000	7000
18000	15000

a) Find s_y^2, $s_{y.x}^2$, and $s_{x.y}^2$
b) Find r^2 and estimate ρ^2 for the universe.
c) Test, at the .05 level of significance, the hypothesis that $\rho = 0$.
d) Discuss factors that would tend to make ρ positive, and factors that would tend to make ρ negative. What does your answer to part (c) suggest are the more prevalent factors?

10–15. A random sample of 20 people who were convicted during March 1978 gave these results:

X = Age
Y = Number of Months of Sentence
$\Sigma X = 1040$
$\Sigma Y = 210$
$\Sigma XY = 12040$
$\Sigma X^2 = 64720$
$\Sigma Y^2 = 2870$

a) Calculate r and r^2.
b) Estimate ρ^2, assuming that X and Y are normally distributed, linearly related, random variables.
c) Test the hypothesis that $\rho = 0$ at the .05 level of significance.
d) Find the linear regression equation $\hat{Y} = a + bX$.
e) Find the linear regression equation $\hat{X} = c + dY$.
f) Multiply the two regression coefficients b and d, computed in parts (d) and (e). Compare the results to r^2, which you calculated in part (a).

***10–16.** The Pearson correlation coefficient of two random variables X and Y will always have the same sign (positive or negative) as the regression coefficient whether you regress Y on X or regress X on Y. Why is this so?

10–17. If two random variables are uncorrelated ($\rho = 0$), does this mean that there is no cause-and-effect relationship between them? Discuss.

10–18. The following data of X and Y values were taken from a table of random numbers. You know, therefore, that $\rho = 0$.

X	Y
7	3
9	25
1	22
8	48
8	36

a) Calculate r.
b) How do you account for the fact that r^2 is not 0, when the population ρ is known to be 0?
c) Given the method by which the data were selected, would you say that a t test for the significance of r is appropriate? Why or why not?

*10.3 Nonlinear and Multiple Regression Analysis

The techniques of regression analysis and correlation analysis discussed so far in this chapter have been involved with only *two* random variables X and Y. This section returns to regression analysis, but we introduce two new situations:

1. when the relationship between two variables X and Y is a *nonlinear* one, and
2. when there is more than one independent variable used to predict Y.

The mathematical expression

$$Y = a + bX + cX^2 \qquad \textbf{10.13}$$

is the equation for a *parabola,* which describes a curvilinear relationship. The constant term a describes where the parabola crosses the Y axis. The coefficient b gives the slope of the parabola at the Y axis. The coefficient c describes the direction of curvature. If c is positive, the curve will bend upward at the ends; if c is negative, the curve will bend downward. Figure 10.2 illustrates several of the possibilities with scatter diagrams.

In most cases, the decision between using a linear (first-degree) equation and a parabolic (second-degree) equation will not be clear-cut. A scatter diagram will generally reveal some degree of a linear component as well as suggestions of curvilinearity. One of the difficulties in interpreting scatter diagrams is that of deciding whether apparent departures of the points from the linear regression line are in fact indicators of a curvilinear relationship or merely random deviation from the linear regression line. There is a general rule of "thrift" which says that, if in doubt, use the simpler model. An elaborate model may be hard to justify or interpret theoretically.

Statistical tests have been developed to test for values of the parameters of the second-degree polynomial regression equation. Computer programs will generally print out the level of significance of the statistics in testing the hypotheses that the corresponding parameters are zero.

Another nonlinear model that is commonly used in regression analysis is the exponential model. If the relationship between X and Y is

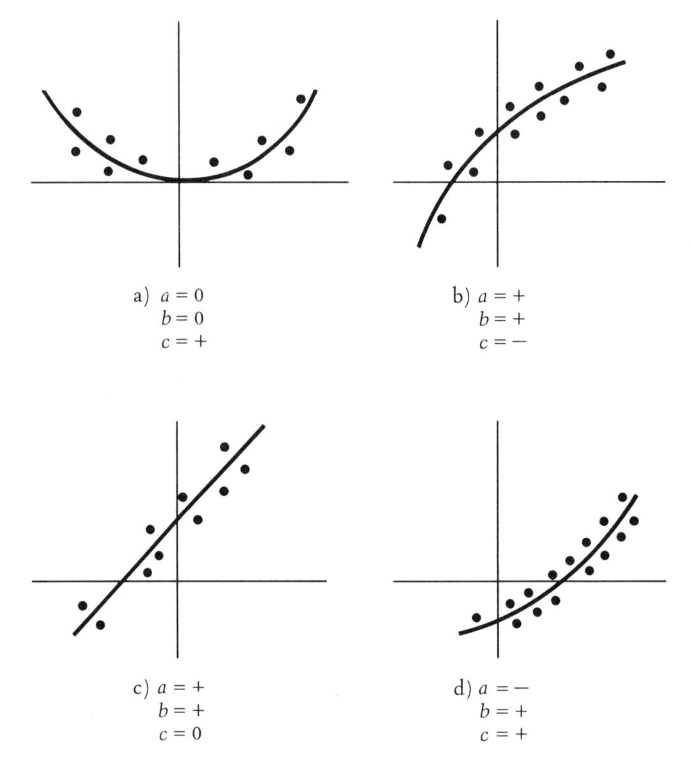

Figure 10.2. *SCATTER DIAGRAMS OF PARABOLIC REGRESSION Several Possible Forms of the Parabolic Regression Equation*

of the form $E(Y|X) = \alpha\beta^X$, the parameters α and β can be estimated by a regression equation of the form

$$\hat{Y} = ab^X \qquad \textbf{10.14}$$

This equation can most easily be solved by transforming the dependent variable into the logarithm of Y. The universe relationship then becomes the linear relationship

$$E(\log Y|X) = \log \alpha + X \log \beta,$$

and the estimating equation becomes the linear equation

$$\log \hat{Y} = \log a + X \log b \qquad \textbf{10.15}$$

The statistics "log a" and "log b" can be computed by the method of linear regression analysis set forth in Section 10.1.

The exponential model is appropriate if you expect Y to change by a constant percentage or proportion as X changes, or if the data in a scatter diagram indicate a tendency for the Y values to change by a constant proportion for unit changes in X. Exponential models are widely used to express changes over time, when constant rates of growth or decline are assumed. Models of economic growth, of population growth, of pollution growth, etc. frequently are based on exponential models.

Multiple Regression Analysis

The regression models discussed so far have considered the effect of only one independent variable X on a dependent random variable Y. In any realistic example, the variable Y will depend on many variables other than X, so that a regression equation expressing $\hat{Y}$ as a function of only one independent variable X will leave out many other factors that might be of considerable importance. The methods of regression analysis permit you to bring more than one independent variable into the model to explain Y.

> A regression equation expressing $\hat{Y}$ as a linear function of two or more independent variables is known as a MULTIPLE LINEAR REGRESSION EQUATION.

By bringing new explanatory variables into the equation you can reduce the random, or unexplained, variation in Y, which is expressed as the standard error of Y, $s_{y.x}$ around the regression line. In addition to reducing the standard error, you can shed new light on the effect of other independent variables that are already in the equation. The following illustration will emphasize this point.

A random sample of 8 employees of the Equal Employment Opportunity Club revealed the statistics on height and income shown in Table 10.4. These statistics were very disturbing to the chairperson of "Equality for Little Folks" (ELF), an organization dedicated to the health, wealth, and welfare of short people. It appeared that the EEOC was pursuing a policy of paying people largely on the basis of their heights. In order to determine whether such a policy was indeed being pursued, a staff person for ELF conducted a linear regression analysis, and calculated the regression equation to be

$$\hat{Y} = -33.63 + .6459X$$

A test of the hypothesis that $B = 0$ was rejected at the .05 level.

There is strong reason to believe that B is positive. Since for every additional inch of height the equation predicts an increase of $645.90 in income, EEOC is obviously discriminating on the basis of height—or is it?

Table 10.4. *HEIGHTS AND INCOMES Random Sample of Eight Employees of Equal Employment Opportunity Club*

INCOME Y	HEIGHT X	SEX
$ 5000	61 inches	Female
6000	59	Female
14000	73	Male
16000	67	Male
4000	65	Female
9000	69	Male
7000	63	Female
11000	71	Male

One of the most important points to emphasize in interpreting regression coefficients is that an association between variables does not imply causation. Even though higher X values may be associated with higher Y values, we cannot infer from that fact that higher X values *cause* higher Y values. There are three possible cause-and-effect relationships to consider.

1. X causes Y.
2. Y causes X.
3. There is a common cause (or several common causes) which make X and Y behave in a related way.

Statistics alone cannot tell you which of these three (or combination of them) is operating. But statistical analysis can give clues to causation by identifying third and fourth variables that may be common causes.

A statistical analyst might have looked for other variables describing the 8 EEOC employees, to determine whether characteristics other than height might be operating here. Suppose the analyst found that 4 of the employees were male and 4 were female, as indicated in Table 10.4. Since sex could very well be related to both height and income, the introduction of the variable "sex" into the model could shed new light on the relationship between height and income.

The multiple regression equation that predicts income on the basis of height and sex is of the form

$$\hat{Y} = a + b_1X_1 + b_2X_2 \qquad \textbf{10.16}$$

where X_1 = height in inches
X_2 = sex

It will be necessary to treat sex as a quantitative variable. You can do this by coding the variable X_2, so that $X_2 = 0$ if male and 1 if female.

The calculation of the statistics a, b_1, and b_2 is lengthy and beyond the scope of this text. Statisticians nearly always rely on a computer to give them a multiple regression equation. The multiple regression equation, obtained from a computer printout, turns out to be

$$\hat{Y} = 26.5 - .2X_1 - 7.6X_2$$

A very different picture appears now that we have introduced a second independent variable. The coefficient of X_1 (Height) is now $-.2$. The effect of height on income has now become negative! And the effect of sex appears to be very great.

How can height be negatively related to income in the multiple regression equation when it was positively related in the simple linear regression equation? The answer is that, in the simple model, taller people earned more because the men earned more than the women, and the men were taller. The multiple regression equation removed the effect of sex from the coefficient of X_1 by holding sex constant. The equation reveals that within one sex the taller people actually earn a little less.

You can interpret the statistics of the multiple regression equation as follows:

1. The value of a is 26.5. According to the model, a person whose X_1 value is 0 (0 inches tall) and whose X_2 value is 0 (male) would be predicted to have an income of $26,500. As discussed in Section 10.1, interpretations of values extrapolated beyond the range of the data are often nonsensical if they are made in terms of the specific problem. You need not worry about the earning capacity of someone who is zero inches tall.
2. The value of b_1 is $-.20$. This means that *holding sex constant,* you would predict a *decrease* of $200 in income for each additional inch of height.
3. The value of b_2 is -7.60. This means that *holding height constant,* you would predict a $7600 decrease in income as the coded "sex" variable X_2 increases from 0 to 1. Since males are 0 and females are 1, this means that for any given height you would predict that males earn $7600 more.

It appears now that EEOC is practicing sex discrimination more than height discrimination. The next task that confronts an analyst is to test for the significance of the b_1 and b_2 coefficients. We will not present the details of tests for the regression coefficients of a multiple regression equation in this book; however, a computer program that gives you these regression coefficients will generally give you, if asked (and sometimes even if not asked) the levels of significance. In the present example, the

b_1 coefficient is not significantly different from 0, but the b_2 coefficient is significant at the .05 level. You therefore have sufficient evidence to claim sex discrimination, but insufficient evidence to claim height discrimination.

The general form of the multiple regression equation for k independent variables is as follows.

Multiple regression equation: *k* independent variables

$$\hat{Y} = a + b_1X_1 + b_2X_2 + b_3X_3 + \cdots + b_kX_k \qquad 10.17$$

The coefficients b_i are known as PARTIAL REGRESSION COEFFICIENTS.

The partial regression coefficient b_1 tells you the amount $\hat{Y}$ changes for a unit change in X_1, while all the other independent variables remain constant, and each of the other partial regression coefficients in a similar fashion tell you the net effect on $\hat{Y}$ of a particular variable while all the other independent variables in the model are held constant.

A multiple regression equation cannot be drawn as a line on a plane surface, since each of the independent variables would require a separate dimension for visual presentation. In the case of two independent variables, the equation can be represented as a plane in three-dimensional space. In the case of more than two independent variables, physical representation of the equation becomes virtually impossible; we can describe the regression equation as a k-dimensional hyperplane in $k + 1$ dimensional space.

You can run statistical tests for the significance of the partial regression coefficients. Such tests assume that the observations are independent observations from a population that has the relationship

$$E(Y|X_1, X_2, \cdots, X_k) = A + B_1X_1 + B_2X_2 + \cdots + B_kX_k$$

The test for each of the B coefficients being zero involves the t statistic. The test is based on the assumption that the Y values are normally distributed around the regression hyperplane, and that the variance $\sigma^2_{Y.123\ldots k}$ is a constant for all $X_1, X_2, \ldots, X_k$ values.

EXERCISES

***10–19.** A random sample of 24 school districts predicted the mathematical achievement level of sixth grade pupils by the multiple regression equation

$$\hat{Y} = a + b_1X_1 + b_2X_2$$

where $\hat{Y}$ is predicted math achievement score
X_1 is mean class size
X_2 is mean family income in the school district

The regression equation was computed to be

$$\hat{Y} = 3.2 - .32X_1 + .0025X_2$$

a) Predict the mean score in a district with a mean class size of 28.2 and a mean income of $9200.
b) Interpret the meaning of the regression coefficients −0.32 and +.0025 in terms of this problem.

***10–20.** The earnings of the employees of the Bivouac Day Care Center are estimated by the multiple regression equation

$$\hat{Y} = 6851 - 1.40X_1 + 3894X_2$$

where X_1 is age in years
X_2 is sex ($X_2 = 0$ if female and 1 if male)

The ages of the employees in the sample ranged from 21 to 48 years.

a) Interpret the meaning of the regression coefficients.
b) Predict the earnings of a 42-year-old male.
c) Predict the earnings of a 0-year-old female, according to the equation. If the answer seems nonsensical, does this mean that the model is useless? Why or why not?
d) Is it possible that in this sample older people generally make more money than younger people? Discuss.

***10–21.** A multiple regression was calculated to be

$$\hat{Y} = 12.4 - 0.11X_1 + 0.03X_2 + 0.04X_3$$

where $\hat{Y}$ is sales of Product A
X_1 is price of Product A
X_2 is price of competing Product B
X_3 is price of competing Product C

a) Interpret the meaning of the coefficient −0.11.
b) If Product A sells for $2.00, Product B sells for $1.90, and Product C sells for $2.10, estimate the sales of Product A.

***10–22.** Although statistical tests for regression coefficients require the assumption that the sample observations are independent, you can fit a regression equation to a set of data that are not independently observed. This is often done in a time-series analysis, in which observations for one year are clearly dependent on observations for previous years. Given the following series on population in a region:

Year	*Population*
1955	100,000
1960	128,000
1965	160,000
1970	200,000
1975	256,000

a) Plot a scatter diagram of these data.
b) Compute a linear regression equation $\hat{Y} = a + bX$.
c) Compute an exponential regression equation of the form

$$Y_{\text{exp}} = cd^X$$

[Suggestions: Make the transformation

$$\log Y_{\text{exp}} = \log c + X \log d$$

where X is year and Y is population. Use a Table of Common Logarithms. It will also help to code the variables. Express Y in thousands. Let $X = 0$ for 1965 and express other X values as deviations from 0.]

Summary of Chapter

A linear regression equation is used to estimate the parameters of the equation $E(Y|X) = A + Bx$. The test of the null hypothesis that $B = 0$ will tell you whether you have evidence that X is a predictor of Y. The parameter ρ^2 can be estimated from the statistics of the regression equation. This parameter is the proportion of the variance of the dependent variable Y that is associated with the independent variable X.

While regression analysis looks at the effect of one variable on a dependent random variable, correlation analysis looks at the degree of relationship between two random variables. The coefficient of linear correlation r is used to test the null hypothesis that there is no linear relationship between the two random variables.

Nonlinear regression analysis looks at the effects that an independent variable has on a dependent variable that cannot be described by a linear equation. Multiple regression analysis looks at the effect that two or more independent variables have on a dependent random variable. Multiple regression analysis also enables you to isolate the effect of each independent variable on the dependent variable, holding all other independent variables constant. No computational procedures were presented for nonlinear and multiple regression analysis; the discussion was confined to interpretation of the coefficients.

REVIEW EXERCISES FOR CHAPTER 10

***10–23.** Given the following statistics on 5 countries:

Country	*Per Capita Income*	*Percent of GNP in Agriculture*
A	$2000	20
B	100	90
C	500	50
D	1000	30
E	1400	10

a) Find the least squares regression line predicting per capita income on the basis of percent of Gross National Product in the Agriculture sector. Compute s_y^2 and $s_{y.x}^2$.
b) Turn the problem around and treat percent of GNP in Agriculture as the dependent variable. Compute the linear regression equation. Compute s_x^2 and $s_{x.y}^2$.
c) Find r^2, the coefficient of determination.
d) Multiply the two regression coefficients you calculated in part (a) and part (b). Compare your answer to your answers in part (c).

***10–24.** Treating the variables in Exercise 10–23 both as random variables, calculate the coefficient of correlation r. Assuming that the two random variables are normally distributed, test the hypothesis at the .05 level of significance that $\rho = 0$.

***10–25.** Given the following multiple regression equation:

$$\hat{Y} = 57.2 - 0.29X_1 - 4.8X_2 + 0.13X_3 + .014X_4 - .047X_5 + 3.2X_6 - 2.9X_7$$

where $\hat{Y}$ is life expectancy of a 40-year-old person
X_1 is number of pounds overweight
X_2 is number of packs of cigarettes smoked in a day
X_3 is number of years mother lives
X_4 is number of years father lives
X_5 is diastolic blood pressure
X_6 is sex ($X_6 = 0$ if male and 1 if female)
X_7 is race ($X_7 = 0$ if white and 1 if nonwhite)

a) What is the life expectancy of a 40-year-old black male, 6 pounds underweight, who smokes half a pack of cigarettes a day, whose diastolic blood pressure is 90, whose mother lived to be 76, and whose father lived to be 64?
b) According to the regression equation, would quitting the half pack of cigarettes a day increase his life expectancy by 2.4 years? Why or why not?
c) Suppose the expected change in diastolic blood pressure induced by smoking a pack of cigarettes a day was +10 points, but that smoking was unrelated to any of the other independent variables. How much difference would there be in the life expectancies of two randomly selected persons, one of whom smoked a pack of cigarettes a day and the other of whom did not smoke? (Assume the individuals were identical to each other in all other characteristics.)
d) Suppose a person looking at the equation decided, since nothing was mentioned about cigar smoking, that cigar smoking is not related to life expectancy. Would such a conclusion be justified?
e) From looking at the equation and from your general knowledge, would you say that cigar smoking is less associated with life expectancy than cigarette smoking?

10–26. Given, the data following on page 340 on vocabulary (number of English words recognized = X) and income (\$ = Y):
Plot a scatter diagram of these data. What mathematical model would you propose for a regression equation that predicts income from vocabulary?

X	Y
10,000	8,000
25,000	13,000
80,000	15,000
50,000	16,000
40,000	18,000
5,000	4,500
12,000	10,500
22,000	15,000
30,000	16,000
16,000	12,000

10–27. Let Y be miles per gallon, X_1 be horsepower, and X_2 be speed driven over a test course. A random sample of 5 cars yielded the following results:

Y	X_1	X_2
30.2	80	64
25.4	100	55
26.8	120	55
30.0	90	46
22.5	110	60

a) Find the regression equation $\hat{Y} = a + bX_1$.
b) Find the regression equation $\hat{Y} = c + dX_2$.
c) Interpret the meaning of the coefficients and b and d that you calculated.

10–28. The ages and salaries of 10 randomly selected nurses at the Big Bertha Health Care Center are:

Age	*Salary*	*Age*	*Salary*
44	12,000	22	16,000
28	9,000	55	27,000
56	15,000	40	20,000
44	12,000	36	21,000
35	21,000	30	17,000

a) Find the linear regression equation predicting salary from age.
b) Test the hypothesis that $B = 0$.

10–29. a) Calculate the proportion of the variability in salaries that can be explained by age, for the 10 nurses in Exercise 10–28.
b) Estimate what this proportion is for the whole population.

10–30. Six randomly selected members of a fraternity had the following statistics on cigarette and liquor (ounces of alcohol) consumption:

Cigarettes	*Liquor*
14	0.4
4	0.2
30	2.5
0	0.0
6	0.4
11	1.0

a) Find the coefficient of linear correlation r.
b) Assuming normal data, test for independence of the two variables.

10–31. Ten randomly selected employees of Dynamic Consultants, Inc., were identified by salary and distance from work, as follows:

Salary	*Distance*
$12,000	2.5 miles
10,000	2.0 miles
9,000	0.4 miles
25,000	2.0 miles
20,000	4.5 miles
33,000	6.8 miles
7,000	2.2 miles
32,000	2.0 miles
19,000	4.2 miles
10,000	1.0 miles

a) Compute the linear correlation coefficient between salary and distance from work.
b) Test the hypothesis that there is no correlation between salary and distance for all employees, using a .10 level of significance.

10–32. Thirteen individuals were assigned randomly to special diets with varying amounts of animal protein. Their daily protein intake (in grams) and systolic blood pressure were as follows:

Protein Intake	*Blood Pressure*
30	140
40	110
50	120
60	120
70	170
80	140
90	160
100	220
110	220

Protein Intake	*Blood Pressure*
120	130
130	240
140	210
150	230

a) Find the linear regression equation which predicts blood pressure on the basis of protein intake.
b) Find the standard error $s_{y.x}$.
c) What proportion of the variance in blood pressure is explained by protein intake?
d) Test the hypothesis that the regression coefficient B is zero. Use a .05 level of significance.
e) Would it be appropriate in this example to test the significance of the linear correlation coefficient? Discuss.

10–33. In an attempt to determine whether a linear relationship existed between a senior citizen's age and that citizen's accident rate, a study was conducted on 128 senior citizens who drive. A linear regression equation was computed to predict accident rate (Y) from age (X.)

The following statistics were computed, using this equation:

$$\Sigma(Y - \hat{Y})^2 = 30.0$$
$$\Sigma(Y - \hat{Y})^2 = 32.0$$

a) What proportion of the variability in accident rate can be attributed to age in this sample?
b) Test the hypothesis that there is no linear relationship between age and accident rate. Use a .01 level of significance.

SUGGESTED READINGS

CLARK, CHARLES T., and LAWRENCE L. SCHKADE. *Statistical Methods for Business Decisions.* Cincinnati, Ohio: Southwestern Publishing Co., 1969. Chapters 16–18.

EZEKIEL, M., and K. FOX. *Methods of Correlation and Regression Analysis,* 3rd ed. New York: John Wiley and Sons, Inc., 1959.

HAMBURG, MORRIS. *Statistical Analysis for Decision-Making,* 2nd ed. New York: Harcourt, Brace, Jovanovich, Inc., 1977. Chapter 9.

MORONEY, M. J. *Facts from Figures.* Harmondsworth, Middlesex: Penguin Books, Ltd., 1956. Chapter 16.

NETER, JOHN, and WILLIAM WASSERMAN. *Fundamental Statistics for Business and Economics,* 3rd ed. Boston: Allyn and Bacon, Inc., 1966. Chapter 15.

O'TOOLE, A. L. *Elementary Practical Statistics.* New York: The McMillan Company, 1964. Chapters 8 and 9.

ELEVEN

Distribution-Free, or Nonparametric, Statistical Tests

Outline of Chapter

11.1 The Uses and Drawbacks of Distribution-Free Statistical Tests

Distribution-free statistical tests can be used more generally than can the tests you studied previously because they make fewer assumptions about the population from which they are drawn. But they use less information and therefore are not so likely to establish your research hypothesis.

11.2 One-Sample Tests

The runs test tests the randomness of a sequence of data.

Goodness of fit tests test the assumptions you may want to make about the probability distribution of the population.

11.3 Two-Sample Tests

The sign test looks at two related samples and compares the numbers of increases and decreases in some quantitative measure.

The median test compares two independent samples by counting the numbers in each that are above and below the median.

The Mann-Whitney *U* test compares two independent samples by ranking all the individuals and summing the ranks in each sample. The Kolmogorov-Smirnov test is a test for differences in averages and dispersions between two independent samples.

11.4 Distribution-Free Tests for Correlation

The Spearman rank correlation coefficient is used to test the correlation between the ranks of paired data. The Kendall coefficient of concordance is used to test the correlation among the ranks of three or more variables.

Objectives for the Student

1. Recognize those situations in which a parametric test is not warranted and a nonparametric test is called for.
2. Run a nonparametric runs test for the randomness of data.
3. Test for goodness of fit using the chi-square statistic.
4. Run nonparametric tests for the differences between two samples.
5. Run a nonparametric test for correlation between two variables.
6. Run a nonparametric test for correlation among three or more variables.

Every statistical test that you have encountered thus far has had at least one disturbing feature: it makes assumptions about the probability distribution of the population which in most cases you cannot be sure are valid. All of the tests involving the t statistic, for example, whether they be tests for means, for differences between means, for the coefficient of correlation, or for regression coefficients B required that you assume that the underlying probability distribution of the random variable is normal. But in Chapter 6 it was pointed out that very few probability distributions in nature are truly normal. It would seem, therefore, that generalizations about populations from samples can seldom be made, and that scientific research based on samples and the theory of samples will seldom have much validity.

However, two other findings by statisticians come to the rescue. One is the fact that larger sample sizes can do wonders in softening the need for strict assumptions about the population. The Central Limit Theorem lets you use the normal probability distribution for means (and other sample statistics as well) in large samples. When the sample size is 30 or more, you generally do not have to worry about the shape of the probability distribution at all; for samples of more than 10, you are fairly safe in using the t statistic if the probability distribution is very roughly normal; but when the sample size is 6 or less, you should not use a t test test unless you have strong reason to believe that the underlying random variable of the population has a normal probability distribution.

The other thing that can come to the rescue is a set of statistical tests that make no assumptions about the probability distribution of the random variable, or at the most fewer assumptions about the distribution than the t test make. Such tests are known as "nonparametric," or (more appropriately) "distribution-free" statistical tests.

11.1 The Uses and Drawbacks of Distribution-Free Statistical Tests

Figure 9.1 is a tree diagram illustrating the situations appropriate for the standard normal statistic z and the t statistic. Two branches of the tree, however, were followed by a question mark (?) to indicate that no tests had yet been discussed that covered these cases. In both of these cases

1. the probability distribution of X was unknown, or at least could not be assumed to be normal, and
2. the sample size was small.

When these conditions occur, you need to use a distribution-free test to test a hypothesis about a parameter.

Suppose that the president of the Orange Cab Company, in seeking legislation favoring his company, claims that the amount of time people have to wait for a taxicab at the corner of 4th and Main Streets averages less than 5 minutes. To test the hypothesis, somebody takes a random sample of the times between cabs (to the nearest minute) and gets the following results.

Observation	*Number of Minutes since Previous Cab*
1	—
2	2
3	6
4	4
5	2
6	4

Should the researcher run a t test to see whether the mean amount of time between cabs is less than 5 minutes?

The answer in this case is strongly no. The sample size is very small, ($n = 5$, since only 5 intervals between cabs were observed), and for such a small sample you should not use a t test unless you know that the probability distribution of times between taxicabs is very close to a normal distribution. Unless you have some *a priori* reason to believe that the time between taxicabs is normally distributed, you should not run a t test. In the case of taxicabs, there is considerable evidence, both theoretical and empirical, that the probability distribution is very unsymmetrical and a far cry from the normal distribution. In the present example, a t test would reject the null hypothesis at the .05 level, but the sign test (a distribution-free test that is suitable for this problem) would not. Although it is tempting to use the t test (especially tempting for the president of Orange Cabs) it is not an appropriate test here. There is simply not enough evidence to reject the null hypothesis at the .05 level.

The major drawback of nonparametric tests is that they do not make as much use of the information in the sample as does a "parametric" test (such as a t test), and therefore you have a lower probability of being able to reject the null hypothesis when it is false.

> The POWER of a statistical test is its ability to reject the null hypothesis, when H_0 is false. For any value of the parameter for which you are testing (other than the value hypothesized under H_0), the power of the test is the probability of rejecting the null hypothesis.

Nonparametric tests suffer from the disadvantage of having less power than corresponding parametric tests. They will therefore require a larger sample size to bring to bear on the problem.

There is some disagreement among statisticians and those involved in statistical research as to the types of situations in which one should use distribution-free methods as opposed to parametric methods. Those inclined toward greater use of parametric tests emphasize the importance of using all of the available information. They point out that t tests use more information from the sample than do tests using distribution-free statistics, and that t tests are not very sensitive to moderate departures of the population from the normal. If the data look reasonably normal, such people say, or if you have reason to believe the population has one mode and is fairly symmetrically distributed about that mode, by all means use a parametric test, unless you are dealing with a very small sample. On the other hand, proponents of nonparametric tests point out that in many cases nonparametric tests are 80% or 90% as powerful as the equivalent parametric tests, and this slight decrease in power is a small price to pay for the insurance that your test is making no unwarranted assumptions.

When you are dealing with nominal or ordinal-level data and small samples, you should not employ tests requiring the assumption of normality; in such cases you should use distribution-free tests. Some of the social sciences are concerned much more with variables that are nominal or ordinal-level than with those that are interval or ratio-level.

EXERCISE

11–1. Given the following variables and sample sizes, identify them as being testable by parametric methods; by distribution-free (nonparametric) methods only; or by neither.

Variable	*Sample Size* n
(1) Heights of adult males	10
(2) Waiting time for taxicabs	5
(3) Waiting time for taxicabs	50
(4) Views about a proposed abortion law	12
(5) A variable whose values are not randomly selected	100

11.2 One-Sample Tests

There is a wide variety of nonparametric tests; this chapter will introduce a few of the more widely used ones in order to give you a feeling for the way they are used.

The Runs Test

All of the statistical tests discussed in this book are based on the assumption that the sample is randomly selected. Sometimes you might want to check on the randomness of the sample by observing the order in which the data were collected or recorded. Suppose, for example, a survey revealed that out of 40 people in a sample, 24 favored a certain proposal. Suppose, however, that further inspection of the data revealed that *the first 24 people* who were interviewed were all in favor of the proposal, and the last 16 were all opposed. You would have good reason to suspect in this case that the method of obtaining the sample was not a purely random one. If the sample were truly random, the chances of getting a "run" of 24 favorable responses followed by a "run" of 16 unfavorable responses is too low to be believable.

In another situation, however, you might find the data to be *too mixed up* to be believable: if, in a sample of 20 people, 10 were Republicans and 10 were Democrats, you should be suspicious of the randomness of the sample if the sequence of individuals, in the order in which they were interviewed was

R D R D R D R D R D R D R D R D R D R D

In both these illustrations we can test for randomness by counting the "runs," or sequences of identical symbols. If the number of runs is either too great or too small, we reject the hypothesis that the sample was randomly selected.

In the runs test you have a sequence of n things of two kinds: n_1 are of one kind and n_2 are of the other kind. You count the number of runs and turn to Table VI to find the critical number of runs, r, at the .05 level. If the number of runs is at or below the *lower* critical value given in Table VI.a, or above the upper critical value given in Table VI.b, then you have either too few or too many runs ro accept the hypothesis of randomness.

EXAMPLE 11.1 The participants at the Southwestern Conference on Organizational Development arrived at the reception desk in the following order (participants were denoted only by sex).

MM	FFFFF	MM	FF	M	FFFF	MMMMMMM
1	2	3	4	5	6	7

Test the hypothesis, at the .05 level of significance, that the arrivals were random with respect to sex.

Solution: The null hypothesis is that the M's and F's are in a random order.

There are 12 M's and 11 F's. Therefore, let $n_1 = 12$ and $n_2 = 11$. There are 7 runs. For $n_1 = 12$ and $n_2 = 11$, Table VI.a tells you that the lower critical number of runs is 7. Thus, if the number of runs ≤ 7, we will reject H_0 at the .05 level of significance. Since there are 7 runs, we reject H_0; there are too few runs to believe in the randomness of the arrivals. The conclusion suggests that there is a tendency for men to bunch up with each other and for women to bunch up with each other in their arrivals.

Even without a formal statistical test, inspection of this sequence might have led you intuitively to believe that the data were not random. But when the number of runs is higher than expected, you can be more easily fooled. For instance, suppose that these same 23 people arrived at the final banquet in the following order:

M M F M F M F F M F M F M M F M F M F F M F M

Again, $n_1 = 12$ and $n_2 = 11$. Table VI.b gives the upper critical value of the number of runs as 18. There are 19 runs in the sequence.

Again, this is more than enough to reject H_0 at the .05 level of significance. The chief difference between this case and the previous one is that you reject H_0 in favor of the alternative hypothesis because there are *too many runs.* As before, the sexes are not randomly distributed, but in this case the men and women mix with each other more than you would expect in a random sequence. Results such as these could have interesting implications about the behavior of the sexes, both in arriving at conferences and arriving at banquets. ■

Tests for Goodness of Fit

Another set of one-sample tests are those that determine whether a set of observed data fits a theoretical probability distribution. Many of these tests use the χ^2 statistic, with which you are already familiar. The test is a nonparametric test inasmuch as it treats the random variable as nominal (the weakest assumption you can make about the level of measurement).

There is an important distinction between the *assumption of* a particular form or shape of a probability distribution (which we make when conducting a parametric test) and a *test for* a particular form or shape of a probability distribution. In the former case, we make the assumption without testing, and therefore run the risk of following an invalid procedure if the assumptions happen to be erroneous. In the latter case, however, we specifically test the assumption we want to make about the form of the distribution. (A hypothesis is simply an assumption that we test.) If we reject the assumption that the data fit the model, then we should not run statistical tests that assume that model for the data.

If we are tempted to run a t test on the assumption that the data come from a normal distribution, we may want to test to see whether the data fit the normal model. The tests for goodness of fit for the normal distribution are a bit lengthy and cumbersome, and the procedure will not be discussed here. One of these involves the χ^2 statistic, and the logic of the procedure is the same as that for the goodness-of-fit test for a uniform probability distribution.

EXERCISES

11–2. On 20 successive trading days, the New York Stock Exchange index showed the following sequence of advances and declines. (Advances are listed as +, declines as −.)

$$+ + - - - + + - - - + - - - + - - - - +$$

Is the sequence a random one? Test at the .05 level of significance.

11–3. The runs test can be used to test for a trend in a sequence of interval or ratio-level data. If the data, arranged in order of time, show a significant upward (or downward) trend, then most of the data in the early period will be below (or above) the median, and most of the later data will be above (or below) the median. In such a case, there will be fewer runs than would be expected if the data were purely random over time.

Using the data below, which give rates in 5-year intervals, use the runs test to test the null hypotheses that

a) there is no upward or downward trend in the marriage rate, and
b) there is no upward or downward trend in the divorce rate in the United States between 1890 and 1975.

Year	*Marriage Rate Per 1000*	*Divorce Rate Per 1000*
1890	9.0	0.5
1895	8.9	0.6
1900	9.3	0.7
1905	10.0	0.8
1910	10.3	0.9
1915	10.0	1.0
1920	12.0	1.6
1925	10.3	1.5
1930	9.2	1.6
1935	10.4	1.7
1940	12.1	2.0
1945	12.2	3.5
1950	11.1	2.6
1955	9.3	2.3
1960	8.5	2.2
1965	9.3	2.5
1970	10.6	3.5
1975	10.0	4.8

Source: Division of Vital Statistics, National Center for Health Statistics

11–4. A number of social workers were asked their preference for a weekday off from work, in exchange for a weekend day of work. It was hypothesized that social workers have no preference for any of the 3 midweek days, Tuesday, Wednesday, and Thursday. The results of the questions were:

Day	*Number Preferring That Day*
Tuesday	4
Wednesday	12
Thursday	8
Total	24

Assuming that these 24 are a random sample of social workers, test the hypothesis that social workers in general have no preference for any of these 3 days over any other.

11.3 Two-Sample Tests

The Sign Test

Section 9.1 discussed tests for two related samples. For example, when we observe a number of individuals before and after a treatment, we can observe the changes in some outcome variable (e.g., weight gain or loss, change in test score, change in accident rate, etc.). In Section 9.1 you ran a t test to test whether the differences were significantly different from zero. The assumption underlying the t test is that the population of differences is normal and that $\mu = 0$. If the assumption of normality is not met, you should not use the t test. However, you can test the null hypothesis that the median change is 0 (or the probability of an increase is .5), by observing the positive changes, indicated by plusses, and negative changes, indicated by minuses.

EXAMPLE 11.2 Suppose a statistician reviewed the research in Example 9.4 and decided that a t test was not justified in that situation, because the extreme smallness of the sample makes the assumption of an underlying normal distribution especially critical. Since the statistician saw insufficient evidence of this normality, he recommended that a distribution-free test be used. The sign test is a test well suited to the problem. This test looks only at the direction of changes (positive or negative) and ignores the magnitude of the changes. Table 11.1 sets up the sign test for the data of Example 9.4.

The null hypothesis is that the probability of an increase in accidents is 1/2, or H_0: $p = .5$. The alternative hypothesis is H_1: $p \neq .5$.

The only assumptions required to run this test are that the obser-

Table 11.1. *THE SIGN TEST*
Data from Table 9.1

Highway Segment	Accidents Before	Accidents After	Difference δ	Sign of Difference
A	6	7	+1	+
B	4	9	+5	+
C	10	21	+11	+
D	6	11	+5	+
E	4	7	+3	+
		No. of + scores = 5		
		No. of − scores = 0		
		No. of ties = 0		
		Total 5		

vations are independent and that the parameter p is constant. These assumptions enable you to use the binomial probability distribution, with $n = .5$. Does the sample result, that $X = 5$ increases, lead you to reject H_0? The binomial probability that $X = 5$, is found from Equation 5.7 as

$$f(5) = \frac{5!}{0!5!}(.5)^5(.5)^0 = .0312$$

Since the probability of 5 "plusses" out of 5 trials is only .0312 under the null hypothesis, it may seem that you can reject the null hypothesis at the .05 level of significance. However, this test was set up as a two-tail test, so that the rejection region of H_0 should consist of x values at the lower end of the distribution as well as the upper end. In a two-tail test, a result of 0 "plusses" is just as far from the expected number of successes under the null hypothesis as the result "5 plusses." The probability of 0 plusses is given by

$$f(0) = \frac{5!}{0!5!}(.5)^0(.5)^5 = .0312$$

so that the probability of *either* 0 or 5 plusses is $.0312 + .0312 = .0624$, which is greater than .05. Therefore, a result of 0 or 5 plusses is not sufficient to reject H_0 at the .05 level in a two-tail test. ■

The example illustrates the comparative "lack of power" of the sign test. In a sample of size 5, you could not have achieved a more decisive indicator that H_0 was false than 5 out of 5 "successes." Yet even such a strong result is not enough to refute H_0 at the .05 level. No matter what the sample result would have been, you could not have rejected H_0. Even if H_0 is false, the probability of rejecting it, using this particular test, is zero. The power of the test is zero. The sign test is therefore a useless test for a sample of size 5, if we set a level of significance at .05. It is necessary to have a larger sample to make effective use of this test.

The sign test ignores ties. That is, if there had been a pair of observations that showed zero change, we would not have included that observation in the sample. For example, if we had made 20 paired observations, of which 12 showed increases, 6 showed decreases, and 2 showed no change, the sign test would have recognized only the 18 observations of increases and decreases, and tested the null hypothesis with a binomial distribution with $n = 18$.

The following example shows how the sign test can be used even when the data are only ordinal-level.

EXAMPLE 11.3 Forty-six randomly selected patients who had undergone at least a year of psychoanalytic therapy were given a battery of personality tests both before and after their treatments. The evaluators of the tests agreed that the scores on these tests meet the requirements of ordinal-level data, inasmuch as one can say that the higher the score the better the indication of a well-adjusted personality; but many of the evaluators were skeptical about the ability of the test scores to meet the requirements of interval-level or ratio-level variables. Because of this skepticism, the numerical changes in patients' scores before and after treatment were not recorded; only the directions of the changes were recorded, with the following results:

Positive change in scores:	22
Zero change in scores:	10
Negative change in scores:	14

Do these results show, at the .05 level of significance, that scores are more likely to increase than decrease after the treatment?

Solution: Let p be the proportion of the population who improve (ignoring those who do not change). Let X be the number in the sample who improved, and n be the number of observations.

The null hypothesis is $H_0: p = .5$.
The alternative hypothesis is $H_1: p > .5$.

There are 22 plusses and 14 minuses, so that $n = 36$. (Ignore the 10 zeros.) For $n = 36$, use the normal approximation to the binomial, then

$$E(X) = np = 36(.5) = 18.0$$

and

$$\sigma X = \sqrt{np(1 - p)} = \sqrt{36(.5)(.5)} = 3.0$$

For the one-tail test at the .05 level of significance, the decision rule is:

If $z \leqslant 1.645$, accept H_0.
If $z > 1.645$, reject H_0.

Calculate

$$z = \frac{X - E(X)}{\sigma X} = \frac{22 - 18}{3} = 1.33$$

Since $z \leqslant 1.645$, we accept H_0. There is insufficient evidence to show that more than 50% of the cases improve. ■

The Median Test

The sign test was a test for paired observations from two *related* samples. When the two samples are independent, you can also test for the differences in the central tendency of the two samples even if the

Table 11.2. *CROSS CLASSIFICATION FOR THE MEDIAN TEST*

	Above Median	Below Median	
Hypoxians	12	8	20
Marsupians	5	9	14
	17	17	34

variables do not meet the criteria for interval or ratio-level variables. Suppose, for example, that you wished to compare two ethnic groups in their degree of "social conformity" as measured by test scores that permitted analysis only up to the ordinal level of measurement.

Assume that 20 Hypoxians and 14 Marsupians were tested, and that their 34 scores were ranked from highest to lowest. The 17 lowest scores are below the median, and the 17 highest scores are above the median. The results are shown in Table 11.2.

The null hypothesis is that the probability of being at or above the median is .5 for members of each group. You can test the hypothesis with the χ^2 statistic by the method presented in Section 9.3. At the .05 level of significance, the critical value of chi-square for 1 degree of freedom is 3.841. Use Equations 3.1 and 3.2 to calculate

$$\chi^2 = \frac{(12 - 10)^2}{10} + \frac{(8 - 10)^2}{10} + \frac{(5 - 7)^2}{7} + \frac{(9 - 7)^2}{7} = 1.943$$

The value of chi-square is insufficient to reject H_0.

This result illustrates how frustrating the use of the more cautious nonparametric techniques can be. The data strongly *suggest* that Hypoxians tend to have higher conformity scores than Marsupians, and it may be difficult to resist the temptation to run a *t* test. Because of the greater power of the *t* test, it would have a better chance of rejecting the null hypothesis than the median test. But hopping from test to test in the hope of coming up with one that gives significant results is not a valid procedure.

A problem in the median test is how to classify data points that fall at the median. One method of dealing with this problem is to classify the data into two categories: (a) values less than or equal to the median and (b) values greater than the median. Unless a high proportion of values fall at the median, this method of classification should be satisfactory for the median test.

The Mann-Whitney *U* Test

The median test has the disadvantage of ignoring a great amount of information. If the data can be quantified beyond mere classification of being above or below the median, then you lose something by classifying the data in this way. A test that uses more information is the Mann-Whitney U test. It is a test designed for two samples of data that can be ranked. The test looks, not on the actual values of the data, but only on the *ranks* of the data. To illustrate the Mann-Whitney U test, consider the same problem that was discussed for the median test. Suppose the raw "social conformity" scores of the 20 Hypoxians and 14 Marsupians are as follows.

Hypoxians	*Marsupians*
62.5	58.2
56.8	49.7
55.2	56.1
56.3	49.1
58.5	57.3
59.6	53.4
63.5	44.5
71.8	45.7
75.2	58.6
66.1	43.8
58.4	58.8
57.2	50.9
54.6	51.8
52.1	43.2
51.4	
48.2	
46.4	
49.8	
53.5	
54.2	

Assuming that these are random samples, can you say, at the .05 level of significance, that Hypoxians and Marsupians differ in their scores?

You cannot use a t test for differences between means unless you assume that the data come from a population that is reasonably close to normal. On the other hand, the median test score is certainly valid; since you have the raw scores and have more information than the median test uses. The Mann-Whitney U test has the advantage of using more information than the median test while not depending on the assumptions that the t test requires.

To use the Mann-Whitney U test, combine the two samples, ranking all 34 individuals by score. Ranking from lowest to highest you have:

Rank	*Score*	*Ethnic Group*	*Rank*	*Score*	*Ethnic Group*
1.	43.2	M	18.	55.2	H
2.	43.8	M	19.	56.1	M
3.	44.5	M	20.	56.3	H
4.	45.7	M	21.	56.8	H
5.	46.4	H	22.	57.2	H
6.	48.2	H	23.	57.3	M
7.	49.1	M	24.	58.2	M
8.	49.7	M	25.	58.4	H
9.	49.8	H	26.	58.5	H
10.	51.4	H	27.	58.6	M
11.	51.8	H	28.	58.8	M
12.	52.1	M	29.	59.6	H
13.	53.4	M	30.	62.5	H
14.	53.5	H	31.	63.5	H
15.	54.2	H	32.	66.1	H
16.	54.6	H	33.	71.8	H
17.	54.9	M	34.	75.2	H

Let n_1 be the number of observations in the 1st sample (Hypoxians).
Let n_2 be the number of observations in the 2nd sample (Marsupians).
Let R_1 be the sum of the ranks in the 1st sample.
Let R_2 be the sum of the ranks in the 2nd sample.

In this example, $n_1 = 20$, $n_2 = 14$.

$$R_1 = 5 + 6 + 9 + 10 + 11 + 14 + 15 + 16 + 18 + 20 + 21 + 22 + 25 + 26 + 29 + 30 + 31 + 32 + 33 + 34 = 407$$

$$R_2 = 1 + 2 + 3 + 4 + 7 + 8 + 12 + 13 + 17 + 19 + 23 + 24 + 27 + 28 = 188$$

The null hypothesis is that the two samples are drawn from the same population, so that the expected value of the mean of the ranks from the two samples would be equal. To test H_0, define the statistic

$$U = n_1 n_2 + \frac{n_1(n_1 + 1)}{2} - R_1 \qquad \textbf{11.1}$$

It can be shown that the expected value of U, under the assumption of the null hypothesis, is

$$E(U) = \frac{n_1 n_2}{2} \qquad \textbf{11.2}$$

and the standard deviation of U is

$$\sigma_U = \sqrt{\frac{n_1 n_2(n_1 + n_2 + 1)}{12}} \qquad \textbf{11.3}$$

When n_1 and n_2 are greater than 10, the sampling distribution of U is approximately normal. You can therefore test H_0 using the standard normal variable

$$z = \frac{U - E(U)}{\sigma_U}$$

Calculate U from Equation 11.1:

$$U = (20)(14) + \frac{20(20 + 1)}{2} - 407 = +83$$

From Equation 11.2

$$E(U) = \frac{(20)(14)}{2} = 140$$

From Equation 11.3

$$\sigma_U = \sqrt{\frac{(20)(14)(20 + 14 + 1)}{12}} = 28.58$$

Then

$$z = \frac{83 - 140}{28.58} = -1.99$$

For a two-tail test, this is significant at the .05 level. There is evidence that the two ethnic groups do differ with respect to the measured characteristic.

Compare this result with that of the previous example, in which the median test failed to reject H_0. By bringing more information to bear upon the problem, the Mann-Whitney U test has a better chance of rejecting H_0 (if it is false) than does the median test.

Tied Observations

If two observations have equal values, they will share the same rank. In the case of tied ranks, each observation should be given a rank equal to the *mean* of the ranks that are shared by the tied values. For example, in the following hypothetical data, 2 observations are tied for 1st place, so that each gets a rank of $(1 + 2)/2 = 1.5$; and 3 are tied for 5th place, so each gets a rank of $(5 + 6 + 7)/3 = 6.0$.

A large number of ties will produce some distortion if the sample

Group A		*Group B*	
Score	*Rank*	*Score*	*Rank*
95	1.5	95	1.5
92	3.0	88	6.0
90	4.0	88	6.0
88	6.0	81	8.0

sizes are small, and in such a case there is a correction factor which can be applied to the calculation of σ_U. However, in most cases the effect of ties will be too negligible to necessitate the use of this correction factor.

The Kolmogorov-Smirnov Test For Differences

Both the median test and the Mann-Whitney U test are designed to determine whether two independent samples come from a population having the same average. Like these tests, the Kolmogorov-Smirnov test compares two independent samples, but it is able to detect not only differences in average but differences in dispersion between the two samples as well.

The Kolmogorov-Smirnov test compares the cumulative frequency distributions of two independent samples. To run the test, record the cumulative frequency distribution of each sample, using the same class intervals for each sample. Let

$$S_1(X) = \frac{K_1}{n_1}$$

where K_1 is the number of observations in the first sample less than or equal to each class X, and n_1 is the number in the sample; and let

$$S_2(X) = \frac{K_2}{n_2}$$

where K_2 is the number of observations in the second sample less than or equal to each X, and n_2 is the number in the second sample.

For each class, compute the difference

$$S_1(X) - S_2(X)$$

Let the maximum of these differences be

$$D = \text{maximum}\,[S_1(X) - S_2(X)]$$

The null hypothesis is that there is no difference between the two populations. The alternative hypothesis is the one-sided alternative that the first sample tends toward lower values than the second sample. According to the null hypothesis, the statistic

$$\frac{4D^2 n_1 n_2}{n_1 + n_2}$$

is distributed approximately as χ^2 with 2 degrees of freedom. If the value of this statistic is greater than or equal to the critical value of χ^2 as designated in Table IV, you can reject H_0 at the designated level of significance. This is a one-tail test; rejection of H_0 means that you assert that the first sample has a lower average, or lower extreme values, than the second sample.

EXAMPLE 11.4 In order to test for differences in tolerance for alcohol between two national groups, suppose an experimenter has 16 Germans and 20 Russians each drink 6 ounces of vodka during a one-hour period, and 20 minutes thereafter administers a "sobriety" test. The test scores and the calculations of $S_1(X)$ for the Germans and $S_2(X)$ for the Russians are shown in Table 11.3. From the table, we can calculate

$$\chi^2 = \frac{4D^2 n_1 n_2}{n_1 + n_2} = \frac{4(.5000)^2(16)(20)}{36} = 8.889$$

Table IV indicates that the critical value of χ^2 for 2 degrees of freedom at the .05 level of significance is 5.99. Since the value of the test statistic exceeds this critical value, reject H_0. You can say that the population from which the Germans were drawn tends to have lower scores than the population from which the Russians were randomly drawn. ■

The test statistic is only approximately distributed as χ^2, though

Table 11.3. *THE KOLMOGOROV-SMIRNOV CALCULATION The Difference Between Two Independent Samples*

SCORES	GERMANS			RUSSIANS			
X	f_1	K_1	$S_1(X)$	f_2	K_2	$S_2(X)$	$S_1(X) - S_2(X)$
60–63	1	1	.0625	0	0	.0000	.0625
64–67	3	4	.2500	1	1	.0500	.2000
68–71	4	8	.5000	2	3	.1500	.3500
72–75	4	12	.7500	2	5	.2500	.5000
76–79	2	14	.8750	5	10	.5000	.3750
80–83	0	14	.8750	5	15	.7500	.1250
84–87	2	16	1.0000	5	20	1.0000	.0000
	16			20			

$n_1 = 16$ $\quad n_2 = 20$

$S_1(X) = K_1/n_1 \quad S_2(X) = K_2/n_2$

$D = \text{maximum}\,[S_1(X) - S_2(X)] = .5000$

the approximation gets better as the sample sizes get larger. For small samples, as in the above example, the test is conservative; the error due to the approximation will tend to underestimate the significance of the test statistic. In the above example, the computed value of χ^2 is significant at the .01 level. Books that go into nonparametric statistics in greater depth will often publish tables showing critical values for the Kolmogorov-Smirnov test for smaller samples. These will tend to show greater significance for small sample results than the χ^2 table.

EXERCISES

11–5. During a political campaign Mr. Sling decided that, in order to improve his chances of winning the election, he should begin to stress the point in his speeches that his opponent had been divorced three times. In order to assess the effectiveness of the emphasis on this campaign issue, his staff took a random sample of 40 people and interviewed them both before and after the first week of intensive campaigning on this issue. Of the 40, 12 showed attitudes more favorable to the candidate, 20 showed attitudes less favorable, and 8 showed little or no change in attitude.

a) Is there evidence, at the .05 level of significance, that there was a shift in voter attitudes toward the candidate during this week of campaigning?

b) Would it be a mistake to assume that a shift in attitudes during the week was *due* to the campaign? Why?

11–6. After a 6-week physical fitness course, 8 participants all had higher test scores in physical fitness than they had at the start of the program. Assume that the 8 participants are a random sample of all people between 20 and 60 years of age who have had no major health disabilities. Does the sample result show, at the .05 level of significance, that people tend to improve their physical fitness during the time that they take such a course?

11–7. A payroll clerk at the 3650th Squadron classified the squadron personnel by their military status and pay as follows.

	Pay Above Median	Pay Below Median
Regular Air Force	67	83
Air Force Reserve	33	17

Test the hypothesis, at the .01 level of significance, that there is no difference between regular Air Force and Air Force Reserve in pay. Assume that the squadron is a random sample of regular and reserve Air Force personnel.

11–8. In a class in International Politics, the students whose native language was English received 4 A's, 5 B's, 3 C's, and 2 D's. Those whose native language was not English received 1 A, 3 B's, 6 C's, and 2 D's.

a) Can you run a parametric test to determine whether students whose native language is English differ in grade expectations from students whose native language is other than English? Discuss.
b) Run two nonparametric tests to determine whether the difference in native language makes a difference in the grade one gets.
c) What level of measurement is warranted for the variable "academic grade"?

11–9. Forty pupils in the fifth grade at PS 149 sold tickets to the school play. The numbers of tickets sold by the boys and girls were:

Boys: 0, 14, 3, 8, 1, 2, 0, 0, 5, 0, 4, 0, 0, 2, 2, 1, 0, 0, 3, 0, 22, 2
Girls: 2, 5, 12, 0, 0, 4, 6, 16, 18, 2, 0, 19, 7, 0, 2, 5, 15, 36

a) Test the hypothesis at the .05 level of significance that boys and girls are equal in the number of tickets they sell, assuming that the probability distribution of tickets sold is normal.
b) Test the same hypothesis as in (a), but not making the assumption that the distribution is normal.
c) Discuss whether the test based on the normal assumption is a valid one to use.

11–10. A random sample of 8 military and 6 civilian employees at Camp Burnside was tested for psychological characteristics conducive to "mobility." The scores were:

Military: 36, 4, 12, 2, 9, 12, 10, 15
Civilian: 1, 5, 11, 14, 8, 3

a) Perform a median test at the .10 level of significance to determine whether military and civilian employees are equal in their scores.
b) Conduct a t test for differences between means at the .10 level of significance. From inspection of the data, do you believe the assumptions necessary for the test are justified?
c) Discuss the influence on both of these test outcomes of the military person whose score was 36.

11–11. In a random sample of 12 unionized textile manufacturing companies and 15 nonunionized companies, the mean weekly earnings of the employees were:

Unionized Companies	*Nonunionized Companies*
$ 99.78	$106.12
102.36	102.14
114.14	128.12
125.10	116.40
131.02	112.60
122.09	115.88
102.68	122.00
105.12	139.06
109.13	122.17

Unionized Companies	*Nonunionized Companies*
102.10	141.00
119.76	138.25
104.10	126.14
	133.71
	125.82
	132.00

Use the Kolmogorov-Smirnov test to determine whether there is evidence of a difference between unionized companies and nonunionized companies in their mean weekly earnings. Use a .05 level of significance.

11–12. In order to determine whether male engineers differ from female engineers in effectiveness, a company studies the effectiveness ratings of 20 male, and 12 female engineers, randomly selected from a large set of ratings. The ratings were on a scale of 1 to 5.

Males: 2, 2, 1, 5, 2, 3, 1, 4, 3, 1, 2, 5, 4, 2, 3, 4, 2, 3, 2, 3
Females: 2, 3, 5, 5, 3, 4, 2, 5, 4, 1, 4, 5

a) Test to determine whether the sexes differ in effectiveness rating in this company. Use either the Mann-Whitney U test or the Kolmogorov-Smirnov test.
b) Why is it particularly difficult to run the median test in this example?
c) Why is it not appropriate to run a parametric (t) test for this type of data?

11.4 Distribution-Free Tests for Correlation

Tests for the significance of the Pearson linear correlation coefficient require the assumption that both variables are normally distributed. If the distribution of the two random variables is unknown, it is still possible to test for the correlation between the *ranks* of the data. The distribution of ranks is a uniform distribution. For instance, if 8 people are ranked by income, the probability distribution of the rank of a randomly selected person is $P(r) = 1/8$ ($r = 1, 2, \ldots, 8$).

Section 3.3 presented a method of measuring the rank correlation between two variables, the Spearman rank correlation coefficient.

EXAMPLE 11.5 Research proposals which are submitted to a government agency are initially screened by two evaluators. The proposals for a particular project are ranked by each of the evaluators. Suppose the two evaluators rank 8 proposals (designated by the letters A through H) as follows:

Evaluator I: B, A, D, C, H, G, F, E
Evaluator II: D, B, F, A, C, H, G, E

Can you say, at the .05 level of significance, that the two evaluators tend to agree (i.e., that there is a positive correlation between their evaluations)?

Discussion: Before running any statistical test, you must be satisfied that the ranking of these proposals is a random sample of these evaluators' rankings. Table 11.4 lists the rankings of the 8 projects by the 2 evaluators and shows the calculation of the Spearman rank correlation coefficient, r_s. The values of r_s that are significant at the .05 and .01 levels of significance for a two-tail test are shown in Table VII. If running a one-tail test, double the level of significance. For a sample of size 8, you need a value of r_s of at least .643 to be significant at the .05 level in a one-tail test. The computed value of .667 is therefore sufficient to reject H_0.

For larger values of $n (n \geq 10)$ you can test the significance of r_s by using the t statistic

$$t = \frac{r_s \sqrt{n - 2}}{1 - r_s^2} \qquad \textbf{11.4}$$

and use Table III to find significant values of t.

Notice that you calculated t from r_s in exactly the same way that you calculate t from r when testing for the significance of the Pearson coefficient of linear correlation. ■

Table 11.4. *CALCULATION OF SPEARMAN RANK CORRELATION COEFFICIENT*

Project	Rank by Evaluator #1	Rank by Evaluator #2	d	d^2
A	2	4	−2	4
B	1	2	−1	1
C	4	5	−1	1
D	3	1	+2	4
E	8	8	0	0
F	7	3	+4	16
G	6	7	−1	1
H	5	6	−1	1
			0	28

$$r_s = 1 - \frac{6\Sigma d^2}{n(n^2 - 1)} = 1 - \frac{6(28)}{8(64 - 1)} = .667$$

Testing for Correlation Among Three or More Sets or Rankings

The methods of rank correlation between two sets of ranks can be extended to any number k of sets of rankings. The Kendall coefficient of concordance W is a measure of the correlation among several sets of ranks. For example, if you want to know whether a set of variables (such as artistic aptitude, mathematical aptitude, verbal aptitude, and structural visualization) tend to vary in the same direction (so that individuals high in one will tend to be high in others), you could rank a set of randomly selected individuals on the basis of these variables and test to see whether the variables are correlated. Another situation where it might be desirable to measure correlations among ranks is one in which professors from different academic disciplines evaluate papers on an interdisciplinary subject. In this case, the coefficient of concordance could give you an indication of the degree of consensus or lack of consensus among the professors.

For an example of a test for the coefficient of concordance, assume that 3 judges rank 5 projects on the basis of their perceptions of the effectiveness of the projects in achieving their goals. You can calculate Kendall's coefficient of concordance by setting up a k-by-n table, in which k is the number of judges ($k = 3$) and n is the number of projects being evaluated ($n = 5$).

Enter the ranks (1 to n) that each judge gives to each project, as illustrated in Table 11.5.

The sum of the ranks (R_j) for each column j is shown in the table. The null hypothesis is that there is no correlation among the judges' rankings. According to H_0, the expected values of R_j would all be equal. The greater the degree of variability among the 5 R_j values, the stronger the evidence we have to reject H_0. We measure this variability by the

Table 11.5. *RANKS ASSIGNED BY THREE JUDGES TO FIVE PROJECTS*

	Projects				
Judge	I	II	III	IV	V
A	4	3	5	1	2
B	4	2	5	1	3
C	5	4	3	2	1
R_j	13	9	13	4	6

statistic SS, which is the sum of the squared deviations of the R_j values around their mean:

$$SS = \Sigma\left(R_j - \frac{\Sigma R_j}{n}\right)^2 \qquad \textbf{11.5}$$

We use SS to calculate W by

$$W = \frac{SS}{1/12k^2(n^3 - n)} \qquad \textbf{11.6}$$

In the present example,

$$\frac{\Sigma R_j}{n} = \frac{13 + 9 + 13 + 4 + 6}{5} = 9.00$$

Then, $SS = \Sigma(R_j - 9)^2 = (13 - 9)^2 + (9 - 9)^2 + (13 - 9)^2 + (4 - 9)^2 + (6 - 9)^2 = 66$. So

$$W = \frac{66}{1/12(3)^2[5^3 - 5]} = .733$$

The coefficient of concordance is .733. This number is a measure of the degree of agreement among the 3 judges. When W is 1, there is complete agreement among the judges; when there is a high degree of disagreement among the judges, W tends to be close to zero.

In order to test for the significance of the coefficient of concordance, use the statistic SS as the test statistic. Table VIII gives the critical values of SS for $n = 3$ to 7 and $k = 3$ to 6, 8, 10, 15, and 20. In the present example, $n = 5$ and $k = 3$. Table VIII gives the critical value of SS at the .05 level of significance as 64.4. The calculated value of SS is 66.0, so that we are able to reject H_0. There is evidence of agreement among the 3 judges.

When the number n of items to be ranked is greater than 7, you can test the significance of the coefficient of concordance using the χ^2 statistic. The statistic

$$\chi^2 = \frac{SS}{1/12k(n^2 + n)} \qquad \textbf{11.7}$$

is approximately distributed as the χ^2 distribution with $n - 1$ degrees of freedom. Table IV gives the significant values of χ^2.

EXAMPLE 11.6 Three department chairpersons were asked to rank 11 proposed projects in order to determine which ones should receive priority in funding. The results of the rankings are shown in Table 11.6. Find the Kendall coefficient of concordance W and test for significance.

Table 11.6. *RANKING BY THREE DEPARTMENT CHAIRPERSONS OF ELEVEN PROPOSALS*

	Proposed Project										
Chairperson	I	II	III	IV	V	VI	VII	VIII	IX	X	XI
A	1	2	3	4	5	6	7	8	9	10	11
B	5	1	4	3	2	7	6	10	8	9	11
C	4	5	2	6	3	10	1	8	7	9	11
R_j	10	8	9	13	10	23	14	26	24	28	33

$$\Sigma R_j = 198$$

$$\frac{\Sigma R_j}{n} = \frac{198}{11} = 18$$

Table 11.7 *CALCULATION OF SS AND W Data from Table 11.6*

R_j	$R_j - 18$	$(R_j - 18)^2$
10	−8	64
8	−10	100
9	−9	81
13	−5	25
10	−8	64
23	5	25
14	−4	16
26	8	64
24	6	36
28	10	100
33	15	225
198		800

$$\frac{\Sigma R_j}{n} = 18$$

$$SS = \Sigma(R_j - 18)^2 = 800$$

$$W = \frac{800}{1/12\ 3^2\ [11^3 - 11]} = .8080$$

Solution: Table 11.7 shows the calculations for *SS* and *W*. To test for the significance of *W*, use Equation 11.7 to calculate

$$\chi^2 = \frac{800}{1/12(3)(11)(11 + 1)} = 24.24$$

Table IV gives $\chi^2_{.05}$ for $11 - 1 = 10$ degrees of freedom as 18.31. Since the calculated value of $\chi^2 \geqslant 18.31$, we reject H_0. You have evidence of a degree of consensus among the 3 chairpersons. ■

EXERCISES

11–13. A consumer testing agency tested the tape recorders of 6 manufacturers for quality. The 6 models were ranked both on the basis of overall quality and on the basis of suggested retail price. The rankings were as follows.

Model	*Rank by Quality*	*Rank by Price*
A	5	1
B	1	5
C	6	4
D	2	6
E	4	2
F	3	3

a) Find the Spearman rank correlation coefficient, r_s. Does there seem to be a positive or a negative relationship between quality and price?
b) Test the significance of r_s at the .05 level.

11–14. Mr. and Mrs. Brown were asked to rank 5 TV programs in order of their preference. Their rankings were:

Program	*Mr. Brown's Ranking*	*Mrs. Brown's Ranking*
TV Wrestling	1	5
Baseball game	2	4
"The Woman in Blue"	5	1
"Let's Make a Million"	4	2
"Barbara Williams and the News"	3	3

a) Calculate the Spearman rank correlation coefficient and test its significance at the .01 level.
b) What policy might you suggest for the Browns as a result of your answers to (a)?

11–15. The girls in the senior class at Agnew High School rated the 20 boys in the class on the basis of "looks, charm, and sex appeal." The ranks of the 20 boys by the average score of the girls' rating and by their academic averages were as follows (rank 1 = best, rank 20 = worst).

Boy	*Girls' Ranking*	*Academic Rank*
A	6	16
B	2	3
C	7	18
D	4	9
E	13	15
F	8	8
G	3	14
H	17	17
I	9	11
J	20	12
K	14	7
L	10	6
M	1	4
N	18	19
O	15	13
P	19	20
Q	16	10
R	11	2
S	12	5
T	5	1

a) Does there seem to be a tendency for boys of higher academic rank to get higher ratings from the girls, or is the tendency the other way around? Calculate the Spearman rank correlation coefficient in answering this question.

b) Test the significance of the coefficient you calculated in part (a) at the .05 level. Is there evidence of a relationship between ratings by girls and academic rank? (Assume that these are a random sample of all high school seniors.)

11–16. The performance of 10 mutual funds in terms of percentage change in market value were recorded for two successive years as shown below. Using a rank correlation technique, test the hypothesis that there is no relationship between a fund's performance in one year and the next.

Fund	*Year 1*	*Year 2*
A	+22%	−15%
B	+64%	−8%
C	+7%	−2%
D	+46%	−32%
E	+35%	−45%
F	+42%	+16%

Fund	*Year 1*	*Year 2*
G	+17%	−4%
H	+9%	+18%
I	+16%	−19%
J	+11%	+5%

11–17. In Exercise 3–6 you calculate the Spearman rank correlation coefficient of the areas and the population of the states of the United States. Test the significance of both these measures at the .05 level of significance.

11–18. Nine students took a battery of 4 aptitude tests. On each test, the students' scores were ranked as follows.

Test	*Student*								
	A	B	C	D	E	F	G	H	I
Inductive Reasoning	5	3	1	2	7	4	6	9	8
Deductive Reasoning	3	5	2	1	4	6	8	7	9
Ideaphoria	4	8	2	5	6	7	1	9	3
Structural Visualization	9	4	7	6	2	5	1	8	3

a) From visual inspection of the data, do you think the scores on the 4 tests are positively related?
b) Calculate the Kendall coefficient of concordance W.
c) Test the significance of W at the .05 level.

11–19. Suppose 3 judges all ranked 5 Olympic divers in the same order. Calculate W and test its significance at the .01 level.

***11–20.** A value of zero for the Kendall coefficient of concordance W indicates a total lack of consensus, or total disagreement, among the rankings. Unfortunately total disagreement among 3 sets of rankings is not easily defined.

Construct a set of rankings A, B, and C of 5 individuals I, II, III, IV, and V so that the statistics SS and W equal zero. From inspection of these rankings, would you say the three sets were in total disagreement? Can you construct another set of rankings that you would say are in greater disagreement?

Summary of Chapter

Distribution-free, or nonparametric, statistics are designed to be used when

1. you cannot assume the probability distribution of the population is normal, and
2. the sample size is too small to make use of the Central Limit Theorem.

In many of the social sciences the data are usually either nominal or ordinal (they can be classified or ranked but not measured quantitatively). Researchers who do much work with small samples involving this type of data should be especially familiar with nonparametric methods.

Nonparametric tests were presented which correspond to the parametric tests (using the z or t statistic) in Chapters 8, 9, and 10. In addition, a test for randomness and a test for goodness of fit were presented.

REVIEW EXERCISES FOR CHAPTER 11

11–21. The weather for 30 successive days was recorded in Slippery Rock. Letting R denote the "rainfall during the day" and N denote the event "no rainfall during the day," the chief meteorologist of Slippery Rock recorded the rainfall as

N N N R R N R N N N N N N N N R R N N N N N N R N N N R R R

Does rainfall come in "spells," or does it come randomly? Test the hypothesis of randomness at the .05 level of significance.

11–22. The new Director of the Department of Highways decided to interview all 6 employees in the department in the first three weeks that he took over the Directorship. In a follow-up study, 40 of the 66 said they felt better about the new director after the interview, 18 of the 66 said they felt worse, and 8 indicated that there was no major change in their feeling.

a) Test the hypothesis at the .05 level of significance that interviewing tends to result in more people feeling better than feeling worse.
b) What assumptions must you make in making this statistical test?

11–23. Ten intersections were ranked by the number of accidents occurring in the previous month. Six of these had traffic lights. The rankings (1 = most accidents) were as follows (L = a traffic light, N = no light).

Rank	
1	L
2	L
3	L
4	N
5	L
6	L
7	N
8	L
9	N
10	N

a) Assuming that these 10 are a random sample of intersections, run a nonparametric test to determine whether the presence of a traffic light is related to the number of accidents.

b) If intersections with traffic lights tend to have more accidents than do intersections without lights, would you conclude that traffic lights cause accidents? Discuss what statistical procedures you might use.

11–24. A survey of 16 alumni of South Hadley University revealed the following figures on income and contributions to the University Fund Drive. (Assume that these 16 are a random sample of alumni.)

X (Income)	*Y (Contribution)*
$ 40,000	$ 800
12,000	30
7,000	0
18,000	50
8,000	10
24,000	150
10,000	8
13,000	25
19,000	20
7,700	655
144,000	0
21,000	27
10,500	15
6,700	0
12,600	45
55,000	1000

a) Compute the Spearman rank correlation coefficient, and test the significance of r_s at the .05 level of significance.

b) Compute the Pearson linear correlation coefficient. Run a t test to test the significance of r, making the assumption that X and Y are normally distributed.

c) Compare the results of the tests you conducted in parts (a) and (b). From inspection of the data, do you think the assumptions underlying the t test are valid?

11–25. Sixteen persons went on Dr. Svelte's reducing regime. The results before and after completion of the six-week regime are given below. (Assume that these 16 people are a random sample of people who take Dr. Svelte's reducing regime.)

Person	*Age*	*Sex*	*Weight*	
			Before	*After*
Howard Jones	52	M	154	146
Jennifer Jones	50	F	135	122
Wilfred Evans	61	M	182	179

Person	*Age*	*Sex*	*Weight*	
			Before	*After*
Mark Taft	22	M	224	226
Valerie Voorhees	18	F	162	138
Heide Alsleben	42	F	144	122
Peter Greene	26	M	129	116
Sigmund Fromme	55	M	175	175
Maxine Morris	43	F	132	121
Bridget O'Conner	24	F	116	94
Sheila Weinberg	52	F	129	122
Jane Cheng	44	F	103	98
Kristine Andersen	25	F	166	142
Alexander Kolmogorov	36	M	182	156
Natasha Smirnov	36	F	194	168
Penelope Plum	32	F	215	195

a) Test the hypothesis, at the .05 level of significance, that the median weight loss is zero. Make no assumptions about the probability distribution of weight losses.

b) Test the hypothesis, at the .05 level of significance, that the median weight loss is not more than 10 pounds.

c) Do women tend to lose more weight from Dr. Svelte's regime than do men? Test at the .05 level of significance, using the median test.

d) Is there a relationship between age and weight loss? Test at the .05 level of significance, using a distribution-free test.

11–26. Thirteen randomly selected faculty members were classified by age and total number of students enrolled in their courses. The results were:

Faculty Member	*Age*	*Number of Students*
A	22	142
B	44	113
C	39	88
D	45	79
E	72	84
F	38	106
G	36	112
H	42	102
I	29	146
J	34	153
K	55	101
L	24	134
M	76	155

Calculate the Spearman rank correlation coefficient r_s and test its significance at the .05 level of significance.

11–27. Five randomly selected families showed the following data on incomes and amount paid in rent.

Family	*Income*	*Rent*
A	$14,000	$1,900
B	26,000	3,900
C	12,000	1,800
D	730,000	400
E	13,000	2,000

a) Calculate the Spearman rank correlation coefficient r_s and test its significance at the .05 level.

b) Calculate the Pearson coefficient of linear correlation. Assuming that the two variables are normally distributed, test the hypothesis that $\rho = 0$ at the .05 level of significance. (It will be easier to do this if you express income in thousands of dollars and rent in hundreds of dollars.)

c) Does this exercise give you a clue as to the kind of distortion that is introduced in testing for ρ under the assumption of normality when that assumption is unjustified? Discuss.

11–28. In order to determine whether a deck of cards was properly shuffled, someone observed the top 16 cards and identified them by color. The order of the cards (R = red, B = black) was

R B R B R B B R B R B R B R R B

Test the hypothesis, at the .05 level of significance, that the sequence is random.

11–29. Seven randomly selected students were ranked by their grades in 4 courses.

	Student						
Subject	I	II	III	IV	V	VI	VII
English	7	4	5	6	1	3	2
Algebra	5	3	7	1	4	2	6
Social Studies	5	3	6	2	4	1	7
Science	6	4	5	1	3	2	7

Is there a tendency for students to have similar rankings in the 4 subjects? Test at the .05 level of significance.

SUGGESTED READINGS

BRADLEY, J. V. *Distribution-Free Statistical Tests.* Englewood Cliffs, N.J.: Prentice-Hall, 1968.

HARNETT, DONALD L. *Introduction to Statistical Methods,* 2nd ed. Reading, Mass.: Addison-Wesley Publishing Co., 1975. Chapter 15.

KRAFT, C. H., and C. VANEDEN. *A Nonparametric Introduction to Statistics.* New York: McMillan Publishing Co., 1968.

OTT, LYMAN. *An Introduction to Statistical Methods and Data Analysis.* North Scituate, Mass.: Duxbury Press, 1977, chapter 11.

SIEGEL, SIDNEY. *Nonparametric Statistics for the Behavioral Sciences.* New York: McGraw-Hill Book Company, 1956.

TWELVE

Experimental Design in the Evaluation of Public Programs

Outline of Chapter

12.1 Designing Research for Evaluation

There are problems in setting up standards for comparison when evaluating programs.

12.2 Experimental Design and Internal Validity

Before getting the information, you need to determine what information you want, how you are going to get it, and how you are going to analyze what you find. You need to set up your experiment or research in a way that identifies the effect of what you are measuring and not some unwanted effects.

12.3 Problems in Developing True Experimental Designs for Evaluation

True experimental designs are often difficult or impossible to achieve in the evaluation of public programs. Regression analysis can be a particularly effective tool when you cannot physically control for all your variables.

12.4 Interactions, Quasi-Experimental Designs, Nonexperimental Designs, and Problems of External Validity

A host of problems in evaluation research can be dealt with by the use of imaginative techniques that come measurably close to ideal experimental designs.

Objectives for the Student

1. Become aware of the need for and dangers of quantitative measures in evaluating public programs.
2. Recognize the factors that can threaten the internal validity of an experiment.
3. Become familiar with methods for controlling for the effect of certain variables.
4. Understand methods of controlling for interactions.
5. Distinguish between experimental designs, quasi-experimental designs, and nonexperimental designs.
6. Distinguish between internal validity and external validity.

Previous chapters have been devoted to presentation of the theory that you need for making inferences and decisions. This chapter discusses the evaluation of public programs, a subject that draws heavily on statistical theory. The chapter will introduce the subject of statistical inference, known as experimental design, in the context of evaluation research. It will not present a comprehensive study of the theory of experimental design or the many types of design that can be used for various purposes. Such a study is the subject matter of a full course in experimental design for those who have completed a basic statistics course.

12.1 Designing Research for Evaluation

Suppose that a job-training program has been set up in an area in which there is a high unemployment rate. The population in the area is heavily represented by groups characterized by high unemployment rates: minority groups, people with few skills or skills that are technologically obsolecent. Suppose also that the program has been in effect for five years, and that there is considerable disagreement about the degree of success of the program. You have been asked to evaluate the performance of the program since its inception. How do you approach your task, and how will you carry it out?

In answering these questions, you will need a criterion of effectiveness that can be measured. The measure that you use should be in accordance with the stated goals of the program. If, in the job program, the goal is to create a low unemployment rate among the participants in the program, you might use percentage unemployed among those completing the program as your measure. In many cases programs will have multiple goals, and measures of their effectiveness will necessarily be more complicated to reflect this mutliplicity of goals. In the present example, however, let us assume for the sake of simplicity that a low unemployment rate is the goal, and the measure of effectiveness will be the percentage unemployed among the completers of the program.

Why do you need a specific quantitative measure? Some may argue that the goals of almost any program are too varied and rich to be reducible to a single number as a measure. There may be some substance to this objection: quantitative measures may fail to do justice to the many possible effects, both desired and undesired, that the program can have on all the people involved. On the other hand, to avoid a quantitative measure may cause you to lose all sense of the effectiveness of the program, and you may even lose sight of the fundamental purposes of the program. Furthermore, such a measure is necessary if you want to achieve some semblance of an objective or unbiased evaluation. The proponents of the program are likely to see all kinds of good results in it that the detractors fail to see. It is up to the evaluator as an outsider to evaluate the program on the basis of criteria that can be measured without reliance on the feelings or prejudices of those intimately involved in it.

Having decided on an operationally defined measure, you still need to set a standard for comparison. Assume that you determine that the unemployment rate among all those who have gone through the program is 18.4%. Is this an indicator of a successful program or an unsuccessful one? With what do you compare the 18.4% figure? If all the people who went through the program were unemployed to start with, then an 18.4% unemployment rate would be a remarkable improvement over what you could interpret as a 100% unemployment rate for those starting the program.

Although a claim of success based on an improvement of the unemployment rate over 100% seems obviously simple minded, the line of reasoning has often been used in reports and claims for the effectiveness of programs. For example, claims have sometimes been made for the benefits of medical treatments on the basis that a certain percentage of patients have improved under treatment. The fallacy of claiming that the treatment was beneficial is, of course, that the patients undergoing treatment were probably all sick at the time they started treatment, and over the course of time a certain percentage would be expected to improve. And a program designed strictly for unemployed persons is almost inevitably going to end with some of the participants getting employed, no matter how ineffective the program is.

The question that you really need to answer is: *How does the unemployment rate of this group of individuals compare with what the unemployment rate would have been had these individuals not participated in the job program?*

The more you examine this question, of course, the clearer it will become that it is an impossible question to answer precisely. The only way you can truly answer it is by sending the individuals through the program, observing their unemployment rate, then turning time back and keeping these same people out of the program, again observing their unemployment rates. Since this procedure involves a few practical difficulties, the next best thing to do is to compare these individuals with a very similar group of individuals who resemble the first group of individuals in every way except for one: they did not participate in the jobs program. The difference between the unemployment rates of the two groups can then be attributed to the program itself, and you have a measure of the effectiveness of the program on the unemployment rate. Even this idea of getting a second group that is identical to the first in every respect other than that it did not go through the program is an impossible ideal to achieve. It is both an art and a science to obtain two or more groups that are nearly identical in all respects except for the treatment they receivc. The goal of isolating the effect of treatment from all other effects is a primary goal of experimental design.

EXERCISES

12–1. Suggest a quantitative measure that can serve as a measure of effectiveness for the following types of programs.

a) A program to increase the reading skills of sixth grade pupils who have given indications of reading ability far below what is expected for their age.

b) A new treatment designed to enable victims of spinal cord injury to lead economically productive lives.
c) A tactical antiaircraft missile system.
d) The development of an oral insulin pill for diabetics (assume that you are doing the evaluation for the pharmaceutical company that is planning to develop it).
e) The development of an oral insulin pill for diabetics (assume that you are doing the evaluation for the Food and Drug Administration).

12.2. Consider an antiballistic missile system that has been installed over a period of years.

a) Is it possible to arrive at any kind of quantitative measure of the effectiveness of the system?
b) Is the fact that there has been no occasion to use the system an indication that its value has been zero?
c) How would you go about putting a measure on the effectiveness of the system?

12–3. A university is considering whether to discontinue its intercollegiate football program. You have been asked to evaluate the effectiveness of the program over the past ten years. Discuss possible measures that you could use to measure the effectiveness of the program.

12–4. An old home remedy for colds consists of honey and vinegar. Suppose that in a large number of cases of bad colds in which this remedy was suggested, 60% of the colds improved within a week's time.

a) What are the fallacies in concluding from this information that vinegar and honey are an effective remedy for colds?
b) What other information would you need before drawing a conclusion about the effectiveness of this remedy?

12.2 Experimental Design and Internal Validity

An experimental design attempts to discern the effectiveness of a treatment or program on a group of individuals. In order to evaluate a program, you should

1. determine a measure of outcome or effectiveness;
2. observe changes in this measure on individuals who have gone through the program (the experimental group);
3. compare the changes in the experimental group with some standard or baseline for comparison; and
4. isolate the effects of the treatment or program from all other factors that could also be creating changes in the experimental group.

Some of the problems encountered in performing the first three tasks were introduced in Section 12.1. But underlying the whole process of observing an experimental group and a baseline for comparison is the need to control for a host of factors other than the treatment that could be causing changes. Campbell and Stanley have identified eight of these factors:

1. Maturation—During the course of the program individuals get older. Any changes in the individuals that are attributable to increased age should not be attributed to the program.
2. History—During the course of the program, changes may occur in the economy, in the level of technology, or in society in general. Any changes in the individuals that are attributable to the effect of these societal changes should not be attributed to the program.
3. Testing—If individuals are tested before undergoing a program, the experience of being tested may affect the results of a test at the end of the program. Also, the fact that they are being tested may affect the behavior of individuals who are involved in the experiment.
4. Instrumentation—If the measuring instruments or the observers or scorers change during the course of the experiment or program, the recorded results can be affected.
5. Statistical regression—If individuals are selected for the program, for an experimental group, or for a comparison group on the basis of their scores on a pretest (particularly extreme scores) then there will be a tendency for their scores to move toward the mean score of the whole group.
6. Selection—If the members of the experimental group or comparison group are selected on the basis of different characteristics (age, location, willingness to participate, etc.) then differential changes in scores could be attributed to differences in these characteristics.
7. Experimental mortality—Typically, some members in both the experimental group and the comparison group "drop out" of the experiment somewhere along the way. If there is a tendency for the experimental group to differ from the comparison group in the mortality rate or type of people who drop out, the experimental results can be affected.
8. Selection-maturation interaction—The combined effect of a different basis of selection for two or more groups and a different rate of maturation due to this selection difference, can create a difference in a result.

As you read this list, you should become acutely aware of how easy it is for experimental results to fool you. These eight factors are threats to what is known as the internal validity of an experiment.

INTERNAL VALIDITY is the characteristic of an experiment that enables you to infer that a change in the experimental group is due to the treatment itself and not to any other factors.

Unwanted effects such as those above are often known as "noise." Noise is unsolicited data that gets in the way of the information you are looking for. You can think of an experimental design as a method of eliminating, or at least reducing, the noise that gets in your way.

How can you reduce the noise resulting from these unwanted effects? To reduce the effect of maturation and history, you might try to reduce the time of a program to such a short period that these effects would be negligible. Unfortunately such a remedy is impractical if the program is by nature long lasting, if the people involved are growing or changing rapidly (as in the case of children), if economic and social conditions are rapidly changing, or if those directing the program are not going to shorten it simply to go along with your research needs.

Also mentioned in Section 12.1 was the need for selecting a second group of individuals as a basis of comparison with those who go through the program.

When two groups of individuals are selected on a basis that insures that they differ from each other only in whether or not they receive the treatment, and all other differences are contained within statistically defined limits, then the group receiving the treatment is known as the EXPERIMENTAL GROUP and the group not receiving the treatment is known as the CONTROL GROUP.

The term "control group" is often misused to describe any group that is selected as a basis for comparison, even though the basis may be mere convenience. Such a group can be called a "comparison group," but not a control group.

It is important that the individuals assigned to the experimental group do not differ in the most important and relevant characteristics from the individuals assigned to the control group. The fundamental method for achieving this comparability is randomization in the selection process. If, for example, you have 20 subjects who have volunteered for the experiment, you would want to use a random method of assigning 10 to the experimental group and 10 to the control group. In so doing you are controlling for all of the factors that threaten the internal validity of the experiment. You protect yourself against the effect of differential rates of maturation, for example, since, insofar as age is not a factor in the selection of individuals for the two groups, the passage of time should affect the maturation of the two groups about equally.

Random assignment of individuals to the two groups protects you from bias in the selection of individuals, but it does not protect you from chance, or random, error. Even though the random assignment of ten individuals to the experimental group and ten to the control group frees you from a *systematic* tendency to select people for one group who are (for example) older than those in the other group, it is still possible that one group is much older on the average than the other simply because of chance. You can protect yourself from this unwanted "age effect" by one of two methods.

One method would be to specifically control for age by assigning equal numbers of older and younger people to each group. For instance, if 12 of the 20 subjects are at least 30 years old, you can randomly assign 6 of these to the experimental group and 6 to the control group, and of the 8 subjects under 30 you can randomly assign 4 to the experimental group and 4 to the control group. In this manner you insure that each group has an equal percentage of individuals who are 30 years old or over.

This procedure, known as matching, has some drawbacks, however. You may match for characteristics that are unimportant while at the same time overlook some characteristic that is more important. For example, it may turn out that older and younger people differ very little from each other in the characteristic you are measuring, so that you gain nothing by matching for age except the false sense of security that you have conducted a "well-controlled" experiment. On the other hand, perhaps people raised in urban environments differ considerably from those raised in rural areas, or males differ considerably from females, or bald-headed people from hairy people, in the characteristic that you are measuring. However, if you had not known that these were the key variables, you would not have matched for these characteristics, and you might have attributed a difference between the two groups to the program when in fact the difference was due to a higher percentage of bald-headed people in the control group. Adequate matching requires that you know what are the important variables to watch for. However, if you knew this you might not have to make the study because you already knew a great deal about the subject you were studying.

The second method of protection from the "age effect" is to increase your sample size in order to reduce the random error. Recall from Equation 9.2 that the standard error of the difference between two means when the universe means are equal is:

$$\sigma_{\bar{X}_1 - \bar{X}_2} = \sqrt{\frac{\sigma_1^2}{n_1} + \frac{\sigma_2^2}{n_2}} \qquad \textbf{9.2}$$

where σ_1^2 is the variance in the universe from which the experimental group is drawn
σ_2^2 is the variance in the universe from which the control group is drawn
n_1 is the number in the experimental group
n_2 is the number in the control group

Equation 9.2 shows that increasing n_1 and n_2 will decrease the standard error of the difference between the two means. Therefore, if you do not control specifically for age, but let the ages of the people in the two groups be selected at random, Equation 9.2 gives you an idea of how much you can expect the means of the ages in the two groups in your sample to vary from each other. If, for example, the variance in ages is 100, and you assign 10 individuals randomly to each group, the standard error of the difference between the mean ages is

$$\sigma_{\bar{X}_1-\bar{X}_2} = \sqrt{\frac{100}{10} + \frac{100}{10}} = \sqrt{20} = 4.5 \text{ years}$$

The 95% confidence interval (from an extension of Equation 7.5 to differences between two means) is

$$\mu_1 - \mu_2 \pm 1.96\sigma_{\bar{X}_1-\bar{X}_2} = 0 \pm 1.96(4.50) = -8.80 \text{ to } +8.80 \text{ years}$$

You have 95% confidence that the mean ages of the two groups will be between 8.80 years of each other. This may be an unacceptably high interval. But by increasing the sample size to 50 in each group, you can reduce the standard error of the difference to

$$\sigma_{\bar{X}_1-\bar{X}_2} = \sqrt{\frac{100}{50} + \frac{100}{50}} = 4 = 2.00$$

The 95% confidence interval will then be reduced to

$$\mu_1 - \mu_2 \pm 1.96\sigma_{\bar{X}_1-\bar{X}_2} = 0 \pm 1.96(2) = -3.92 \text{ to } +3.92$$

With 95% confidence that the means of the two groups will differ by no more than 3.92 years, you can feel more confident that differences in age between the two groups are not strongly influencing any differences in your sample results.

A similar kind of analysis shows the effect of increasing sample size on reducing the effect of a nominal variable. Suppose that 60% of the individuals in the population are male. Then, if you specifically control for sex, you can have exactly 60% males in both the experimental and the control groups. If you do not control for sex, then the differences in proportion of males between the two groups would have an expected value (from Equation 9.8) of

$$\mu_{\hat{p}_1-\hat{p}_2} = 0 \qquad \textbf{9.8}$$

and a standard deviation (from Equation 9.9) of

$$\sigma_{\hat{p}_1-\hat{p}_2} = \sqrt{\frac{p(1-p)}{n_1} + \frac{p(1-p)}{n_2}} \qquad \textbf{9.9}$$

where p is the proportion of males in the whole population
$\hat{p}_1$ is the proportion of males in the experimental group
$\hat{p}_2$ is the proportion of males in the control group

If $p = .60$ and there are 10 individuals in each group, then

$$\mu_{\hat{p}_1-\hat{p}_2} = 0$$

and

$$\sigma_{\hat{p}_1-\hat{p}_2} = \sqrt{\frac{(.6)(.4)}{10} + \frac{(.6)(.4)}{10}} = .219$$

The 95% confidence interval is approximately (from an extension of Equation 7.7 to differences between two proportions)

$$\mu_1 - \mu_2 \pm 1.96\sigma_{\hat{p}_1-\hat{p}_2} = 0 \pm 1.96(.219) = -.429 \text{ to } +.429$$

This appears to be a rather big interval. A difference of over 40% in the male–female percentage between the two groups would undoubtedly be too much, if you had much reason to believe that sex was an important factor in explaining the variable you are measuring. Differences between the groups that you would like to attribute to the program might really be due to the differences in the male–female ratios in the two groups.

If, however, you increase the size of each group to 50, then the standard error of the difference between proportions will be

$$\sigma_{\hat{p}_1-\hat{p}_2} = \sqrt{\frac{(.6)(.4)}{50} + \frac{(.6)(.4)}{50}} = .098$$

and the 95% confidence interval will be

$$0 \pm 1.96(.098) = -.192 \text{ to } +.192$$

With 95% confidence that the male–female ratio of the two groups will not differ by more than 19.2 percentage points, you may feel that it is not necessary to match for sex if you have 50 individuals in each group.

You might well ask why bother to randomize when you can directly control for other variables (e.g., age and sex) by specifying the proportions in each group? The advantage of randomization combined with a large sample is that it reduces the effect of *all* other variables, whether you are aware of their effect or not. Randomization selects people for the two groups on a basis that is totally independent of any of the characteristics that could affect the measurements. Any differences in mean age, sex ratio, mean income, educational attainment, etc., between two randomly selected groups is entirely due to chance, and the larger the sample size in both groups, the smaller the error due to chance is expected to become. For this reason, an experimental design that combines randomization with a large sample is a powerful tool for measuring the effect of a program or treatment.

Unfortunately, most evaluation research is not able to make use of both random selection and large samples. The rest of this chapter will be concerned primarily with attempts to reconcile the requirements of an ideal design with the realities of getting information on public programs.

EXERCISES

12–5. In an experiment to determine the effectiveness of an authoritarian approach vs. a democratic approach in a group's performance of a task, 30 undergraduates were randomly assigned to two groups with the identical task of answering a series of questions involving field research, library research, and use of a computer. A boss is assigned to one group, who is given complete authority to direct the activities of the group members. The other group is assigned one leader, but it is up to the members of the group to determine how they will perform the task. The groups are graded both by the correctness of their answers and the time required to complete the task.

a) In assigning the volunteers for the experiment to the two groups, the experimenters wish to avoid imbalances between the two groups with respect to characteristics that could affect the performance of the groups. List some of the characteristics that you think are most important to control for.

b) Suppose someone wanted to control for the two characteristics race and sex. Given the following individuals classified by race and sex, assign them to the two groups (experimental and control) in a way that would eliminate the effects of these two variables on the outcome of the experiment.

Individual	*Race*	*Sex*	*Individual*	*Race*	*Sex*
Albrecht	White	Male	Philips	White	Male
Bell	Black	Female	Quinn	White	Male
Clark	Black	Female	Roberts	White	Female
Darlington	White	Female	Semple	White	Female
Ellsworth	White	Male	Thomas	Black	Female
Forbes	White	Female	Unger	Black	Male
Glidden	Black	Male	Vining	White	Female
Hobart	White	Female	Wilkins	Black	Male
Inverness	Black	Male	Xavier	White	Male

Individual	*Race*	*Sex*	*Individual*	*Race*	*Sex*
Jackson	Black	Female	Young	White	Male
Kinley	White	Female	Zorn	White	Female
Lothrop	White	Female	Albemarle	Black	Female
McKee	White	Female	Bellefonte	Black	Female
Neville	Black	Male	Craig	White	Male
O'Hara	White	Male	Dithridge	Black	Male

***12–6.** A random sample of 50 individuals was divided into two equal-sized groups in order to test the degree of intellectual deterioration caused by a continuous exposure to TV commercials. The experimental group was exposed to 4 sessions of commercials of 6 hours each; the control group was given similar exposure to nonmalignant programming. Suppose a critical variable affecting the results is literacy. If the selection of the individuals for the two groups is random, set up a 95% confidence interval for the difference between percent of literates in the two groups. (Assume that among the population the literacy rate is 50%.)

***12–7.** In the experiment of Exercise 12–6, how large a sample will you need in order to give you a 95% degree of confidence that the difference in the percent literate in the two groups is not more than 10 percentage points?

12–8. A member of a Commission on Physical Fitness has proposed an exercise program for people of all ages, which involves running, jogging, and various calisthenics. Skeptical members of the commission raise the question of whether or not such a regime is beneficial at all, or whether it may even be harmful.

Propose a general design of an experiment that could help determine the efficacy of such an exercise program. Assume that you have sufficient financial resources to obtain a sample large enough for your needs. Discuss the strengths and weaknesses of the design that you propose. To simplify the discussion, assume that you need to consider the program only for people under 40 years of age who have recently passed an examination which certified that they are in good physical condition.

***12–9.** An experimenter wished to divide 4 subjects into 2 groups, an experimental group and a control group. He wanted to control for 4 variables: educational level, marital status, religion, and political party affiliation. It seemed feasible to do this, since

2 subjects had completed college and 2 had not.
2 subjects were married and 2 were single.
2 subjects were Protestant and 2 were Roman Catholic.
2 subjects were Republicans and 2 were Democrats.

a) Show that it is impossible to control for all 4 variables if the subjects have the following characteristics. (Table p.388)

b) What does this impossibility suggest about the feasibility of controlling for many variables when the sample size is small?

Subject	Educational Level	Marital Status	Religion	Party
A	No college degree	Married	Protestant	Republican
B	No college degree	Single	Protestant	Democratic
C	College degree	Single	Catholic	Republican
D	College degree	Married	Catholic	Democratic

12–10. Outline an approach of determining what harmful or beneficial effects a birth control pill can have on women who take it for a period of at least 5 years. What "noise" variables would you particularly want to control for?

12.3 Problems in Developing True Experimental Designs for Evaluation

Randomization is an effective method of assigning subjects in an experimental design because the basis for selection is completely independent of any variable that you wish to control for. Some of the methods that do *not* guarantee this independence are:

1. self-selection—by which individuals have some influence on whether they are to be in the experimental or the control group.
2. subjective selection—by which the experimenter selects individuals for the groups on the basis of who she or he thinks would make good subjects for each group.
3. selection by geography—by which individuals in one geographical area are selected as experimental subjects, while individuals in another area are selected as controls.
4. selection on the basis of convenience—whether this is by location of an individual's home or place of work, work schedule, ability, etc.
5. selection on the basis of need—by which individuals are placed in the experimental group because someone determined that their need for the treatment or program was great.

This is not to say that selection on the basis of any of these methods will necessarily bias the results of the experiment, but on the other hand they may, and you need to watch like a hawk for such biases. A true experimental design does not use such methods for selection. In an ideally designed experiment, the selection of individuals for the experimental and control groups is done by use of a table of random numbers or random selection using a computer.

But a true experimental design is not often feasible in evaluating programs. The situation most ripe for an experimental design is in formative research. Formative research looks ahead at what a future program might accomplish, and may use randomly selected individuals for a simulated program that you hope resembles the real program you plan to carry out. But in summative research, you will almost never be able to employ an experimental design. Summative research looks back at a program that has been completed, and assesses its effectiveness.

> A research design on a program that has been completed before the research is carried out is known as an EX POST FACTO RESEARCH DESIGN.

An *ex post facto* research design is by its very nature not a true experimental design. You cannot randomly assign individuals to groups when the individuals were assigned to the program long before you came along, and by methods over which you had no control. Unless the program itself was set up purely for experimental purposes, you can be quite sure that individuals were selected for the program for reasons of choice, of convenience, of need, etc., but certainly not on the basis of random selection.

What all this adds up to is not that the discussion of experimental design in the previous section is irrelevant to the evaluation of public programs that have been completed. Far from it! Simply because an evaluation study is not a perfect experimental design does not mean that it is useless. Even when the study falls short of being an ideal "experiment" it is important to be able to determine how far from the ideal the study is. When you can do this you are in a position to determine how much or how little weight to give to a study, the kinds of uncertainty you will have after you know the results of the study, and ways to make better studies in the future.

Go back to the problem of evaluating a program whose participants showed an 18.4% unemployment rate after the program had been in effect for 5 years. The people going through the program, unfortunately for your purposes, were not randomly selected, nor was a control group selected for comparison purposes. You would want to compare the unemployment rate of the people who came through the program with the unemployment rate of the same people had they not gone through the program. Since you cannot do this, the next best thing to do is to compare the unemployment rate with a group of people as similar as possible to those who went through the program. Lacking a randomly selected experimental or control group, you look for a nonrandomly selected group who appear to resemble the experimental group in every important respect.

But how do you find such a group? You want the comparison group to be similar to the group that went through the program in all those characteristics that could affect the unemployment rate. It would be easy to start a list of such characteristics but difficult to complete it. The list would be a long one that would probably include age, color, sex, educational attainment, skills, marital status, etc. The list must be confined to traits that are identifiable or measurable. Such variables as ambition, degree of extroversion, mental stability, etc. may be important factors in one's getting employed, but they would probably not be included in your list because it may be impossible to identify or measure people by such variables. You can control only for variables that are operationally defined.

The comparison group must be not only a group that resembles the experimental group in the most relevant characteristics, but it must also be one for which you have sufficient data for comparison. There is no formal method that ensures that you can find such a group; you simply do the best you can. But in so doing you must be aware of the shortcomings of the comparison group. If the program was carried out in one city, then there are going to be problems in comparing the results with a group in a different city, even though that group may impress you as being strikingly similar to the program group. You cannot get around the fact that any two cities are different in many ways, and some of these differences are going to create noise that can affect the validity of your conclusions. One answer to this difficulty is to compare the experimental group with another group in the same city. However, the group selected for the program might have been selected because it came from the highest unemployment area in the city; because of this fact, no other area in the city will be strictly comparable.

These warnings should not prevent you from carrying out the study; they should merely keep you on guard against overlooking some pitfalls or jumping to unwarranted conclusions. In spite of the difficulties, you may be able to find a group that is similar to the program group in many respects; a careful evaluation taking cognizance of the difficulties of comparison is far better than no study at all.

In doing evaluation research you must steer a middle course between excessive perfectionism and too little concern for rigor and the need to protect the validity of the results. The first of these will result in your doing no evaluations at all; the second will result in your doing something worse than no evaluation at all. Examples abound of excessive carelessness or sloppiness; conclusions based on inadequate sample sizes, inadequate controls, subjective evaluations of the results by individuals with vested interests in the results, etc. When you cannot achieve adequate sample size or controls, you still should control for as many variables as you can; in your report make clear the difficulties and impossibilities that you were up against; and interpret your findings with a degree of caution and restraint that is commensurate with the weaknesses in your study. Good research is characterized not only by its ability to overcome obstacles but also by its openness and honesty about those obstacles and the degree to which the researcher was able or not able to overcome them.

Regression Analysis

One of the major advantages of an understanding of statistics in doing research is that you can use various methods of statistical analysis to control for the effect of some variables when physical control is difficult or impossible. In a classical experiment, if you want to observe the effect that X has on Y, you hold all variables constant except X and Y. In an experimental design, "holding variables constant" may mean insuring that both the experimental and control group are equal with respect to those variables. But if you leave the laboratory and do your research in the outside world, these variables will not stand still for you, and you cannot find groups that are exactly comparable with respect to all the variables you want to control for. But statistical techniques can enable you to control effectively for these variables; one of the most useful of these techniques is multiple regression analysis, a subject that was introduced in Chapter 10.

Suppose that in studying the effect of a job-training program you came up against the problem of not being able to find a group of individuals that was truly comparable to the group that went through the program. You may have access, however, to a large subset of the population that includes the group you are studying, or at least includes individuals who represent the characteristics of the group you are studying. If your information includes data on unemployment rate, age, race, educational level, and other characteristics that you suspect may affect the unemployment rate, then you may be in a good position to use a technique such as multiple regression, even though you are not able to use a true experimental design. Using the methods presented in Chapter 10, you can come up with a multiple regression equation showing the effects of a large number of variables on the unemployment rate. Suppose you obtain the following multiple regression equation.

$$\hat{Y} = .0462 + .0141X_1 - .0060X_2 - .0031X_3 + .0038X_4$$

where $\hat{Y}$ is the predicted unemployment rate given X_1, X_2, X_3, X_4
X_1 is the percent minority
X_2 is the mean age
X_3 is the mean number of years of education
X_4 is the percent living in rural areas

Suppose that the group that had gone through the program has the following characteristics.

36% minority
mean age of 44 years
mean educational level of 11.4 years
0% living in rural areas

Then the predicted unemployment rate for this group would be

$$\hat{Y} = .0462 + .0141(36) - .0080(44) = .0031(11.4) + .016(0) = .25446$$

This calculation says that you would "expect" a group with the indicated characteristics to have an unemployment rate of 25.446%. The fact that the actual unemployment rate was only 18.4% suggests that the program was effective in reducing the unemployment rate.

It would be premature to claim from this analysis that the program reduced the unemployment rate by 25.446 − 18.400 = 7.046 percentage points. For one thing, you should be concerned with the question of whether the multiple regression equation took into consideration all the important explanatory variables. Perhaps there is some other variable that influences unemployment (call it X_5) in which the experimental group differs considerably from those who supplied the data for the multiple regression equation. In this case, the difference between the program group's unemployment rate (18.4%) and the predicted unemployment rate (25.446%) might be due to the difference in X_5 and not to the program itself. The same danger exists for all other variables (call them X_6, X_7, etc.) that were not brought explicitly into the multiple regression equation but might be part of the explanation for differences in the unemployment rates which you might naively be attributing to the program.

In theory, there are two fundamental ways to get around this difficulty:

1. Control for *all* relevant variables (through either a perfectly matching control group or a comprehensive multiple regression equation that brings into the equation all variables that have much impact on the unemployment rate.
2. Take a larger sample of groups that have gone through the program, ideally selected at random.

The first of these conditions calls for either a very fortunate occurrence of a matching group or a knowledge of *all* the relevant variables. You will almost never succeed totally in achieving either of these; through good fortune or great effort you may come reasonably close.

The second of these methods has the advantages of randomization mentioned earlier. In this case, however, you need to take a random sample of groups, not merely individuals. The reason for this is that in *ex post facto* research the individuals within a group are not generally randomly or independently selected. All of the individuals in one group are going to share some characteristics that none of the individuals in another group share (e.g., the members of one group may all be residents of Cleveland, and the members of the other group residents of Cincinnati).

Evaluation of programs becomes more reliable as the sample size increases; therefore your reliability is better if your program is conducted with more than one set of individuals. In that case, you may be able to compare several program groups with several similar groups of people who have not gone through the programs. You may be able to compare the mean unemployment rate of the groups that completed the program with the mean unemployment rate of groups that did not complete the program, or with the unemployment rate predicted by a multiple regression equation based on data obtained from those who have not completed the program. If statistical tests show that the differences between the observed and predicted unemployment rates are statistically significant, then you have strong reason to believe that the program was effective.

A basic problem in an *ex post facto* research design is that most programs were set up to serve a clientele, not to do research on them. The problem is similar to that of much medical research, where the data must often come from clinical practice rather than from experimental research. And the goals of a public program or clinical practice are plainly different and sometimes incompatible with those of research. If research were the primary goal, the directors of programs would select areas, groups, and individuals within groups using randomization techniques, and evaluation of the effectiveness of the programs would be very much facilitated. But alas for the evaluator, most public programs were not conceived and developed for purposes of evaluation.

Because of these inherent difficulties in evaluation, it will be necessary in authorizing and funding evaluation studies to provide funds sufficient for doing a job that will cope with the obstacles. Inadequate funding will support only superficial studies, which may be worse than useless.

If adequately supported evaluation is accepted as a way of life, there is some hope that programs in the future may be set up in a manner that is a little bit more conducive to their evaluation. Although there may be ethical or political objections to employing randomization techniques in selecting individuals or groups, there may be opportunities to use such techniques at times. For example, if a program involves two or more groups, it may be possible to assign people to the groups randomly, vary some procedures between the two groups, and use each of the groups as a control for some experimental techniques in the other. Such devices take imagination and skill. If the director of the program is also concerned with the needs of research, then there will be opportunities to use this imagination and skill.

EXERCISES

12–11. What are the main advantages of a research study designed before the execution of a program over an *ex post facto* research design?

12–12. A number of states and localities in the United States have passed legislation restricting in some manner the purchase or use of firearms. If one of the goals of such legislation is the reduction of the number of deaths from firearms, how might you go about evaluating the effectiveness of such legislation? Discuss the design of your research.

12–13. There is still disagreement among economists about the factors causing the economic collapse in the United States and the world in the years 1929–1933. Some attribute the collapse primarily to monetary policy; such people say that the Federal Reserve Board restricted the money supply and raised interest rates at a time when such action resulted in a severe contraction of economic output. These analysts claim that the crisis could have been avoided or more quickly cured by a more judicious handling of the money supply. Others feel that the cause of the crisis had its roots in fiscal policy, that the crisis was brought on and aggravated by the failure of government to offset taxation by sufficient spending to stimulate economic activity.

We have plenty of data on the supply of money, interest rates, government spending and taxation, and levels of economic activity during these years. Why is it, therefore, so difficult to assess the effects of monetary and fiscal policy that different experts still disagree as to what was the cause and possible cure of the crisis? Discuss in general terms, not necessarily as an economist, but from the point of view of the general problems of evaluating in retrospect the effect of various policies.

12–14. A study of the graduates of three colleges in a metropolitan area revealed that, among the graduates of the classes of 15 years ago,

the mean income of graduates of College A was $23,500
the mean income of graduates of College B was $14,600
the mean income of graduates of College C was $17,300

The result of this study was used by a representative of College A to show that College A helps you to earn more money. What other possible explanations could be offered for the differences in mean incomes among the 15-year classes of the 3 colleges? You should be able to identify at least 6 of them.

12–15. Given the fact that there are other possible explanations for the differences in mean income among the 15-year graduates of the 3 colleges in Exercise 12–14, discuss how you might conduct an evaluation of the 3 colleges with respect to their ability to increase the incomes of their 15-year graduates.

12–16. In a certain locality, a random sample of 200 people who smoked over 20 cigarettes a day showed a mean of 14.2 pounds overweight and a standard deviation of 10 pounds. A random sample of 250 nonsmokers showed a mean of 6.8 pounds overweight and a standard deviation of 10 pounds.

a) Test the hypothesis, at the .01 level of significance, that heavy smokers and nonsmokers do not differ in their mean excess weight.

b) Can you conclude, from your answer to (a), that smoking tends to cause excess weight?
c) What reasons can you think of for smokers to carry more excess weight on the average than nonsmokers, other than the effect of smoking on weight?
d) How would you design a study to determine the effect of smoking on weight, which controlled for the factors you mention in part (c)?

12.4 Interactions, Quasi-Experimental Designs, Nonexperimental Designs, and Problems of External Validity

INTERACTIONS

The discussion on controlling for extraneous variables has been based on the assumption that the effect of each variable on the sample result was independent of every other variable. However, sometimes two or more variables work together in influencing a result.

> The combined effect of two or more variables is known as an INTERACTION.

Suppose, for example, you wanted to measure the effect of some program on income. Since you have seen studies that showed that males earn more income on the average than females, and that married people tend to earn more income than single people, you want to control for the variables "sex" and "marital status." Suppose you have 20 subjects, and

14 are male and 6 are female
12 are married and 8 are single

If you assign 10 people to an experimental group and 10 to a control group, you could control for sex and marital status by assigning 7 males and 3 females to each group, and by assigning 6 married people and 4 single people to each group. In this way you would control for the effect of each of the two variables sex and marital status. Such control would be adequate if the effect of each of these variables could be totally isolated from the effect of the other, i.e., if there is no interaction between the two. But here is a case of two variables that have a high degree of interaction with each other. The effect of being married is much different on male incomes than it is on female incomes.

In order to control for interactions you need to set up your experimental and control groups so that they have equal proportions, not merely of male and female, and married and single, but also of married males, single males, married females, and single females. For example if the 20 individuals were classified as shown in Table 12.1, then to control for the interactions you would need to assign 4 of the 8 married males randomly to the experimental group and 4 to the control group, 2 of the 4 married females to the experimental group and 2 to the control group, etc. In this way the two groups would be equal not only with respect to the main variables, sex and marital status, but to the interactions of sex and marital status.

In order to control for all possible interactions among 2, 3, or more variables, experimenters can insure equal representation of all of the combinations only by painstaking design.

> A design that apportions all combinations of the control variables equally to the experimental and the control groups is known as a COMPLETE FACTORIAL DESIGN.

Unfortunately, complete factorial designs are generally feasible only in ideal or nearly ideal experimental situations. Suppose, for example, that you wanted to control for four variables: Sex (male and female), Race (white, black, and other), Age (under 20, 20 to 30, 30 to 40, 40 to 50, 50 and over), and Marital Status (married, single, widowed, divorced). To have just *one* subject in each combination of categories or cell, would require 2 × 3 × 5 × 4 = 120 subjects. Such a sample would give no indication of the variance of individuals within each category. If you wanted to have 10 observations in each category, you would need a total sample of 1200 subjects. Not only is it likely to be costly to get a sample of 1200 subjects, but it will be even more difficult and costly to find 10 subjects in each cell.

Because of the great difficulties inherent in doing complete factorial designs on many variables, researchers must often content them-

Table 12.1. *CROSS CLASSIFICATION OF TWENTY INDIVIDUALS BY AGE AND MARITAL STATUS*

	MARRIED	SINGLE	
MALE	8	6	14
FEMALE	4	2	6
	12	8	

selves with something less than a complete factorial design. An incomplete factorial design generally controls for the main effects (i.e., the effect of each variable in isolation from the others) but not all of the interaction effects. In evaluating programs, you will seldom have the opportunity to attempt such ambitious analysis as a complete factorial design controlling for a large number of variables and their interactions.

Another disadvantage of attempting to match for too many variables and their combinations is that you lose some of the benefits of randomization. The benefits of randomization increase with the sample size. If you have 100 individuals, a random assignment of 50 to each group will control fairly effectively for all variables. But if you try to apportion all the combinations equally, such as black females with fallen arches, then you may end up assigning the 2 individuals in this category equally to the experimental and the control group. Although you do this randomly, there is little gain from random assignment of such a small group of people.

The essence of a true experimental design is randomization. If the subjects are selected for the experimental group and the control group in a random manner, you can control for the effects of history, maturation, and all the other variables that threaten the internal validity of the experiment.

Quasi-Experimental Designs

If it is totally impractical to select subjects randomly, as in *ex post facto* research designs, you should at least make a serious attempt to develop a design that controls for as many of the threats to internal validity as is practical. A design that controls for many of these factors but nevertheless falls short of a true experimental design is a quasi-experimental design.

> A QUASI-EXPERIMENTAL DESIGN involves a comparison of at least two groups, an experimental group and a comparison group. The comparison group is not a true control group inasmuch as it is not randomly selected. It is selected by methods that control for the effect of a number of those factors that threaten the internal validity of the design.

An example of a quasi-experimental design is the time-series design. Consider a study of a highway-safety campaign which involved lowering the speed limit on a highway during the years 1975 and 1976. The number of fatalities on the highway for a 6-year period were:

Year	*No. of Fatalities*
1973	14
1974	16
1975	12
1976	13
1977	18
1978	21

A graph of this record is shown in Figure 12.1.

The "experimental" years are 1975 and 1976. The "comparison" years are 1973, –74, –77, and –78. In no sense can you say that the experimental years and control years are randomly selected. Nevertheless, the use of the comparison years may effectively control for most of the unwanted effects other than history. If you investigate historical trends and find no reason to believe that there is anything about 1975 and 1976 that would lead you to expect fewer fatalities than in the other years *other than the effect of the lowered speed limit,* then you have some valid basis for inferring that a decrease in fatalities in those 2 years from what you would have predicted on the basis of other years is due to the lowered speed limit.

NONEXPERIMENTAL DESIGNS

Sometimes even quasi-experimental designs are very difficult to achieve. In such cases, researchers have often abandoned comparison groups altogether.

> Research on a treatment or program that does not use a carefully selected, or randomly selected, comparison group is sometimes called a NONEXPERIMENTAL DESIGN.

Figure 12.1. *FATALITIES ON A HIGHWAY (FIVE YEARS)*

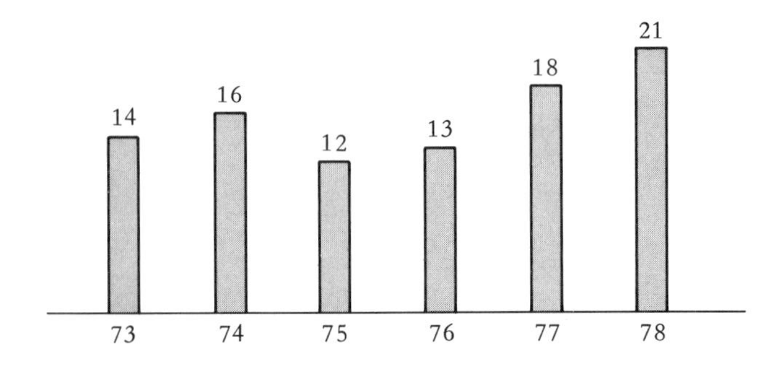

While you should not consider all nonexperimental designs as "bad," you should nevertheless conduct them and interpret them only with the greatest caution. Their chief use is to give insight into a program and provide suggestions for improvement. The great danger of nonexperimental designs is that they will be used as definitive evaluations, and that conclusions about the worthwhileness of a program will be made that could affect funding or even decisions to instigate or terminate a program. If evaluation is to be used for that purpose, it is probably better not to attempt to do a study until you have the means to conduct at least a quasi-experimental design.

One of the commonest forms of nonexperimental design is the *before/after* study. You are probably familiar with many reports, as well as advertisements, reporting a change in some measure (number of cavities, weight, whiteness of laundry, etc.) between the beginning of some program and the end. The problem with this design is that it controls for none of the unwanted influences on internal validity. You have no basis for saying what would have happened to the group if there had been no program. Another common nonexperimental design is the *after-only* study. This is an *ex post facto* design that looks only at the group that has undergone the program and does not look at a control or comparison group. Like the before/after study, it can be useful for preliminary purposes and in giving insight or clues, but not for summative purposes.

When used inappropriately nonexperimental designs can be worse than no research at all. The reason I say this is that there is danger that the study may get published. And a published study, whether good or bad, will get quoted and cited; its findings will be accepted and may even be used as the basis for future decisions, and even its methodology might get accepted or imitated. The fact that nonexperimental designs have been used in the past for certain evaluations may serve as a precedent for using similar designs in the future. As long as journals publish results based on such studies, as long as agencies fund studies that cannot be adequately designed, and as long as colleges and universities place emphasis on publication rather than quality of research, we are likely to be inundated by inadequately designed studies that can lead only to suspicion about research in program evaluation and in the social sciences in general.

External Validity

The discussion thus far has been concerned with problems affecting the results of the experiment itself. Beyond such problems are those of generalizing the experimental results to a wider universe.

> EXTERNAL VALIDITY characterizes the situation in which the experimental results can be applied to a larger population or universe.

The problem of external validity is the same one with which we are concerned in all statistical inference—that of generalizing the sample results. If the evaluation of a program is going to be used to decide whether similar programs are to be instigated, you will need to decide how far the results of the evaluation can be generalized—what is the universe of "similar programs"?

To illustrate the nature of external validity, suppose that a school district installed an experimental remedial reading program in 5 randomly selected eighth grade classes in the district, while all other classes, which used the existing remedial reading program, served as controls. Because the experimental classes were randomly selected, you are able to conduct a true experimental design, and the problems of internal validity should be well taken care of. The problem of external validity arises when you ask "to what population can the result be generalized?" If you use the results to make inferences about similar programs in other eighth grade classes in other school districts, you should ask whether this district is representative of other school districts. If you want to make inferences about the effect of such a program on sixth or seventh grade pupils, or ninth or tenth grade pupils, you should ask whether the effects of a program on eighth grade pupils is going to be the same as the effects on pupils in other grades. It is tempting to make the assumption that it will be the same, and it may even be reasonable to believe that it will be, but it is not *valid* to make the assumption without some experimental evidence on pupils in other grades.

Another problem in achieving external validity is that of measurement bias. You need to ask whether the measuring instrument that you are using is measuring the thing you say it is measuring. You want to measure the effectiveness or success of a program; your problem is, what measure is really a valid measure of effectiveness. You cannot measure effectiveness until you have clearly and thoroughly defined the goals of the program in operationally defined terms. Many programs are multipurpose, and a measure of effectiveness is going to have to take these multiple purposes into consideration and give a proper weighting to them.

Consider a program to train people to get jobs. You can consider several possible measures of effectiveness, for example,

1. the proportion of those entering the program who complete it
2. the proportion of those completing the program who get placed in jobs
3. the proportion of those who get placed who are in the same job 6 months later
4. the proportion of those who get placed who are employed in any job 6 months later.

You could use one of these measures, a combination of them, or some other measure. What you use depends on the stated goal of the program. If both immediate placement and retention in the job are goals, a measure that combines numbers 2 and 3 might be most appropriate.

EXERCISES

12–17. Suppose you want to conduct an experiment on 8 subjects, 4 of whom you would assign to an experimental group and 4 to a control group. Suppose 4 of the subjects are white and 4 are black; 4 are male and 4 are female; 4 are over 30 years old and 4 are 30 years old or under.

a) Would it be possible to set up a complete factorial design controlling for race, sex, age, and all the interactions? Why or why not?

b) Is it possible to set up an experiment controlling for the main effects of race, sex, and age?

12–18. A city added fluoride to its water supply 25 years ago and has continued to maintain the fluoride content ever since. The stated purpose of the fluoride program was to reduce tooth decay, especially among persons under 20 years of age. Suppose you have been asked to evaluate the effectiveness of the fluoride program. Describe how you would approach the problem. Assume that you have identified a similar city in another state about 500 miles to the west that has kept its water completely free of fluoride.

12–19. In what ways can an evaluation study that has made no effort to set up a control or comparison group be better than no study at all? In what ways can such a study be worse than no study at all?

***12–20.** For years the citizens living near the corner of Hatfield and McCoy Streets had been petitioning to get a traffic light installed. Finally on April 1, 1975, the light was installed. The monthly accident record over a period of time covering both the period before and the period after installation is given:

Months	*No. of Accidents*	*Months*	*No. of Accidents*
April 1974	5	April '75	4
May '74	8	May '75	2
June '74	4	June '75	5
July '74	7	July '75	7
August '74	3	August '75	4
Sept. '74	9	Sept. '75	3
Oct. '74	4	Oct. '75	2
Nov. '74	11	Nov. '75	7
Dec. '74	10	Dec. '75	3
Jan. '75	14	Jan. '76	8
Feb. '75	9	Feb. '76	10
March '75	8	March '76	5

a) Graph the monthly accident totals over the 24-month period. Does the graph indicate a substantial decrease in the number of accidents after the installation of the light?

b) From the data, describe the extent of change in the number of accidents before and after the installation of the light. What descriptive measures are appropriate?

c) Run a statistical test to determine whether there has been a significant decrease in the accident rate since the installation of the light, using the t statistic. What assumptions are necessary in running this test?

d) Run a nonparametric test to determine whether there has been a significant decrease in the accident rate since the installation of the light. What assumptions do you need to make in running this test?

e) Is it more appropriate in this case to run a test that considers the data a set of independent observations, or one that treats the data as a set of related, or paired, observations? Discuss the merits and problems in each approach.

f) Would this study be properly considered an experimental design, a quasi-experimental design, or a nonexperimental design? What is the experimental group and what is the control, or comparison, group, if such a group exists?

12–21. A Day Care Center has been set up in East Bilgewater in order to enable mothers and fathers of children 2 to 5 years old to continue working in full-time jobs. In order to evaluate the effectiveness of the center, suggest a more specific statement of the goals of the program.

Summary of Chapter

As the spending of public money becomes subjected to more and more critical scrutiny, it is going to become increasingly important to justify expenditures on proposed and on-going programs through careful evaluation procedures. The ideal procedure is an experimental design, which has the advantage of insuring the internal validity of the procedure. Since experimental designs are often difficult or impossible to achieve in evaluating public programs, it is often necessary to resort to quasi-experimental designs, which can control for many of the unwanted variables.

Even the best-controlled evaluating procedures can have a problem of external validity, that is, can the results be generalized to new situations or new programs?

The evaluation of programs is both an art and a science. It is important to understand statistical theory and experimental design, but it is equally important to have the imagination and skill to apply this theory in areas where tradition, politics, and human emotion are not always conducive to analytical study.

REVIEW EXERCISES FOR CHAPTER 12

12–22. In evaluating a program of a school of Public Affairs, an evaluator has defined the purpose of the school as "training people to perform jobs in the public sector." His measures of the effectiveness of the program are the number of graduates placed in public-sector jobs, and number of graduates still in public-sector jobs 5 years after graduation.

a) Is this statement of goals or his measure too narrow?
b) Is there any way you can broaden the statement of the goal while still maintaining an operationally defined measure of the effectiveness of the program?

12–23. In an attempt to determine the effectiveness of training in psychotherapy, a team of researchers conducted a study on 3 randomly selected groups of volunteers: one group received psychotherapeutic treatment from trained therapists; one group received treatment from university professors who were highly rated by their students as teachers but who had received no psychotherapeutic training; and a third group received no treatment. After 6 months, the subjects in the 3 groups were scored by measures that comprised several indicators of psychological health. The scores were:

Group A (Psychotherapists)	*Group B (Professors)*	*Group C (Control)*
67	60	47
58	72	59
70	58	43
57	70	58
63	65	53

a) Run a t test to determine whether those treated by the psychotherapists have significantly higher scores than the untreated persons.
b) Run a t test to determine whether those treated by the professors have significantly higher scores than those receiving no treatment.
c) Run a t test to determine whether there is a significant difference in scores of those treated by the psychotherapists and those treated by the professors.

12–24. What is the difference between a comparison group and a true control group? In what ways is the latter superior for doing research? What are some of the reasons why it is a very difficult thing to get a true control group?

12–25. Suppose that you have 4 subjects, whom you wish to assign randomly to 2 groups. Assume the ages of the subjects are 20, 30, 40, and 50.

a) List all 6 ways you could assign the 4 subjects to the 2 groups, and compute the differences between the mean ages of the 2 groups.
b) If you want to control specifically for age, how can you make the assignment?
c) If you believe that age is an important factor affecting the results of your experiment, do you think matching by age is a better method of assigning the individuals than a purely random method?

12–26. Suppose that you have 4 subjects whom you wish to assign randomly to 2 groups. Assume that 2 of the subjects are male and 2 are female.

a) List all 6 ways you could assign the 4 subjects to the 2 groups, and compute the differences in proportion of females between the 2 groups in each case.
b) If you want to control specifically for sex, how can you make the assignment?
c) If you believe that sex is an important factor affecting the results of your experiment, do you think that matching by sex is a better method of assigning the individuals than a purely random method?

12–27. A "cloud seeder" claims that he can increase the monthly rainfall in Horsechester by a program of continuous seeding of clouds. In the first 3 months of his cloud-seeding efforts, monthly rainfall averaged 6.4 centimeters, up from the previous year's mean of 4.2 centimeters. The cloud-seeder claims that his program is a success. Discuss some of the problems in assessing the benefits of the cloud-seeding program. What questions would you want answered, and what other information might you want to have?

12–28. In an attempt to determine the effect of class size on educational achievement, a study looked at the mean class sizes of a random selection of schools throughout a state, and recorded the mean scores of a battery of educational achievement tests of eleventh grade pupils. The results of the study included the regression equation

$$\hat{Y} = 84.5 - 0.73X$$

where $\hat{Y}$ is the predicted score on the battery of tests and X is the mean class size. From the equation, it appears that an increase of 1 in mean class size is associated with a decrease of 0.73 in test score.

Assume, as you realistically would in situations like this, that the assignment to the classes was *not* random. What other variables besides the class size might be affecting the achievement test scores? Do you have reason to suspect that some of these variables might be affecting the size of class that the individuals select? Discuss, with consideration of the eight factors that were listed as affecting internal validity.

12–29. A program has been set up to provide loans to members of minority groups who have gone into business for themselves. It has been proposed that the measure of effectiveness of the program be the total amount of money that is paid back. What are the advantages and disadvantages of using this as a measure of effectiveness? What other measures can you suggest be used? To what extent does the measure of effectiveness depend on the type of organization that is providing the loans?

12–30. Out of a class of 74 tenth grade pupils, 44 volunteered to participate in a physical fitness program developed by the State Board of Health. The 44 volunteers (24 boys; 20 girls) and the 30 who did not volunteer (16 boys; 14 girls) were given several tests, the volunteers being tested both before and after the 3-month program. Of the 44 volunteers, 36 completed the program (18 boys; 18 girls). The results are tabulated below.

a) List 6 or 8 conclusions that are *suggested* by these data. What are some of the threats to internal validity that you must contend with in coming to these conclusions?

	Volunteers (Before)		Volunteers (After)		Nonvolunteers	
	Mean	*Stand. Dev.*	*Mean*	*Stand. Dev.*	*Mean*	*Stand. Dev.*
Boys						
Situps	23.2	2.4	29.6	2.2	13.4	3.8
Chinups	7.8	.97	9.4	1.4	6.5	2.0
60-yd.dash	7.2″	0.8″	7.0″	0.7″	7.1″	0.9″
Girls						
Situps	10.5	2.0	24.1	3.8	4.2	0.5
Chinups	0.6	0.1	3.2	0.3	0.2	0.1
60-yd.dash	7.4″	0.9″	6.9″	0.8″	9.4″	1.2″

b) In drawing conclusions about the effectiveness of the program in promoting physical fitness, what are some of the problems of external validity?

12–31. The following conclusions reflect factors that could affect the internal validity of the experiments. In each case, identify at least one factor other than the treatment or program that could be operating.

a) After 2 years of a diet consisting of large quantities of yogurt and wheat germ, individuals had significantly more gray hair than before.

b) The class was given a pretest of mathematical skills. The 20 students who scored lowest on the pretest were assigned to Mr. Snark's mathematics section; the other students were assigned to Miss Treblecief's section. After the course the students were given a post test. The mean score of Mr. Snark's students improved by 22.4 points; the mean score of Miss Treblecief's students improved by only 14.4 points. The gap between the two groups thus narrowed by 8.0 points.

SUGGESTED READINGS

CAMPBELL, DONALD T., and JULIAN STANLEY. *Experimental and Quasi-Experimental Designs for Research.* Chicago, Ill.: Rand McNally, and Company, 1966.

CLARK, CHARLES T., and LAWRENCE L. SCHKADE. *Statistical Methods for Business Decisions.* Cincinnati, Ohio: Southwestern Publishing Company, 1969. Chapter 15.

EYSENCK, H. J. "The Effects of Psychotherapy: An Evaluation," *Journal of Consulting Psychology, XVI* (1952), 319–24.

FISHER, SIR RONALD A. *The Design of Experiments.* Edinburgh: Oliver and Boyd, 1949.

MENDENHALL, WILLIAM. *Introduction to Linear Models and the Design and Analysis of Experiments.* Belmont, California: Wadsworth Publishing Company, 1968.

OTT, LYMAN. *An Introduction to Statistical Methods and Data Analysis.* North Scituate, Mass.: Duxbury Press, 1977. Chapter 9.

ROSSI, PETER H. "Boobytraps and Pitfalls in the Evaluation of Social Action Programs," *Proceedings of the Social Science Section.* Washington, D.C.: American Statistical Association, 1966. Pp. 127–32.

WEISS, CAROL H. *Evaluation Research.* Englewood Cliffs, N.J.: Prentice-Hall, Inc., 1972.

Appendix

APPENDIX A
STATISTICAL TABLES

Table I. CUMULATIVE BINOMIAL PROBABILITIES

$$F(c) = P(X \leqslant c) = \sum_{x=0}^{c} \binom{n}{x} (1-p)^{n-x} p^x$$

Example: If $p = 0.20$, $n = 7$, $c = 2$, then $F(2) = P(X \leqslant 2) = 0.8520$.

		p									
n	x	0.05	0.10	0.15	0.20	0.25	0.30	0.35	0.40	0.45	0.50
2	0	0.9025	0.8100	0.7225	0.6400	0.5625	0.4900	0.4225	0.3600	0.3025	0.2500
	1	0.9975	0.9900	0.9775	0.9600	0.9375	0.9100	0.8775	0.8400	0.7975	0.7500
3	0	0.8574	0.7290	0.6141	0.5120	0.4219	0.3430	0.2746	0.2160	0.1664	0.1250
	1	0.9928	0.9720	0.9392	0.8960	0.8438	0.7840	0.7182	0.6480	0.5748	0.5000
	2	0.9999	0.9990	0.9966	0.9920	0.9844	0.9730	0.9571	0.9360	0.9089	0.8750
4	0	0.8145	0.6561	0.5220	0.4096	0.3164	0.2401	0.1785	0.1296	0.0915	0.0625
	1	0.9860	0.9477	0.8905	0.8192	0.7383	0.6517	0.5630	0.4752	0.3910	0.3125
	2	0.9995	0.9963	0.9880	0.9728	0.9492	0.9163	0.8735	0.8208	0.7585	0.6875
	3	1.0000	0.9999	0.9995	0.9984	0.9961	0.9919	0.9850	0.9744	0.9590	0.9375
5	0	0.7738	0.5905	0.4437	0.3277	0.2373	0.1681	0.1160	0.0778	0.0503	0.0312
	1	0.9774	0.9185	0.8352	0.7373	0.6328	0.5282	0.4284	0.3370	0.2562	0.1875
	2	0.9988	0.9914	0.9734	0.9421	0.8965	0.8369	0.7648	0.6826	0.5931	0.5000
	3	1.0000	0.9995	0.9978	0.9933	0.9844	0.9692	0.9460	0.9130	0.8688	0.8125
	4	1.0000	1.0000	0.9999	0.9997	0.9990	0.9976	0.9947	0.9898	0.9815	0.9688
6	0	0.7351	0.5314	0.3771	0.2621	0.1780	0.1176	0.0754	0.0467	0.0277	0.0156
	1	0.9672	0.8857	0.7765	0.6554	0.5339	0.4202	0.3191	0.2333	0.1636	0.1094
	2	0.9978	0.9842	0.9527	0.9011	0.8306	0.7443	0.6471	0.5443	0.4415	0.3438
	3	0.9999	0.9987	0.9941	0.9830	0.9624	0.9295	0.8826	0.8208	0.7447	0.6562
	4	1.0000	0.9999	0.9996	0.9984	0.9954	0.9891	0.9777	0.9590	0.9308	0.8906
	5	1.0000	1.0000	1.0000	0.9999	0.9998	0.9993	0.9982	0.9959	0.9917	0.9844
7	0	0.6983	0.4783	0.3206	0.2097	0.1335	0.0824	0.0490	0.0280	0.0152	0.0078
	1	0.9556	0.8503	0.7166	0.5767	0.4449	0.3294	0.2338	0.1586	0.1024	0.0625
	2	0.9962	0.9743	0.9262	0.8520	0.7564	0.6471	0.5323	0.4199	0.3164	0.2266
	3	0.9998	0.9973	0.9879	0.9667	0.9294	0.8740	0.8002	0.7102	0.6083	0.5000
	4	1.0000	0.9998	0.9988	0.9953	0.9871	0.9712	0.9444	0.9037	0.8471	0.7734
	5	1.0000	1.0000	0.9999	0.9996	0.9987	0.9962	0.9910	0.9812	0.9643	0.9375
	6	1.0000	1.0000	1.0000	1.0000	0.9999	0.9998	0.9994	0.9984	0.9963	0.9922

Source: From Irwin Miller and John E. Freund, *Probability and Statistics for Engineers*, © 1965 by Prentice-Hall, Inc.

Table I *Continued*

n	x	p = 0.05	0.10	0.15	0.20	0.25	0.30	0.35	0.40	0.45	0.50
8	0	0.6634	0.4305	0.2725	0.1678	0.1001	0.0576	0.0319	0.0168	0.0084	0.0039
	1	0.9428	0.8131	0.6572	0.5033	0.3671	0.2553	0.1691	0.1064	0.0632	0.0352
	2	0.9942	0.9619	0.8948	0.7969	0.6785	0.5518	0.4278	0.3154	0.2201	0.1445
	3	0.9996	0.9950	0.9786	0.9437	0.8862	0.8059	0.7064	0.5941	0.4770	0.3633
	4	1.0000	0.9996	0.9971	0.9896	0.9727	0.9420	0.8939	0.8263	0.7396	0.6367
	5	1.0000	1.0000	0.9998	0.9988	0.9958	0.9887	0.9747	0.9502	0.9115	0.8555
	6	1.0000	1.0000	1.0000	0.9999	0.9996	0.9987	0.9964	0.9915	0.9819	0.9648
	7	1.0000	1.0000	1.0000	1.0000	1.0000	0.9999	0.9998	0.9993	0.9983	0.9961
9	0	0.6302	0.3874	0.2316	0.1342	0.0751	0.0404	0.0207	0.0101	0.0046	0.0020
	1	0.9288	0.7748	0.5995	0.4362	0.3003	0.1960	0.1211	0.0705	0.0385	0.0195
	2	0.9916	0.9470	0.8591	0.7382	0.6007	0.4628	0.3373	0.2318	0.1495	0.0898
	3	0.9994	0.9917	0.9661	0.9144	0.8343	0.7297	0.6089	0.4826	0.3614	0.2539
	4	1.0000	0.9991	0.9944	0.9804	0.9511	0.9012	0.8283	0.7334	0.6214	0.5000
	5	1.0000	0.9999	0.9994	0.9969	0.9900	0.9747	0.9464	0.9006	0.8342	0.7461
	6	1.0000	1.0000	1.0000	0.9997	0.9987	0.9957	0.9888	0.9750	0.9502	0.9102
	7	1.0000	1.0000	1.0000	1.0000	0.9996	0.9996	0.9986	0.9962	0.9909	0.9805
	8	1.0000	1.0000	1.0000	1.0000	1.0000	1.0000	0.9999	0.9997	0.9992	0.9980
10	0	0.5987	0.3487	0.1969	0.1074	0.0563	0.0282	0.0135	0.0060	0.0025	0.0010
	1	0.9139	0.7361	0.5443	0.3758	0.2440	0.1493	0.0860	0.0464	0.0232	0.0107
	2	0.9885	0.9298	0.8202	0.6778	0.5256	0.3828	0.2616	0.1673	0.0996	0.0547
	3	0.9990	0.9872	0.9500	0.8791	0.7759	0.6496	0.5138	0.3823	0.2660	0.1719
	4	0.9999	0.9984	0.9901	0.9672	0.9219	0.8497	0.7515	0.6331	0.5044	0.3770
	5	1.0000	0.9999	0.9986	0.9936	0.9803	0.9527	0.9051	0.8338	0.7384	0.6230
	6	1.0000	1.0000	0.9999	0.9991	0.9965	0.9894	0.9740	0.9452	0.8980	0.8281
	7	1.0000	1.0000	1.0000	0.9999	0.9996	0.9984	0.9952	0.9877	0.9726	0.9453
	8	1.0000	1.0000	1.0000	1.0000	1.0000	0.9999	0.9995	0.9983	0.9955	0.9893
	9	1.0000	1.0000	1.0000	1.0000	1.0000	1.0000	1.0000	0.9999	0.9997	0.9990

11	0	0.5688	0.3138	0.1673	0.0859	0.0422	0.0198	0.0088	0.0036	0.0014	0.0005
	1	0.8981	0.6974	0.4922	0.3221	0.1971	0.1130	0.0606	0.0302	0.0139	0.0059
	2	0.9848	0.9104	0.7788	0.6174	0.4552	0.3127	0.2001	0.1189	0.0652	0.0327
	3	0.9984	0.9815	0.9306	0.8389	0.7133	0.5696	0.4256	0.2963	0.1911	0.1133
	4	0.9999	0.9972	0.9841	0.9496	0.8854	0.7897	0.6683	0.5328	0.3971	0.2744
	5	1.0000	0.9997	0.9973	0.9883	0.9657	0.9218	0.8513	0.7535	0.6331	0.5000
	6	1.0000	1.0000	0.9997	0.9980	0.9924	0.9784	0.9499	0.9006	0.8262	0.7256
	7	1.0000	1.0000	1.0000	0.9998	0.9988	0.9957	0.9878	0.9707	0.9390	0.8867
	8	1.0000	1.0000	1.0000	1.0000	0.9999	0.9994	0.9980	0.9941	0.9852	0.9673
	9	1.0000	1.0000	1.0000	1.0000	1.0000	1.0000	0.9998	0.9993	0.9978	0.9941
	10	1.0000	1.0000	1.0000	1.0000	1.0000	1.0000	1.0000	1.0000	0.9998	0.9995
12	0	0.5404	0.2824	0.1422	0.0687	0.0317	0.0138	0.0057	0.0022	0.0008	0.0002
	1	0.8816	0.6590	0.4435	0.2749	0.1584	0.0850	0.0424	0.0196	0.0083	0.0032
	2	0.9804	0.8891	0.7358	0.5583	0.3907	0.2528	0.1513	0.0834	0.0421	0.0193
	3	0.9978	0.9744	0.9078	0.7946	0.6488	0.4925	0.3467	0.2253	0.1345	0.0730
	4	0.9998	0.9957	0.9761	0.9274	0.8424	0.7237	0.5833	0.4382	0.3044	0.1938
	5	1.0000	0.9995	0.9954	0.9806	0.9456	0.8822	0.7873	0.6652	0.5269	0.3872
	6	1.0000	0.9999	0.9993	0.9961	0.9857	0.9614	0.9154	0.8418	0.7393	0.6128
	7	1.0000	1.0000	0.9999	0.9994	0.9972	0.9905	0.9745	0.9427	0.8883	0.8062
	8	1.0000	1.0000	1.0000	0.9999	0.9996	0.9983	0.9944	0.9847	0.9644	0.9270
	9	1.0000	1.0000	1.0000	1.0000	1.0000	0.9998	0.9992	0.9972	0.9921	0.9807
	10	1.0000	1.0000	1.0000	1.0000	1.0000	1.0000	0.9999	0.9997	0.9989	0.9968
	11	1.0000	1.0000	1.0000	1.0000	1.0000	1.0000	1.0000	1.0000	0.9999	0.9998
13	0	0.5133	0.2542	0.1209	0.0550	0.0238	0.0097	0.0037	0.0013	0.0004	0.0001
	1	0.8646	0.6213	0.3983	0.2336	0.1267	0.0637	0.0296	0.0126	0.0049	0.0017
	2	0.9755	0.8661	0.6920	0.5017	0.3326	0.2025	0.1132	0.0579	0.0269	0.0112
	3	0.9969	0.9658	0.8820	0.7473	0.5843	0.4206	0.2783	0.1686	0.0929	0.0461
	4	0.9997	0.9935	0.9658	0.9009	0.7940	0.6543	0.5005	0.3530	0.2279	0.1334
	5	1.0000	0.9991	0.9925	0.9700	0.9198	0.8346	0.7159	0.5744	0.4268	0.2905
	6	1.0000	0.9999	0.9987	0.9930	0.9757	0.9376	0.8705	0.7712	0.6437	0.5000
	7	1.0000	1.0000	0.9998	0.9988	0.9944	0.9818	0.9538	0.9023	0.8212	0.7095
	8	1.0000	1.0000	1.0000	0.9998	0.9990	0.9960	0.9874	0.9679	0.9302	0.8666
	9	1.0000	1.0000	1.0000	1.0000	0.9999	0.9993	0.9975	0.9922	0.9797	0.9539

Table I *Continued*

n	x	p = 0.05	0.10	0.15	0.20	0.25	0.30	0.35	0.40	0.45	0.50
13	10	1.0000	1.0000	1.0000	1.0000	1.0000	0.9999	0.9997	0.9987	0.9959	0.9888
	11	1.0000	1.0000	1.0000	1.0000	1.0000	1.0000	1.0000	0.9999	0.9995	0.9983
	12	1.0000	1.0000	1.0000	1.0000	1.0000	1.0000	1.0000	1.0000	1.0000	0.9999
14	0	0.4877	0.2288	0.1028	0.0440	0.0178	0.0068	0.0024	0.0008	0.0002	0.0001
	1	0.8470	0.5846	0.3567	0.1979	0.1010	0.0475	0.0205	0.0081	0.0029	0.0009
	2	0.9699	0.8416	0.6479	0.4481	0.2811	0.1608	0.0839	0.0398	0.0170	0.0063
	3	0.9958	0.9559	0.8535	0.6982	0.5213	0.3552	0.2205	0.1243	0.0632	0.0287
	4	0.9996	0.9908	0.9533	0.8702	0.7415	0.5842	0.4227	0.2793	0.1672	0.0873
	5	1.0000	0.9985	0.9885	0.9561	0.8883	0.7805	0.6405	0.4859	0.3373	0.2120
	6	1.0000	0.9998	0.9978	0.9884	0.9617	0.9067	0.8164	0.6925	0.5461	0.3953
	7	1.0000	1.0000	0.9997	0.9976	0.9897	0.9685	0.9247	0.8499	0.7414	0.6047
	8	1.0000	1.0000	1.0000	0.9996	0.9978	0.9917	0.9757	0.9417	0.8811	0.7880
	9	1.0000	1.0000	1.0000	1.0000	0.9997	0.9983	0.9940	0.9825	0.9574	0.9102
	10	1.0000	1.0000	1.0000	1.0000	1.0000	0.9998	0.9989	0.9961	0.9886	0.9713
	11	1.0000	1.0000	1.0000	1.0000	1.0000	1.0000	0.9999	0.9994	0.9978	0.9935
	12	1.0000	1.0000	1.0000	1.0000	1.0000	1.0000	1.0000	0.9999	0.9997	0.9991
	13	1.0000	1.0000	1.0000	1.0000	1.0000	1.0000	1.0000	1.0000	1.0000	0.9999
15	0	0.4633	0.2059	0.0874	0.0352	0.0134	0.0047	0.0016	0.0005	0.0001	0.0000
	1	0.8290	0.5490	0.3186	0.1671	0.0802	0.0353	0.0142	0.0052	0.0017	0.0005
	2	0.9638	0.8159	0.6042	0.3980	0.2361	0.1268	0.0617	0.0271	0.0107	0.0037
	3	0.9945	0.9444	0.8227	0.6482	0.4613	0.2969	0.1727	0.0905	0.0424	0.0176
	4	0.9994	0.9873	0.9383	0.8358	0.6865	0.5155	0.3519	0.2173	0.1204	0.0592
	5	0.9999	0.9978	0.9832	0.9389	0.8516	0.7216	0.5643	0.4032	0.2608	0.1509
	6	1.0000	0.9997	0.9964	0.9819	0.9434	0.8689	0.7548	0.6098	0.4522	0.3036
	7	1.0000	1.0000	0.9996	0.9958	0.9827	0.9500	0.8868	0.7869	0.6535	0.5000
	8	1.0000	1.0000	0.9999	0.9992	0.9958	0.9848	0.9578	0.9050	0.8182	0.6964
	9	1.0000	1.0000	1.0000	0.9999	0.9992	0.9963	0.9876	0.9662	0.9231	0.8491

	10	1.0000	1.0000	1.0000	1.0000	0.9999	0.9993	0.9972	0.9907	0.9745	0.9408
	11	1.0000	1.0000	1.0000	1.0000	1.0000	0.9999	0.9995	0.9981	0.9937	0.9824
	12	1.0000	1.0000	1.0000	1.0000	1.0000	1.0000	0.9999	0.9997	0.9989	0.9963
	13	1.0000	1.0000	1.0000	1.0000	1.0000	1.0000	1.0000	1.0000	0.9999	0.9995
	14	1.0000	1.0000	1.0000	1.0000	1.0000	1.0000	1.0000	1.0000	1.0000	1.0000
16	0	0.4401	0.1853	0.0743	0.0281	0.0100	0.0033	0.0010	0.0003	0.0001	0.0000
	1	0.8108	0.5147	0.2839	0.1407	0.0635	0.0261	0.0098	0.0033	0.0010	0.0003
	2	0.9571	0.7892	0.5614	0.3518	0.1971	0.0994	0.0451	0.0183	0.0066	0.0021
	3	0.9930	0.9316	0.7899	0.5981	0.4050	0.2459	0.1339	0.0651	0.0281	0.0106
	4	0.9991	0.9830	0.9209	0.7982	0.6302	0.4499	0.2892	0.1666	0.0853	0.0384
	5	0.9999	0.9967	0.9765	0.9183	0.8103	0.6598	0.4900	0.3288	0.1976	0.1051
	6	1.0000	0.9995	0.9944	0.9733	0.9204	0.8247	0.6881	0.5272	0.3660	0.2272
	7	1.0000	0.9999	0.9989	0.9930	0.9729	0.9256	0.8406	0.7161	0.5629	0.4018
	8	1.0000	1.0000	0.9998	0.9985	0.9925	0.9743	0.9329	0.8577	0.7441	0.5982
	9	1.0000	1.0000	1.0000	0.9998	0.9984	0.9929	0.9771	0.9417	0.8759	0.7728
	10	1.0000	1.0000	1.0000	1.0000	0.9997	0.9984	0.9938	0.9809	0.9514	0.8949
	11	1.0000	1.0000	1.0000	1.0000	1.0000	0.9997	0.9987	0.9951	0.9851	0.9616
	12	1.0000	1.0000	1.0000	1.0000	1.0000	1.0000	0.9998	0.9991	0.9965	0.9894
	13	1.0000	1.0000	1.0000	1.0000	1.0000	1.0000	1.0000	0.9999	0.9994	0.9979
	14	1.0000	1.0000	1.0000	1.0000	1.0000	1.0000	1.0000	1.0000	1.0000	0.9997
	15	1.0000	1.0000	1.0000	1.0000	1.0000	1.0000	1.0000	1.0000	1.0000	1.0000
17	0	0.4181	0.1668	0.0631	0.0225	0.0075	0.0023	0.0007	0.0002	0.0000	0.0000
	1	0.7922	0.4818	0.2525	0.1182	0.0501	0.0193	0.0067	0.0021	0.0006	0.0001
	2	0.9497	0.7618	0.5198	0.3096	0.1637	0.0774	0.0327	0.0123	0.0041	0.0012
	3	0.9912	0.9174	0.7556	0.5489	0.3530	0.2019	0.1028	0.0464	0.0184	0.0064
	4	0.9988	0.9779	0.9013	0.7582	0.5739	0.3887	0.2348	0.1260	0.0596	0.0245
	5	0.9999	0.9953	0.9681	0.8943	0.7653	0.5968	0.4197	0.2639	0.1471	0.0717
	6	1.0000	0.9992	0.9917	0.9623	0.8929	0.7752	0.6188	0.4478	0.2902	0.1662
	7	1.0000	0.9999	0.9983	0.9891	0.9598	0.8954	0.7872	0.6405	0.4743	0.3145
	8	1.0000	1.0000	0.9997	0.9974	0.9876	0.9597	0.9006	0.8011	0.6626	0.5000
	9	1.0000	1.0000	1.0000	0.9995	0.9969	0.9873	0.9617	0.9081	0.8166	0.6855

Table I *Continued*

n	x	0.05	0.10	0.15	0.20	0.25	0.30	0.35	0.40	0.45	0.50
						p					
17	10	1.0000	1.0000	1.0000	0.9999	0.9994	0.9968	0.9880	0.9652	0.9174	0.8338
	11	1.0000	1.0000	1.0000	1.0000	0.9999	0.9993	0.9970	0.9894	0.9699	0.9283
	12	1.0000	1.0000	1.0000	1.0000	1.0000	0.9999	0.9994	0.9975	0.9914	0.9755
	13	1.0000	1.0000	1.0000	1.0000	1.0000	1.0000	0.9999	0.9995	0.9981	0.9936
	14	1.0000	1.0000	1.0000	1.0000	1.0000	1.0000	1.0000	0.9999	0.9997	0.9988
	15	1.0000	1.0000	1.0000	1.0000	1.0000	1.0000	1.0000	1.0000	1.0000	0.9999
	16	1.0000	1.0000	1.0000	1.0000	1.0000	1.0000	1.0000	1.0000	1.0000	1.0000
18	0	0.3972	0.1501	0.0536	0.0180	0.0056	0.0016	0.0004	0.0001	0.0000	0.0000
	1	0.7735	0.4503	0.2241	0.0991	0.0395	0.0142	0.0046	0.0013	0.0003	0.0001
	2	0.9419	0.7338	0.4797	0.2713	0.1353	0.0600	0.0236	0.0082	0.0025	0.0007
	3	0.9891	0.9018	0.7202	0.5010	0.3057	0.1646	0.0783	0.0328	0.0120	0.0038
	4	0.9985	0.9718	0.8794	0.7164	0.5187	0.3327	0.1886	0.0942	0.0411	0.0154
	5	0.9998	0.9936	0.9581	0.8671	0.7175	0.5344	0.3550	0.2088	0.1077	0.0481
	6	1.0000	0.9988	0.9882	0.9487	0.8610	0.7217	0.5491	0.3743	0.2258	0.1189
	7	1.0000	0.9998	0.9973	0.9837	0.9431	0.8593	0.7283	0.5634	0.3915	0.2403
	8	1.0000	1.0000	0.9995	0.9957	0.9807	0.9404	0.8609	0.7368	0.5778	0.4073
	9	1.0000	1.0000	0.9999	0.9991	0.9946	0.9790	0.9403	0.8653	0.7473	0.5927
	10	1.0000	1.0000	1.0000	0.9998	0.9988	0.9939	0.9788	0.9424	0.8720	0.7597
	11	1.0000	1.0000	1.0000	1.0000	0.9998	0.9986	0.9938	0.9797	0.9463	0.8811
	12	1.0000	1.0000	1.0000	1.0000	1.0000	0.9997	0.9986	0.9942	0.9817	0.9519
	13	1.0000	1.0000	1.0000	1.0000	1.0000	1.0000	0.9997	0.9987	0.9951	0.9846
	14	1.0000	1.0000	1.0000	1.0000	1.0000	1.0000	1.0000	0.9998	0.9990	0.9962
	15	1.0000	1.0000	1.0000	1.0000	1.0000	1.0000	1.0000	1.0000	0.9999	0.9993
	16	1.0000	1.0000	1.0000	1.0000	1.0000	1.0000	1.0000	1.0000	1.0000	0.9999
19	0	0.3774	0.1351	0.0456	0.0144	0.0042	0.0011	0.0003	0.0001	0.0000	0.0000
	1	0.7547	0.4203	0.1985	0.0829	0.0310	0.0104	0.0031	0.0008	0.0002	0.0000
	2	0.9335	0.7054	0.4413	0.2369	0.1113	0.0462	0.0170	0.0055	0.0015	0.0004
	3	0.9868	0.8850	0.6841	0.4551	0.2630	0.1332	0.0591	0.0230	0.0077	0.0022
	4	0.9980	0.9648	0.8556	0.6733	0.4654	0.2822	0.1500	0.0696	0.0280	0.0096

	5	0.9998	0.9914	0.9463	0.8369	0.6678	0.4739	0.2968	0.1629	0.0777	0.0318
	6	1.0000	0.9983	0.9837	0.9324	0.8251	0.6655	0.4812	0.3081	0.1727	0.0835
	7	1.0000	0.9997	0.9959	0.9767	0.9225	0.8180	0.6656	0.4878	0.3169	0.1796
	8	1.0000	1.0000	0.9992	0.9933	0.9713	0.9161	0.8145	0.6675	0.4940	0.3238
	9	1.0000	1.0000	0.9999	0.9984	0.9911	0.9674	0.9125	0.8139	0.6710	0.5000
	10	1.0000	1.0000	1.0000	0.9997	0.9977	0.9895	0.9653	0.9115	0.8159	0.6762
	11	1.0000	1.0000	1.0000	1.0000	0.9995	0.9972	0.9886	0.9648	0.9129	0.8204
	12	1.0000	1.0000	1.0000	1.0000	0.9999	0.9994	0.9969	0.9884	0.9658	0.9165
	13	1.0000	1.0000	1.0000	1.0000	1.0000	0.9999	0.9993	0.9969	0.9891	0.9682
	14	1.0000	1.0000	1.0000	1.0000	1.0000	1.0000	0.9999	0.9994	0.9972	0.9904
	15	1.0000	1.0000	1.0000	1.0000	1.0000	1.0000	1.0000	0.9999	0.9995	0.9978
	16	1.0000	1.0000	1.0000	1.0000	1.0000	1.0000	1.0000	1.0000	0.9999	0.9996
	17	1.0000	1.0000	1.0000	1.0000	1.0000	1.0000	1.0000	1.0000	1.0000	1.0000
20	0	0.3585	0.1216	0.0388	0.0115	0.0032	0.0008	0.0002	0.0000	0.0000	0.0000
	1	0.7358	0.3917	0.1756	0.0692	0.0243	0.0076	0.0021	0.0005	0.0001	0.0000
	2	0.9245	0.6769	0.4049	0.2061	0.0913	0.0355	0.0121	0.0036	0.0009	0.0002
	3	0.9811	0.8670	0.6477	0.4114	0.2252	0.1071	0.0444	0.0160	0.0049	0.0013
	4	0.9974	0.9568	0.8298	0.6296	0.4148	0.2375	0.1182	0.0510	0.0189	0.0059
	5	0.9997	0.9887	0.9327	0.8042	0.6172	0.4164	0.2454	0.1256	0.0553	0.0207
	6	1.0000	0.9976	0.9781	0.9133	0.7858	0.6080	0.4166	0.2500	0.1299	0.0577
	7	1.0000	0.9996	0.9941	0.9679	0.8982	0.7723	0.6010	0.4159	0.2520	0.1316
	8	1.0000	0.9999	0.9987	0.9900	0.9591	0.8867	0.7624	0.5956	0.4143	0.2517
	9	1.0000	1.0000	0.9998	0.9974	0.9861	0.9520	0.8782	0.7553	0.5914	0.4119
	10	1.0000	1.0000	1.0000	0.9994	0.9961	0.9829	0.9468	0.8725	0.7507	0.5881
	11	1.0000	1.0000	1.0000	0.9999	0.9991	0.9949	0.9804	0.9435	0.8692	0.7483
	12	1.0000	1.0000	1.0000	1.0000	0.9998	0.9987	0.9940	0.9790	0.9420	0.8684
	13	1.0000	1.0000	1.0000	1.0000	1.0000	0.9997	0.9985	0.9935	0.9786	0.9423
	14	1.0000	1.0000	1.0000	1.0000	1.0000	1.0000	0.9997	0.9984	0.9936	0.9793
	15	1.0000	1.0000	1.0000	1.0000	1.0000	1.0000	1.0000	0.9997	0.9985	0.9941
	16	1.0000	1.0000	1.0000	1.0000	1.0000	1.0000	1.0000	1.0000	0.9997	0.9987
	17	1.0000	1.0000	1.0000	1.0000	1.0000	1.0000	1.0000	1.0000	1.0000	0.9998
	18	1.0000	1.0000	1.0000	1.0000	1.0000	1.0000	1.0000	1.0000	1.0000	1.0000

Table II. *AREAS UNDER THE STANDARD NORMAL CURVE*

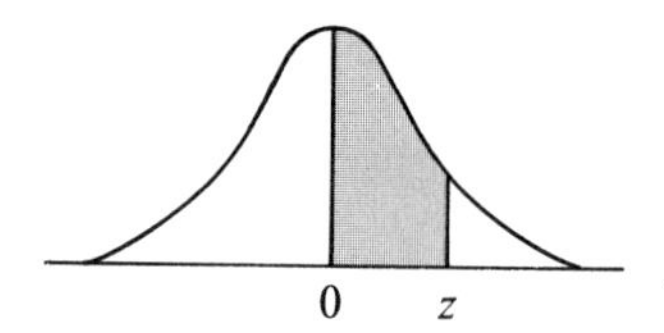

Probabilities that a random variable having the standard normal probability distribution assumes a value between 0 and z.

z	.00	.01	.02	.03	.04	.05	.06	.07	.08	.09
0.0	.0000	.0040	.0080	.0120	.0160	.0199	.0239	.0279	.0319	.0359
0.1	.0398	.0438	.0478	.0517	.0557	.0596	.0636	.0675	.0714	.0753
0.2	.0793	.0832	.0871	.0910	.0948	.0987	.1026	.1064	.1103	.1141
0.3	.1179	.1217	.1255	.1293	.1331	.1368	.1406	.1443	.1480	.1517
0.4	.1554	.1591	.1628	.1664	.1700	.1736	.1772	.1808	.1844	.1879
0.5	.1915	.1950	.1985	.2019	.2054	.2088	.2123	.2157	.2190	.2224
0.6	.2257	.2291	.2324	.2357	.2389	.2422	.2454	.2486	.2517	.2549
0.7	.2580	.2611	.2642	.2673	.2704	.2734	.2764	.2794	.2823	.2852
0.8	.2881	.2910	.2939	.2967	.2995	.3023	.3051	.3078	.3106	.3133
0.9	.3159	.3186	.3212	.3238	.3264	.3289	.3315	.3340	.3365	.3389
1.0	.3413	.3438	.3461	.3485	.3508	.3531	.3554	.3577	.3599	.3621
1.1	.3643	.3665	.3686	.3708	.3729	.3749	.3770	.3790	.3810	.3830
1.2	.3849	.3869	.3888	.3907	.3925	.3944	.3962	.3980	.3997	.4015
1.3	.4032	.4049	.4066	.4082	.4099	.4115	.4131	.4147	.4162	.4177
1.4	.4192	.4207	.4222	.4236	.4251	.4265	.4279	.4292	.4306	.4319
1.5	.4332	.4345	.4357	.4370	.4382	.4394	.4406	.4418	.4429	.4441
1.6	.4452	.4463	.4474	.4484	.4495	.4505	.4515	.4525	.4535	.4545
1.7	.4554	.4564	.4573	.4582	.4591	.4599	.4608	.4616	.4625	.4633
1.8	.4641	.4649	.4656	.4664	.4671	.4678	.4686	.4693	.4699	.4706
1.9	.4713	.4719	.4726	.4732	.4738	.4744	.4750	.4756	.4761	.4767
2.0	.4772	.4778	.4783	.4788	.4793	.4798	.4803	.4808	.4812	.4817
2.1	.4821	.4826	.4830	.4834	.4838	.4842	.4846	.4850	.4854	.4857
2.2	.4861	.4864	.4868	.4871	.4875	.4878	.4881	.4884	.4887	.4890
2.3	.4893	.4896	.4898	.4901	.4904	.4906	.4909	.4911	.4913	.4916
2.4	.4918	.4920	.4922	.4925	.4927	.4929	.4931	.4932	.4934	.4936
2.5	.4938	.4940	.4941	.4943	.4945	.4946	.4948	.4949	.4951	.4952
2.6	.4953	.4955	.4956	.4957	.4959	.4960	.4961	.4962	.4963	.4964
2.7	.4965	.4966	.4967	.4968	.4969	.4970	.4971	.4972	.4973	.4974
2.8	.4974	.4975	.4976	.4977	.4977	.4978	.4979	.4979	.4980	.4981
2.9	.4981	.4982	.4982	.4983	.4984	.4984	.4985	.4985	.4986	.4986
3.0	.4987	.4987	.4987	.4988	.4988	.4989	.4989	.4989	.4990	.4990

Source: Adapted by William Mendenhall and Lyman Ott, *Understanding Statistics*, 2nd ed. (North Scituate, Mass.: Duxbury Press, 1976), p. 348, from Table II of A. Hald, *Statistical Tables and Formulas*.

Table III. *CRITICAL VALUES OF STUDENT'S t DISTRIBUTION*

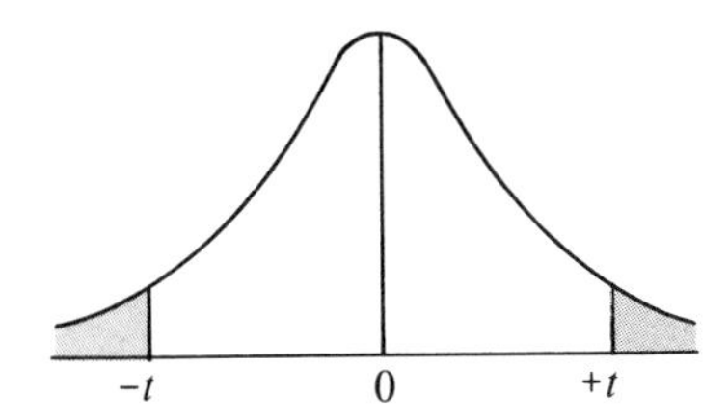

Example: For 15 degrees of freedom, the t-value which corresponds to an area of 0.05 in both tails combined is 2.131.

Degrees of Freedom	Area in Both Tails Combined			
	0.10	0.05	0.02	0.01
1	6.314	12.706	31.821	63.657
2	2.920	4.303	6.965	9.925
3	2.353	3.182	4.541	5.841
4	2.132	2.776	3.747	4.604
5	2.015	2.571	3.365	4.032
6	1.943	2.447	3.143	3.707
7	1.895	2.365	2.998	3.499
8	1.860	2.306	2.896	3.355
9	1.833	2.262	2.821	3.250
10	1.812	2.228	2.764	3.169
11	1.796	2.201	2.718	3.106
12	1.782	2.179	2.681	3.055
13	1.771	2.160	2.650	3.012
14	1.761	2.145	2.624	2.977
15	1.753	2.131	2.602	2.947
16	1.746	2.120	2.583	2.921
17	1.740	2.110	2.567	2.898
18	1.734	2.101	2.552	2.878
19	1.729	2.093	2.539	2.861
20	1.725	2.086	2.528	2.845
21	1.721	2.080	2.518	2.831
22	1.717	2.074	2.508	2.819
23	1.714	2.069	2.500	2.807
24	1.711	2.064	2.492	2.797
25	1.708	2.060	2.485	2.787
26	1.706	2.056	2.479	2.779
27	1.703	2.052	2.473	2.771
28	1.701	2.048	2.467	2.763
29	1.699	2.045	2.462	2.756
30	1.697	2.042	2.457	2.750
40	1.684	2.021	2.423	2.704
60	1.671	2.000	2.390	2.660
120	1.658	1.980	2.358	2.617
Normal Distribution	1.645	1.960	2.326	2.576

Table IV. *CRITICAL VALUES OF THE χ^2 DISTRIBUTION*

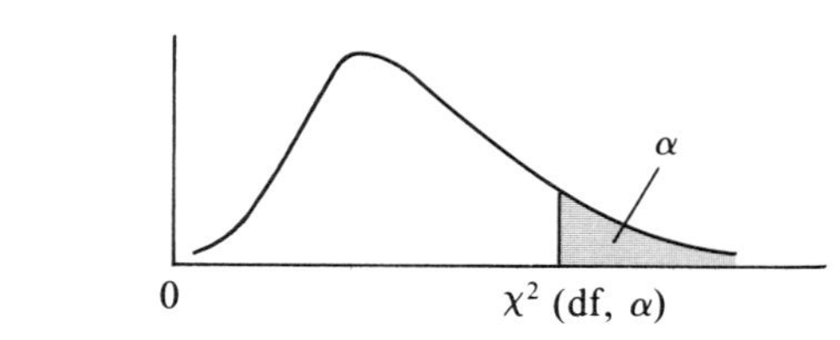

The entries in this table are the critical values for chi square for which the area to the right under the curve is equal to α.

df	α									
	0.995	0.990	0.975	0.950	0.900	0.100	0.050	0.025	0.010	0.005
1	0.0000393	0.000157	0.000982	0.00393	0.0158	2.71	3.84	5.02	6.64	7.88
2	0.0100	0.0201	0.0506	0.103	0.211	4.61	6.00	7.38	9.21	10.6
3	0.0717	0.115	0.216	0.352	0.584	6.25	7.82	9.35	11.4	12.9
4	0.207	0.297	0.484	0.711	1.0636	7.78	9.50	11.1	13.3	14.9
5	0.412	0.554	0.831	1.15	1.61	9.24	11.1	12.8	15.1	16.8
6	0.676	0.872	1.24	1.64	2.20	10.6	12.6	14.5	16.8	18.6
7	0.990	1.24	1.69	2.17	2.83	12.0	14.1	16.0	18.5	20.3
8	1.34	1.65	2.18	2.73	3.49	13.4	15.5	17.5	20.1	22.0
9	1.73	2.09	2.70	3.33	4.17	14.7	17.0	19.0	21.7	23.6
10	2.16	2.56	3.25	3.94	4.87	16.0	18.3	20.5	23.2	25.2
11	2.60	3.05	3.82	4.58	5.58	17.2	19.7	21.9	24.7	26.8
12	3.07	3.57	4.40	5.23	6.30	18.6	21.0	23.3	26.2	28.3

13	3.57	4.11	5.01	5.90	7.04	19.8	22.4	24.7	27.7	29.8
14	4.07	4.66	5.63	6.57	7.79	21.1	23.7	26.1	29.1	31.3
15	4.60	5.23	6.26	7.26	8.55	22.3	25.0	27.5	30.6	32.8
16	5.14	5.81	6.91	7.96	9.31	23.5	26.3	28.9	32.0	34.3
17	5.70	6.41	7.56	8.67	10.1	24.8	27.6	30.2	33.4	35.7
18	6.26	7.01	8.23	9.39	10.9	26.0	28.9	31.5	34.8	37.2
19	6.84	7.63	8.91	10.1	11.7	27.2	30.1	32.9	36.2	38.6
20	7.43	8.26	9.59	10.9	12.4	28.4	31.4	34.2	37.6	40.0
21	8.03	8.90	10.3	11.6	13.2	29.6	32.7	35.5	39.0	41.4
22	8.64	9.54	11.0	12.3	14.0	30.8	33.9	36.8	40.3	42.8
23	9.26	10.2	11.0	13.1	14.9	32.0	35.2	38.1	41.6	44.2
24	9.89	10.9	12.4	13.9	15.7	33.2	36.4	39.4	43.0	45.6
25	10.5	11.5	13.1	14.6	16.5	34.4	37.7	40.7	44.3	46.9
26	11.2	12.2	13.8	15.4	17.3	35.6	38.9	41.9	45.6	48.3
27	11.8	12.9	14.6	16.2	18.1	36.7	40.1	43.2	47.0	49.7
28	12.5	13.6	15.3	16.9	18.9	37.9	41.3	44.5	48.3	51.0
29	13.1	14.3	16.1	17.7	19.8	39.1	42.6	45.7	49.6	52.3
30	13.8	15.0	16.8	18.5	20.6	40.3	43.8	47.0	50.9	53.7
40	20.7	22.2	24.4	26.5	29.1	51.8	55.8	59.3	63.7	66.8
50	28.0	29.7	32.4	34.8	37.7	63.2	67.5	71.4	76.2	79.5
60	35.5	37.5	40.5	43.2	46.5	74.4	79.1	83.3	88.4	92.0
70	43.3	45.4	48.8	51.8	55.3	85.5	90.5	95.0	100.0	104.0
80	51.2	53.5	57.2	60.4	64.3	96.6	102.0	107.0	112.0	116.0
90	59.2	61.8	65.7	69.1	73.3	108.0	113.0	118.0	124.0	128.0
100	67.3	70.1	74.2	77.9	82.4	114.0	124.0	130.0	136.0	140.0

Source: Adapted by Robert Johnson, *Elementary Statistics* (North Scituate, Mass.: Duxbury Press, 1973) from E. S. Pearson and H. O. Hartley, *Biometrika Tables for Statisticians*, Vol. 1 (1962), pp. 130–131.

Table V. *RANDOM NUMBERS*

09188	20097	32825	39527	04220	86304	83389	87374	64278	58044
90045	85497	51981	50654	94938	81997	91870	76150	68476	64659
73189	50207	47677	26269	62290	64464	27124	67018	41361	82760
75768	76490	20971	87749	90429	12272	95375	05871	93823	43178
54016	44056	66281	31003	00682	27398	20714	53295	07706	17813
08358	69910	78542	42785	13661	58873	04618	97553	31223	08420
28306	03264	81333	10591	40510	07893	32604	60475	94119	01840
53840	86233	81594	13628	51215	90290	28466	68795	77762	20791
91757	53741	61613	62269	50263	90212	55781	76514	83483	47055
89415	92694	00397	58391	12607	17646	48949	72306	94541	37408
77513	03820	86864	29901	68414	82774	51908	13980	72893	55507
19502	37174	69979	20288	55210	29773	74287	75251	65344	67415
21818	59313	93278	81757	05686	73156	07082	85046	31853	38452
51474	66499	68107	23621	94049	91345	42836	09191	08007	45449
99559	68331	62535	24170	69777	12830	74819	78142	43860	72834
33713	48007	93584	72869	51926	64721	58303	29822	93174	93972
85274	86893	11303	22970	28834	34137	73515	90400	71148	43643
84133	89640	44035	52166	73852	70091	61222	60561	62327	18423
56732	16234	17395	96131	10123	91622	85496	57560	81604	18880
65138	56806	87648	85261	34313	65861	45875	21069	85644	47277
38001	02176	81719	11711	71602	92937	74219	64049	65584	49698
37402	96397	01304	77586	56271	10086	47324	62605	40030	37438
97125	40348	87083	31417	21815	39250	75237	62047	15501	29578
21826	41134	47143	34072	64638	85902	49139	06441	03856	54552
73135	42742	95719	09035	85794	74296	08789	88156	64691	19202
07638	77929	03061	18072	96207	44156	23821	99538	04713	66994
60528	83441	07954	19814	59175	20695	05533	52139	61212	06455
83596	35655	06958	92983	05128	09719	77433	53783	92301	50498
10850	62746	99599	10507	13499	06319	53075	71839	06410	19362
39820	98952	43622	63147	64421	80814	43800	09351	31024	73167
59580	06478	75569	78800	88835	54486	23768	06156	04111	08408
38508	07341	23793	48763	90822	97022	17719	04207	95954	49953
30692	70668	94688	16127	56196	80091	82067	63400	05462	69200
65443	95659	18288	27437	49632	24041	08337	65676	96299	90836
27267	50264	13192	72294	07477	44606	17985	48911	97341	30358
91307	06991	19072	24210	36699	53728	28825	35793	28976	66252
68434	94688	84473	13622	62126	98408	12843	82590	09815	93146
48908	15877	54745	24591	35700	04754	83824	52692	54130	55160
06913	45197	42672	78601	11883	09528	63011	98901	14974	40344
10455	16019	14210	33712	91342	37821	88325	80851	43667	70883
12883	97343	65027	61184	04285	01392	17974	15077	90712	26769
21778	30976	38807	36961	31649	42096	63281	02023	08816	47449
19523	59515	65122	59659	86283	68258	69572	13798	16435	91529
67245	52670	35583	16563	79246	86686	76463	34222	26655	90802
60584	47377	07500	37992	45134	26529	26760	83637	41326	44344
53853	41377	36066	94850	58838	73859	49364	73331	96240	43642
24637	38736	74384	89342	52623	07992	12369	18601	03742	83873
83080	12451	38992	22815	07759	51777	97377	27585	51972	37867
16444	24334	36151	99073	27493	70939	85130	32552	54864	54759
60790	18157	57178	65762	11161	78576	45819	52979	65130	04860

Source: The RAND Corporation, A Million Random Digits with 100,000 Normal Deviates (New York, The Free Press).

Table VI. *A. LOWER CRITICAL VALUES OF r FOR THE RUNS TEST*

n_1 \ n_2	2	3	4	5	6	7	8	9	10	11	12	13	14	15	16	17	18	19	20
2											2	2	2	2	2	2	2	2	2
3					2	2	2	2	2	2	2	2	2	3	3	3	3	3	3
4				2	2	2	3	3	3	3	3	3	3	3	4	4	4	4	4
5			2	2	3	3	3	3	3	4	4	4	4	4	4	4	5	5	5
6		2	2	3	3	3	3	4	4	4	4	5	5	5	5	5	5	6	6
7		2	2	3	3	3	4	4	5	5	5	5	5	6	6	6	6	6	6
8		2	3	3	3	4	4	5	5	5	6	6	6	6	6	7	7	7	7
9		2	3	3	4	4	5	5	5	6	6	6	7	7	7	7	8	8	8
10		2	3	3	4	5	5	5	6	6	7	7	7	7	8	8	8	8	9
11		2	3	4	4	5	5	6	6	7	7	7	8	8	8	9	9	9	9
12	2	2	3	4	4	5	6	6	7	7	7	8	8	8	9	9	9	10	10
13	2	2	3	4	5	5	6	6	7	7	8	8	9	9	9	10	10	10	10
14	2	2	3	4	5	5	6	7	7	8	8	9	9	9	10	10	10	11	11
15	2	3	3	4	5	6	6	7	7	8	8	9	9	10	10	11	11	11	12
16	2	3	4	4	5	6	6	7	8	8	9	9	10	10	11	11	11	12	12
17	2	3	4	4	5	6	7	7	8	9	9	10	10	11	11	11	12	12	13
18	2	3	4	5	5	6	7	8	8	9	9	10	10	11	11	12	12	13	13
19	2	3	4	5	6	6	7	8	8	9	10	10	11	11	12	12	13	13	13
20	2	3	4	5	6	6	7	8	9	9	10	10	11	12	12	13	13	13	14

Source: From Frieda S. Swed and C. Eisenhart. "Tables for Testing Randomness of Grouping in a Sequence of Alternatives," *Annals of Mathematical Statistics*, 14 (1943):83–86. Reprinted by permisssion.

Table VI. *B. UPPER CRITICAL VALUES OF r FOR THE RUNS TEST*

n_1 \ n_2	2	3	4	5	6	7	8	9	10	11	12	13	14	15	16	17	18	19	20
2																			
3																			
4				9	9														
5			9	10	10	11	11												
6			9	10	11	12	12	13	13	13	13								
7				11	12	13	13	14	14	14	14	15	15	15					
8				11	12	13	14	14	15	15	16	16	16	16	17	17	17	17	17
9					13	14	14	15	16	16	16	17	17	18	18	18	18	18	18
10					13	14	15	16	16	17	17	18	18	18	19	19	19	20	20
11					13	14	15	16	17	17	18	19	19	19	20	20	20	21	21
12					13	14	16	16	17	18	19	19	20	20	21	21	21	22	22
13						15	16	17	18	19	19	20	20	21	21	22	22	23	23
14						15	16	17	18	19	20	20	21	22	22	23	23	23	24
15						15	16	18	18	19	20	21	22	22	23	23	24	24	25
16							17	18	19	20	21	21	22	23	23	24	25	25	25
17							17	18	19	20	21	22	23	23	24	25	25	26	26
18							17	18	19	20	21	22	23	24	25	25	26	26	27
19							17	18	20	21	22	23	23	24	25	26	26	27	27
20							17	18	20	21	22	23	24	25	25	26	27	27	28

Source: From Frieda S. Swed and C. Eisenhart, "Tables for Testing Randomness of Grouping in a Sequence of Alternatives," *Annals of Mathematical Statistics*, 14 (1943):83–86. Reprinted by permission.

Table VII. *CRITICAL VALUES OF SPEARMAN RANK CORRELATION COEFFICIENT*

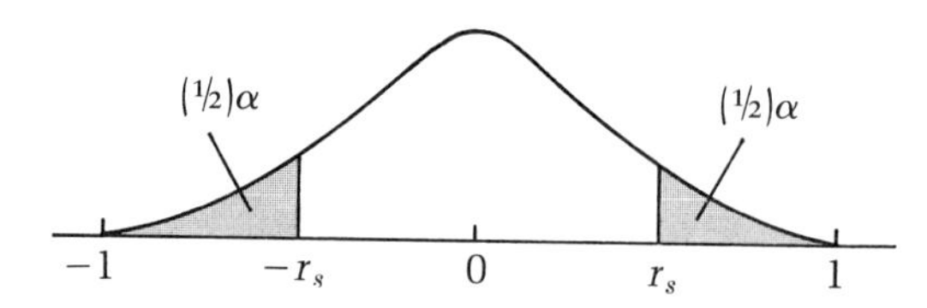

The entries in this table are the critical values of r_s for a two-tailed test at α.

n	$\alpha = 0.10$	$\alpha = 0.05$	$\alpha = 0.02$	$\alpha = 0.01$
5	0.900	—	—	—
6	0.829	0.886	0.943	—
7	0.714	0.786	0.893	—
8	0.643	0.738	0.833	0.881
9	0.600	0.683	0.783	0.833
10	0.564	0.648	0.745	0.794
11	0.523	0.623	0.736	0.818
12	0.497	0.591	0.703	0.780
13	0.475	0.566	0.673	0.745
14	0.457	0.545	0.646	0.716
15	0.441	0.525	0.623	0.689
16	0.425	0.507	0.601	0.666
17	0.412	0.490	0.582	0.645
18	0.399	0.476	0.564	0.625
19	0.388	0.462	0.549	0.608
20	0.377	0.450	0.534	0.591
21	0.368	0.438	0.521	0.576
22	0.359	0.428	0.508	0.562
23	0.351	0.418	0.496	0.549
24	0.343	0.409	0.485	0.537
25	0.336	0.400	0.475	0.526
26	0.329	0.392	0.465	0.515
27	0.323	0.385	0.456	0.505
28	0.317	0.377	0.448	0.496
29	0.311	0.370	0.440	0.487
30	0.305	0.364	0.432	0.478

Note: For a one-tailed test, the value of α shown at the top of Table VII is double the value of α being used in the hypothesis test.

Source: From E.G. Olds, "Distributions of Sums of Squares of Rank Differences for Small Numbers of Individuals," *Annals of Statistics,* 9 (1938):138–148, and amended, 20 (1949):117–118. Reprinted by permission.

Table VIII. *CRITICAL VALUES OF SS IN THE KENDALL COEFFICIENT OF CONCORDANCE*

	N					Additional Values for *N* = 3	
k	3*	4	5	6	7	*k*	*s*
			Values at the .05 Level of Significance				
3			64.4	103.9	157.3	9	54.0
4		49.5	88.4	143.3	217.0	12	71.9
5		62.6	112.3	182.4	276.2	14	83.8
6		75.7	136.1	221.4	335.2	16	95.8
8	48.1	101.7	183.7	299.0	453.1	18	107.7
10	60.0	127.8	231.2	376.7	571.0		
15	89.8	192.9	349.8	570.5	864.9		
20	119.7	258.0	468.5	764.4	1,158.7		
			Values at the .01 Level of Significance				
3			75.6	122.8	185.6	9	75.9
4		61.4	109.3	176.2	265.0	12	103.5
5		80.5	142.8	229.4	343.8	14	121.9
6		99.5	176.1	282.4	422.6	16	140.2
8	66.8	137.4	242.7	388.3	579.9	18	158.6
10	85.1	175.3	309.1	494.0	737.0		
15	131.0	269.8	475.2	758.2	1,129.5		
20	177.0	364.2	641.2	1,022.2	1,521.9		

*Notice that additional critical values of *SS* for *N* = 3 are given in the right-hand column of this table.

Source: From M. Friedman. "A Comparison of Alternative Tests of Significance for the Problem of *m* Rankings," *Annals of Mathematical Statistics,* 11 (1940):86–92. Reprinted by permission.

APPENDIX B.
THE SUMMATION NOTATION AND OTHER SYMBOLS

Capital letters are generally used in this text as the names of variables. For instance, you can let X be age, Y be earnings, and Z be sex. Particular values of these variables are denoted by the lower-case letters x, y, and z. The expression $P(X = x)$ means "the probability that the random variable X takes on a value x. More generally, the function $f(x)$ means the value of the function f when the variable X takes on the value x.

We can be more specific about denoting particular values or observations of a variable by using subscripts. For instance, if we observe n individuals, we can let

x_1 be the first individual's age
x_2 be the second individual's age
x_n be the nth individual's age
y_1 be the first individual's earnings
etc.

The letters i and j are used as "counters," or a generalized number for a subscript, so that

x_i is the ith individual's age, where i can be any integer from 1 to n
y_j is the jth individual's earnings
z_i is the ith individual's sex
etc.

The total age of n individuals can be expressed as

$$x_1 + x_2 + x_3 + \ldots + x_i + \ldots + x_n$$

This can be written more concisely by using the summation sign:

$$\sum_{i=1}^{n} x_i$$

This notation states that the variable being summed has the subscript i, and that you should add up all values of the variable starting with $i = 1$ (the number below the Σ), up to $i = n$ (the number above the Σ).

If it is very obvious which variable is to be summed, we often leave off the subscripts and write the summation simply as Σx.

EXAMPLE 1 If $x_1 = 40$, $x_2 = 20$, and $x_3 = 30$, find $\sum_{i=1}^{3} x_i$.

Solution: $\sum_{i=1}^{3} x_i = x_1 + x_2 + x_3 = 40 + 20 + 30 = 90$

Now find $\sum_{i=1}^{3} \frac{x_i}{3}$.

Solution: $\sum_{i=1}^{3} \frac{x_i}{3} = \frac{x_1}{3} + \frac{x_2}{3} + \frac{x_3}{3} = \frac{40 + 20 + 30}{3} = 30$

EXAMPLE 2 Show that $\sum_{i=1}^{n} cx_i = c\Sigma x_i$ (the sum of a constant times a sum of random variables is equal to the constant times the sum of the random variables).

Solution: The theorem can be proven by expanding $\sum_{i=1}^{n} cx_i$ to read

$$cx_1 + cx_2 + cx_3 + \ldots + cx_n$$

Factoring out the constant makes this equal to

$$c(x_1 + x_2 + x_3 + \ldots + x_n) = c\sum_{i=1}^{n} x_i$$

EXAMPLE 3 Show that $\sum_{i=1}^{n} c = nc$.

Solution:

$$\sum_{i=1}^{n} c = \underbrace{c + c + c + \ldots + c}_{n \text{ terms}}$$

$$= c(\underbrace{1 + 1 + 1 + \ldots + 1}_{n \text{ terms}}) = nc.$$

EXAMPLE 4 Show that $\sum_{i=1}^{n}(x_i + y_i) = \sum_{i=1}^{n} x_i + \sum_{i=1}^{n} y_i$.

Solution:

$$\sum_{i=1}^{n}(x_i + y_i) = (x_1 + y_1) + (x_2 + y_2) + \ldots + (x_n + y_n)$$
$$= (x_1 + x_2 + x_3 + \ldots + x_n) + (y_1 + y_2 + \ldots + y_n)$$
$$= \sum_{i=1}^{n} x_1 + \sum_{i=1}^{n} y_1$$

EXAMPLE 5 Show that $\sum_{i=1}^{n} (x_i - a) = \sum_{i=1}^{n} x_i - na$.

Solution: Use the result of Example 4. Therefore,

$$\sum_{i=1}^{n} (x_i - a) = \sum_{i=1}^{n} x_i - \sum_{i=1}^{n} a = \sum_{i=1}^{n} x_i - na$$

ORDER OF OPERATIONS

It is a convention in mathematical notation that the order of operations in any expression, unless specified to be otherwise, is

1. raising to powers and taking roots
2. multiplication and division
3. addition and subtraction

If you wish to indicate that the operations should be performed in some other order, you show this by placing parentheses around the operations you wish to perform first.

EXAMPLE 6 Distinguish between $\sum_{i=1}^{n} x_i y_i^2$ and $\sum_{i=1}^{n} (x_i y_i)^2$.

Solution:

$$\sum_{i=1}^{n} x_i y_i^2 = x_1 y_1^2 + x_2 y_2^2 + \ldots + x_n y_n^2$$

$$\sum_{i=1}^{n} (x_i y_i)^2 = (x_1 y_1)^2 + (x_2 y_2)^2 + \ldots + (x_n y_n)^2$$
$$= x_1^2 y_1^2 + x_2^2 y_2^2 + \ldots + x_n^2 y_n^2$$

EXAMPLE 7 Rewrite $\sum_{i=1}^{n} (x - \overline{x})^2$ in expanded form.

Solution: Use the formula for squaring a binomial to get

$$
\begin{aligned}
\sum_{i=1}^{n} (x_i - \overline{x})^2 &= \sum_{i=1}^{n} (x_i^2 - 2x_i\overline{x} + \overline{x}_2) \\
&= \sum_{i=1}^{n} x_i^2 - 2\sum_{i=1}^{n} x_i\overline{x} + \sum_{i=1}^{n} \overline{x}^2 \\
&= \sum_{i=1}^{n} x_i^2 - 2\overline{x}\sum_{i=1}^{n} x_i + n\overline{x}^2 \\
&= \sum_{i=1}^{n} x_i^2 - 2\overline{x}n\overline{x} + n\overline{x}^2 \\
&= \sum_{i=1}^{n} x_i^2 - 2n\overline{x}^2 + n\overline{x}^2 \\
&= \sum_{i=1}^{n} x_i^2 - n\overline{x}^2
\end{aligned}
$$

You can use this result to obtain a shortcut formula for the variance s^2, making use of the fact that $\overline{x} = \Sigma x/n$ and $\overline{x}^2 = (\Sigma x)^2/n^2$

$$
\begin{aligned}
s^2 &= \frac{\Sigma(x - \overline{x})^2}{n - 1} = \frac{\Sigma x^2 - n\overline{x}^2}{n - 1} = \frac{\Sigma x^2}{n - 1} - \frac{n\overline{x}^2}{n - 1} \\
&= \frac{\Sigma x^2}{n - 1} - \frac{(\Sigma x)^2}{n(n - 1)}
\end{aligned}
$$

EXERCISES

B.1. Expand the following expressions.

a) $\sum_{i=1}^{3} x_i$

b) $\sum_{i=1}^{4} 3x_i$

c) $\sum_{j=1}^{3} (x_j - 3)$

d) $\sum_{j=1}^{5} y_j^2$

e) $\sum_{i=1}^{3} \frac{x_i}{3}$

f) $\sum_{j=1}^{2} (z_j - k)^2$

g) $\sum_{i=1}^{4} x_i y_i$

B.2. Express each of the following expanded expressions using a summation sign.

a) $x_1 + x_2 + x_3$
b) $(x_1 + 2) + (x_2 + 2) + (x_3 + 2)(x_4 + 2)$
c) $2y_1 + 2y_2 + 2y_3 + 2y_4 + 2y_5$
d) $x_1^2 + x_1^2 + x_3^2$
e) $x_1y_1 + x_2y_2$
f) $3x_1y_1 + 3x_2y_2$

B.3. Given: $x_1 = 0$, $x_2 = 2$, $x_3 = 4$, $x_4 = 6$, and $x_5 = 8$, find

a) $\sum_{i=1}^{5} x_i$

b) $\sum_{i=1}^{5} x_i^2$

c) $\sum_{i=1}^{5} \frac{x_i}{5}$

d) $\sum_{i=1}^{5} \frac{x_i^2}{5} - \left(\sum_{i=1}^{5} \frac{x_i}{5}\right)^2$

e) $\sum_{i=1}^{5} \frac{(x_i - \bar{x})^2}{5}$

f) $\sum_{i=2}^{4} x_i^2$

APPENDIX C. EXPECTED VALUES AND VARIANCES

The expected value of a discrete random variable X is defined to be

$$E(X) = \Sigma x f(x)$$

where $f(x)$ is the probability that the random variable X takes on the value x, and the summation is taken over all values of the random variable.

The expected value of a function g of any random variable X is defined to be

$$E(g(X)) = \Sigma g(x) f(x)$$

where the summation is taken over all values of the random variable.

From these two definitions, you can derive the following theorems.

THEOREM 1 The expected value of a constant is the constant, or

$$E(c) = c$$

Proof: Let $x = c$. Then

$$E(X) = \Sigma x f(x) = \Sigma c f(x) = c \Sigma f(c)$$

and, since $\Sigma f(x) = 1$, $E(c) = c$.

THEOREM 2 $E(cX) = cE(X)$.
Proof: $E(cX) = \Sigma cx f(x) = c \Sigma x f(x) = cE(X)$.

THEOREM 3 $E(a + bX) = \Sigma(a + bx)f(x)$
$$= \Sigma(af(x) + bxf(x))$$
$$= \Sigma af(x) + \Sigma bxf(x)$$
$$= a\Sigma f(x) + b\Sigma xf(x)$$
$$= a + bE(x)$$

This result says that the expected value of a linear function of a random variable is a linear function of the expected value of the random variable. For example, if profit is a linear function of sales, then knowledge of the expected sales will tell you the expected profit.

THEOREM 4* The expected value of a sum of random variables is equal to the sum of the expected values of the random variable, or

$$E(X_1 + X_2 + \ldots + X_n) = E(X_1) + E(X_2) + \ldots + E(X_n)$$

THEOREM 5 The variance of the sum of *independent* random variables equals the sum of the variances of the random variables, or

$$\sigma^2_{X_1+X_2+\ldots+X_n} = \sigma^2_{X_1} + \sigma^2_{X_2} + \ldots + \sigma^2_{X_n}$$

where $X_1, X_2, \ldots, X_n$ are independent random variables.

THEOREM 6 $\sigma^2_c = 0$, where c is a constant.
Proof: Let $x = c$. From Theorem 1, $E(X) = c$. Therefore

$$\sigma^2_X = \Sigma(x - E(X))^2 f(x) = \Sigma(c - c)^2 = 0$$

THEOREM 7 $\sigma^2_{cX} = c^2\sigma^2_X$ where c is a constant.
Proof: $\sigma^2_{cX} = \Sigma(cX - E(cX))^2 f(x)$. From Theorem 2, $E(cX) = cE(X)$. Therefore

$$\sigma^2_{cX} = \Sigma(cx - cE(X))^2 f(x)$$
$$= \Sigma[c^2x^2 - 2c^2xE(X) + c^2(E(X))^2]f(x)$$
$$= c^2\Sigma x^2 f(x) - 2c^2E(X)\Sigma xf(x) + c^2(E(X))^2\Sigma f(x)$$
$$= c^2\Sigma x^2 f(x) - 2c^2(E(X))^2 + c^2(E(X))^2$$
$$= c^2[\Sigma x^2 f(x) - (E(X))^2] = c^2\sigma^2_X$$

THEOREM 8 $\sigma^2_{a+bX} = b^2\sigma^2_X$, where a and b are constants.
Proof: By Theorem 5, $\sigma^2_{a+bX} = \sigma^2_a + \sigma^2_{bX}$; by Theorem 6, $\sigma^2_a = 0$, and by Theorem 7, $\sigma^2_{bX} = b^2\sigma^2_X$. Therefore

$$\sigma^2_{a+bX} = 0 + b^2\sigma^2_X$$

THEOREM 9 $\sigma^2_{A_1X_1+A_2X_2+\ldots+A_kX_k} = A_1^2\sigma^2_{X_1} + A_2^2\sigma^2_{X_2} + \ldots + A_k^2\sigma^2_{X_k}$, **if** $X_1, X_2, \ldots, X_k$ are independent random variables.

*The proofs of Theorems 4 and 5 are omitted.

Proof: By Theorem 5, $\sigma^2_{A_1X_1+A_2X_2+\ldots+A_kX_k} = \sigma^2_{A_1X_1} + \sigma^2_{A_2X_2} + \ldots + \sigma^2_{A_kX_k}$. By Theorem 7, this equals

$$A_1\sigma^2_{X_1} + A_2^2\sigma^2_{X_2} + \ldots + A_k^2\sigma^2_{X_k}$$

THEOREM 10 $E(\overline{X}) = E(X)$, where $\overline{X}$ is the mean of a sample of n independent observations of the random variable X.

Proof: $E(\overline{X}) = E\left(\frac{\Sigma X}{n}\right) = \frac{1}{n} E(\Sigma X)$. By Theorem 2, this equals

$$\frac{1}{n}\Sigma(\mathrm{E(X)}) = \frac{1}{n} nE(X) = E(X)$$

THEOREM 11 $\sigma^2_{\overline{X}} = \sigma^2_X/n$, where X is a random variable and $\overline{X}$ is the mean of a random sample of n independent observations.

Proof:
$$\begin{aligned}\sigma^2_{\overline{X}} &= \sigma^2\left(\frac{\Sigma X}{n}\right)\\ &= \frac{1}{n^2}\sigma^2(\Sigma X) \qquad \text{by Theorem 7}\\ &= \frac{1}{n^2}\Sigma\sigma^2_X \qquad \text{by Theorem 5}\end{aligned}$$

This equals

$$\frac{1}{n^2} n\sigma^2_X = \frac{\sigma^2_X}{n}$$

THEOREM 12 If a random variable X has a binomial distribution with parameters n and p, then $E(X) = np$.

Proof: Equation 5.8 defines the expected value of a discrete random variable as

$$E(X) = \Sigma xP(X = x)$$

For a binomial random variable this becomes

$$E(X) = \sum_{x=0}^{n} x \frac{n!}{x!(n-x)!} p^x(1-p)^{n-x}$$

The first term of this summation (when $x = 0$) is equal to 0. Therefore we can eliminate the first term without changing the value of $E(X)$, so that

$$\begin{aligned}E(X) &= \sum_{x=1}^{n} x \frac{n!}{x!(n-x)!} p^x(1-p)^{n-x}\\ &= \sum_{x=1}^{n} \frac{n!}{(x-1)!(n-x)!} p^x(1-p)^{n-x}\end{aligned}$$

Let $y = x - 1$; let $m = n - 1$. Rewrite $E(X)$ as

$$E(X) = \sum_{y=0}^{m} \frac{(m+1)!}{y!(m-y)!} p^{y+1}(1-p)^{m-y}$$

Rewrite $(m + 1)!$ as $(m + 1)m!$ Rewrite p^{y+1} as $p(p^y)$.

$$E(X) = \sum_{y=0}^{m} (m+1) \frac{m!}{y!(m-y)!} p(p^y)(1-p)^{m-y}$$

Factor the constants $(m + 1)$ and p out of the summation sign.

$$E(X) = (m+1) \sum_{y=0}^{m} \frac{m!}{y!(m-y)!} p^y(1-p)^{m-y}$$

The summation is the sum of the probabilities of the binomial random variable Y with parameters p and m. This expression equals 1, since the sum of the probabilities of any probability distribution is equal to 1. Therefore,

$$E(X) = (m+1)p = np$$

THEOREM 13 If a random variable X has a binomial probability distribution with parameters n and p, then

$$\sigma_X^2 = np(1-p)$$

Proof: Equation 5.9 defines

$$\sigma^2 = \Sigma(x - \mu)^2 f(x)$$

for any discrete random variable X, where μ is the expected value of X. According to Equation 5.10, this is equivalent to

$$\sigma^2 = \Sigma x^2 f(x) - \mu^2$$

In the case of the binomial random variable, therefore

$$\sigma^2 = \sum_{x=0}^{n} (x - np)^2 = \sum_{x=0}^{n} x^2 \frac{n!}{x!(n-x)!} p^x(1-p)^{n-x} - (np)^2$$

The first term of the summation equals 0, so that

$$\sigma^2 = \sum_{x=1}^{n} x^2 \frac{n!}{x!(n-x)!} p^x(1-p)^{n-x} - n^2p^2$$

$$= \sum_{x=1}^{n} x \frac{n!}{(x-1)!(n-x)!} p^x(1-p)^{n-x} - n^2p^2$$

Let $y = x - 1$; let $m = n - 1$. Then

$$\sigma^2 = \sum_{y=0}^{m} (y + 1)\frac{(m + 1)!}{y!(m - y)!} p^{y+1} - n^2p^2$$

Expand the expression under the summation sign by multiplying the binomial $(y + 1)$ with the rest of the expression.

$$\begin{aligned}
\sigma^2 &= \sum_{y=0}^{m} y \frac{(m + 1)!}{y!(m - y)!} p^{y+1}(1 - p)^{m-y} \\
&\quad + \sum_{y=0}^{m} \frac{(m + 1)!}{y!(m - y)!} p^{y+1}(1 - p)^{m-y} - n^2p^2 \\
&= (m + 1)p \sum_{y=0}^{m} y \frac{m!}{y!(m - y)!} p(1 - p)^{m-y} \\
&\quad + (m + 1)p \sum_{y=0}^{m} \frac{m!}{y!(m - y)!} p(1 - p)^{m-y} - n^2p^2 \\
&= (m + 1)pE(Y) + (m + 1)p - n^2p^2
\end{aligned}$$

The first summation is the expected value of the binomial random variable Y with parameters m and p, and the second summation is the sum of the probabilities of Y, which equals 1. Therefore, rewriting $E(Y)$ as $mp = (n - 1)p$, we have

$$\begin{aligned}
\sigma_X^2 &= (m + 1)p(n - 1)p + (m + 1)p - n^2p^2 \\
&= np(n - 1)p + np - n^2p^2 \\
&= (n^2p - np)p + np - n^2p^2 \\
&= np(1 - p)
\end{aligned}$$

APPENDIX D. THE USE OF THE COMPUTER AND PACKAGED PROGRAMS IN STATISTICAL RESEARCH

Possibly no technological innovation in human history has been as rapid as the electronic computer in the quarter-century lasting from the early 1950s to the late 1970s. The impact of the computer has spread from major corporations to small retail businesses, from the defense department and space agency to virtually all branches of federal government, as well as to state and local government; from the natural sciences to the social sciences and the humanities. The computer has greatly expanded the applicability of statistical methods, and in so doing has revolutionized the teaching of statistics.

Until the late 1950s, most statistical analyses of social phenomena were limited to comparatively few variables. This restriction was due not to a lack of understanding of the theories of multivariate analysis; it was due to constraints on the amount of time and money that could be devoted to the heavy calculations necessarily involved when dealing with complex phenomena. Multiple regression methods and other multivariate techniques might take weeks or months of calculating time when the calculations were performed using the desk calculators that had long been featured in statistical laboratories. Since calculation was so often the bottleneck in statistical work, statistics courses placed great emphasis on calculation and short-cut formulas that could moderately reduce computational times.

Communicating with the Computer

It is small wonder that statisticians welcomed an innovation that could perform in seconds calculations that would previously have taken days or weeks. But using the computer involved a major problem in communications. Computers can do only what they are explicitly told to do, and the language they understand is totally unlike the language in which you or I ordinarily express ourselves. Machine language requires that highly skilled programmers are necessary to give instructions to the computer in its own language. Unfortunately, these programmers are not likely to have much understanding of the problems of the statistical researcher, so that communication between programmer and researcher is likely to be difficult.

A major breakthrough was the development of intermediary languages. Intermediary languages enable a researcher to communicate with the computer without having to learn the intricacies of machine language. The best-known of these is FORTRAN. FORTRAN is a general language, usable by many fields and disciplines; other intermediary languages have been developed to be more efficient for one or another specialty, such as business, scientific research, etc.

The benefits of learning an intermediary language such as FORTRAN are very great. With FORTRAN, not only is a person able to instruct the computer to do what she or he wants it to do, but also a person is often able to see much more clearly the logic and the steps necessary in solving a problem. In addition, the user of FORTRAN is less dependent on existing packaged programs, which may not always be able to do exactly what the researcher wants to do.

Packaged Programs

As you have undoubtedly noticed, many statistical formulas are complicated and involve quite a number of steps. This is even more true for multivariate procedures that are beyond the scope of this text. Writing FORTRAN programs for regressions, correlations, and other more advanced procedures is no easy task and may require skills beyond those you possess or plan to acquire. Fortunately, a variety of packaged programs have been developed that enable a person with virtually no programming skill to use a computer. Most computer centers can offer the use of one or more of these packaged programs.

The list of statistical packages is changing so rapidly that any attempt to codify them would rapidly become out-dated. Most statistical programs are *batch* programs; they require you to submit your instructions and data to the computer in one batch and then wait for the result, which will come in the form of a printout that will arrive 5, 10, or more minutes later. A few of the programs are *interactive* programs; they enable you to give instructions and receive almost immediate responses from the computer, so that you are giving and receiving information almost simultaneously.

Since terminal facilities are typically more limited than key punch facilities and card readers, it is probably just as well to use a batch program unless there is a particular benefit to be gained from a conversational feedback. Most of the statistical methods presented in this book can be easily and perhaps most efficiently handled in batch.

There are many statistical packages serving a wide variety of specialized uses, and a few widely used general packages. If you have access to the services of a computer center, you can check to see what general statistical packages are available to you. One package used by many university and business computers is SPSS, which was developed for statistical work in the social sciences. SPSS can be used to calculate not only every statistical measure mentioned in this book, but a wide variety of other measures which can involve many variables. A similar package, OSIRIS, has been linked to SPSS so that it is possible, to take data stored in an OSIRIS file and use it in an SPSS system. This linkage of two systems is an important step toward increasing the flexibility and general usefulness of computer packages.

Another widely used system is SAS. SAS is something between a language and a packaged program. In using SAS, you would in effect be writing a program, but its specialized orientation toward statistical work makes SAS much easier for a statistician to use than a general language such as FORTRAN.

An example of a more specialized program is TROLL, which was developed by the National Bureau of Economic Research. This is especially well adapted to econometric studies, which combine statistical and economic theory.

For certain types of statistical work you could benefit from the use of an interactive program. Such a program is SCSS, which is the interactive counterpart of SPSS. FAKAD, a program developed in Great Britain, is also designed for use on a terminal.

Limitations of Statistical Packages

Learning to use a statistical package is no substitute for an understanding of statistics or statistical inference. Once you know how to use a statistical package, you can feed the computer any kind of data that you want, and the computer will feed back all the means, variances, correlation coefficients, and multiple regression equations you could ask for.

It is another matter, however, to be able to say what all of these numbers mean. For all their amazing features, computers will not interpret their output, nor tell you whether the assumptions underlying the data are valid. If you tell the computer to run a t test even though the data do not warrant the assumptions necessary for a t test, the computer will nevertheless faithfully carry out the test. It will not think for you, caution you, or advise you. It will obediently do everything you tell it to do in a highly efficient way.

Used properly, computer packages can take the drudgery out of statistical work and free peoples' time and energy for creative work rather than calculation. Problems that once upon a time were solvable only in theory can now be solved in practice because calculation times have been reduced to less than a millionth of what they once were. In so doing, computers have opened up new frontiers in the field of statistics.

Glossary of Symbols

A' complement of the set A

$A \subseteq B$ the set A is a subset of B (4.1)

$A \cup B$ the union of A and B; the occurrence of either A or B (or both)

$A \cap B$ the intersection of A and B; the occurrence of both A and B

a_1, a_2, etc. the available courses of action in a payoff table

A a parameter in a simple regression model; the Y intercept (10.2)

a a statistic computed in a simple linear regression equation; the Y intercept (3.7)

B a parameter in a simple regression model; the regression coefficient indicating the change in $E(Y|X)$ for a unit change in X (10.2)

B_1, B_2, etc. parameters of a multiple regression model; the regression coefficients indicating the change in $E(Y|X)$; for a unit change in X; when the other X values are held constant

b a statistic computed in a simple linear regression equation; the regression coefficient (3.7)

b_1, b_2, etc. partial regression coefficients computed in a multiple regression equation; also can be written as $b_{1.23}$, $b_{2.13}$, etc. (3.12, 10.16)

C the coefficient of contingency (3.3)

c a constant

$\text{cov}(X, Y)$ the population covariance of the variables X and Y

d_i the difference between the 2 rankings of the ith observation, in calculating the Spearman rank correlation coefficient (3.4)

$E(X)$ the expected value of a random variable (5.8)

$E(Y|X)$ the expected value of the random variable Y, given that $X = x$ (10.2)

$F(x)$ the cumulative probability that the random variable X is less than or equal to a value x (5.4)

f frequency of occurrences in a class, used in grouped data; the number of observed cases in a class or cell

$f(x)$ the probability that the random variable X takes on

$g(x)$ the value of x, if X is a discrete random variable or $h(x)$ the value x, if X is a discrete random variable; the value of the probability density of the random variable X at x, if X is a continuous random variable

f_e The expected frequency of occurrence in a cell (3.1)

H_0 the null hypothesis

H_1 the alternative hypothesis which you accept only if you reject H_0

m_i the midpoint of the ith class (2.3)

Md the sample median

N the number in a population or universe

n the number of observations in a sample

n a parameter of the binomial distribution; the number of independent trials (5.7)

$n(A)$ the number of outcomes in an experiment contained in the event A (4.7)

$P(A)$ the probability of an event A (4.3)

$P(A \cup B)$ the probability that either A or B (or both) occurs (4.10)

$P(A \cap B)$ the joint probability of A and B; the probability that A and B occur (4.12)

$P(B|A)$ the conditional probability that B occurs, given that A occurs (4.11)

p a parameter of the binomial distribution; the probability of a success in each trial (5.2)

$\hat{p}$ the proportion of successes observed in a sample of size n

$\hat{p}$ the weighted mean of 2 sample proportions (9.10)

$P_0(E)$ the prior probability of an event E (prior to any sample information) (8.2)

$P_1(E|x)$ the posterior probability of an event E, after observing information x (8.2)

r coefficient of linear correlation (Pearson r) computed from a sample (3.6)

r^2 coefficient of determination computed from a sample (3.11)

R_1, R_2 the sum of the ranks in the 2 samples (used in the Mann-Whitney U test)

R_j the sum of the ranks in the jth class (used in calculating Kendall's coefficient of concordance) (11.4)

r_s the Spearman rank correlation coefficient (3.4)

S the sample space; the set of possible outcomes in an experiment

s the standard deviation of a set of n observations (2.6)

s^2 the variance of a set of n observations, given by $\Sigma(X - \overline{X})^2/(n - 1)$

SS the test statistic used to test the significance of Kendall's coefficient of concordance (11.4)

$s_{\overline{X}_1 - \overline{X}_2}$ the estimated standard error of the difference between 2 means (9.4)

$s_{y \cdot x}$ the standard error of estimate, a measure of the dispersion of points around the regression line (10.5) (10.8)

t the statistic $(x - \mu)/(s/\sqrt{n})$, or any statistic distributed as the t distribution (8.1)

U a statistic that measures the difference between the ranks of subsets of a sample, use in the Mann-Whitney U test (11.1)

$u(x)$ the utility of x, where x is a situation, especially an amount of money one has

W Kendall's coefficient of concordance (11.5)

w_i the weight given to the ith observed value (2.4)

X, Y, etc. the names of random variables

x, y, etc. the values that the random variables X, Y, etc. take on

$\overline{X}$, $\overline{Y}$, etc. the arithmetic mean of the sample X values, Y values, etc. (2.1)

$\overline{X}_f$ the arithmetic mean of grouped data (2.3)

$\overline{X}_w$ the weighted arithmetic mean (2.4)

$|X|$ the absolute value of a number X

$\hat{Y}$ the computed ordinate of the regression line in a simple linear regression equation (3.7)

z the standardized normal variable; the value of a normally distributed random variable expressed in number of standard deviations from its mean (6.1)

α (alpha) the probability of a Type I Error if H_0 is true; the level of significance of a test

β (beta) the probability of a Type II Error if H_0 is false

θ (theta) the symbol often given to an unspecified parameter

μ (mu) the expected value or mean of a probability distribution or a population (5.8); the mean or expected value of the normal probability distribution

$\mu_{\overline{X}}$ the expected value of the sample mean

$\mu_{\overline{X}_1-\overline{X}_2}$ the expected value of the difference between 2 sample means (9.1)

ρ (rho) the coefficient of linear correlation in a population or probability distribution (10.11)

ρ^2 the coefficient of determination in a population (10.10)

$\sum_{i=1}^{n} x_i$ the sum of all the x_i, starting with $i = 1$ and going up to $i = n$

σ (sigma) the standard deviation of a population or probability distribution (5.11); the standard deviation of the normal distribution

σ^2 the variance of a population or probability distribution (5.9)

$\sigma_{\overline{X}}$ the standard error of the mean (7.2, 7.3)

$\sigma_{\overline{X}_1-\overline{X}_2}$ the standard error of the difference between 2 means (9.2)

$\sigma^2_{Y \cdot X}$ the variance of Y around $E(Y|X)$ in a simple linear regression model

ϕ (phi) the empty set; an impossible event

χ^2 (chi square) a statistic used to describe the relationship between nominal variables, and to test for independence and goodness of fit (3.2)

Answers to Selected Exercises

CHAPTER 1

1–2. That the sample was randomly selected; that the sample result was not selected in preference to other results that were less favorable; that the sample was sufficiently large to make the "3 out of 5" statement meaningful.

1–5. The governor's position implies that saving a human life is worth at least $500,000. The minority leader's position does not necessarily set a lower value on a life saved, if she truly believes the number of lives saved could be less than half the governor's estimate.

1–6. The population is the set of people who have undergone a surgical procedure. This researcher is comparing two subpopulations: (1) those who undertake heavy exercise routine after their surgery, and (2) those who have had little exercise after their surgery.

The sample consists of two subsamples: the 20 who were subjected to the heavy exercise routine, and the 25 who were kept in bed with little exercise.

1–7. There is a selection bias; those who did not recover are less likely to be interviewed a year later.

1–8. The results might have been much different if someone other than the supervisor had done the interviewing.

1–10. a) The population is all those who actually vote on election day (more precisely it is the actual votes cast).

b) The sample is the set of 1762 voters polled (more precisely, the 1762 responses).

c) Possible biases: Those polled might not all vote, or vote in the same proportion as their responses; there may be a switch in voter opinion between the time of the poll and the time of the election; respondents may vote differently from the way they say they will.

d) Sampling error will result if by chance the 1762 responses differ from the responses if all the population had been sampled.

1–11. Bias can result if Tom tends to generalize faults in his own boss to the set of women. Sampling error will generally be very large if the sample size is only 1.

CHAPTER 2

2–1. a) Variables: Age as of Nearest Birthday; Sex; Occupation.
b) Values: 20, 21, 22, etc.; Male, Female; Low Risk, High Risk.

2–2. Some of the variables and sets of values might be:

Variable	*Values*
Intelligence	Scores on an aptitude test
Age	18, 19, 20, etc.
Sex	Male, Female
Height	Below 66 inches; at least 66 inches

2–3. Among the variables that are irrelevant to the performance on many jobs (which nevertheless may often form the basis for decisions on hiring) are: race, religion, sex, sexual preference, height, physical handicap, weight, etc.

2–4. Possible procedures are:
a) Have the individual stand next to a wall in bare feet, place a shoe box on his (her) head horizontally so that it also touches the wall, mark the wall at the bottom of the shoe box where it touches the wall, and measure the distance from the floor to the mark with a standard tape measure.
b) Ask the individual to identify all sources of personal income.
c) Administer a psychological test which is designed to measure "satisfaction."
d) Obtain effectiveness reports on employees made out by supervisors.
e) Compute mean scores on examination (or administer a self-evaluative questionnaire).

2–5. July, August
July, December
April, August
August, September
August, October

2–7. a) A commonly used indicator of standard of living is per capita income.
b) A primary source of this information is the *Survey of Current Business,* published by the Department of Commerce, Office of Business Economics. A good secondary source is *Statistical Abstract of the United States* published by the Department of Commerce.

2–8. "Quality of life" is a very inclusive concept. It might include economic variables such as per capita income, vital statistics, such as life expectancy or infant mortality, and environmental data such as air and water pollution. Since combining these variables can be done in a variety of ways that might reflect the values of the measurer, you might have difficulty obtaining widely accepted figures on "quality of life" in any locality.

2–10. Such a measure might be an indicator of bone structure and muscular development, as well as excess fat.

2–11. a) Discrete d) Continuous
b) Continuous e) Discrete
c) Continuous

2–12. a) Ratio d) Interval
b) Ordinal e) Nominal
c) Nominal

2–14. a) Less than 1 mile
At least 1 mile but less than 2 miles
At least 2 miles but less than 3 miles
etc.

c) At least 0 but less than 1
At least 1 but less than 2
At least 2 but less than 3
etc.

b) All right as is.

2–15. a) If you use a class interval of $50 and start at $200, you would obtain the following table:

Taxes	*Number of States*
$200.00 to $249.99	4
$250.00 to $299.99	7
$300.00 to $349.99	18
$350.00 to $399.99	9
$400.00 to $449.99	5
$450.00 to $499.99	4
$500.00 to $549.99	1
$550.00 to $599.99	2

b) Grouped in this manner, the data reveal the heavy concentration of states between $300.00 and $350.00.
c) But by grouping the data, you lose information on individual states.

2–17. The width of the bars of a histogram indicate the interval over which the data lie. A histogram is therefore a valid method of presentation only if the data are at least interval level.

2–18. a) The mode is 21 years of age.
The median is 19 years of age.
The mean is 19.78 years of age.
b) More states had a 21-year minimum than any other age. If the states were ranked in order of minimum age, the middle state would have a minimum age of 19. The total of all minimum ages divided by 51 is 19.78.
c) The mode would switch from 21 to 18.
The median would be unchanged.
The mean would be reduced from 19.78 to 19.35.

2–19. a) The mean, if you are interested in population trends.
b) (1) Possibly the median. She would have a 50–50 chance of living at least this long. (2) The mean, since it would determine the premium rate that would cover claims in the long run.
c) The mode.

2–20. The mode is 7.7 minutes.
The median is 3.7 minutes.
The mean is 4.51 minutes.

2–21. Murphy's mode was 100; Schwartz's mode was 80.
Murphy's median was 100; Schwartz's median was 80.
Murphy's mean was 64; Schwartz's mean was 74.

2–22. The mode was 6 hours.
The median was 8 hours.
The mean was 10.2 hours.

2–23. a) No averages c) The median and mean
b) The mode d) The mode

2–24. The mean.

2–25. a) The median is 53,639 people.
The mean is 73,768.1 people.

2–26. a) 4.92 weeks b) 5.057 weeks

2–27. 46.5 teachers. (You must assume that the mean number of teachers within each class is the class mid-point.)

2–29. a) About $750,000, the approximate mid-point of the class with the greatest frequency.
b) The second class ($1,000,000 to 1,499,999.99)
c) $1,910,000.00

2–30. The arithmetic mean. The daily mean will tell him how many words he will write over a period of many days.

2–31. The variance is 1.967.
The standard deviation is 1.40 years.

2–32. a) The variance is 6.00.
The standard deviation is 2.45.
b) The variance is .500.
The standard deviation is .707.

a) City A: Variance = 4.70
Standard deviation = 2.17 accidents
City B: Variance = 3.50
Standard deviation = 1.87 accidents
b) No. If $X^2 > Y^2$, then $\sqrt{X^2} > \sqrt{Y^2}$

2–35. The proposal would tend to decrease the standard deviation of after-tax income.

2–37. a) The mode is 38 mph.
The median is 39 mph.
The mean is 39.42 mph.

2–38. The standard deviation is 6.48 mph.

2–39. a) The class "20 to 29"
b) No. The 4 items in that class are simply identified as "above the median."

2–40. a) 30.269 patients b) 12.872 patients

CHAPTER 3

3–2. a) 60 males; 40 females; 80 employed; 20 unemployed
b) Males: 86.7% employed
Females: 70% employed
c) Employed males: 48
Unemployed males: 12
Employed females: 32
Unemployed females: 8
d) $\chi^2 = 4.17$
e) $C = .200$

3–3. a) Males with male spouses = 14.4
Males with female spouses = 21.6
Females with male spouses = 9.6
Females with female spouses = 14.4
b) $\chi^2 = 60.00$
c) $C = .707$
d) Yes, given the marginal totals, there is no way that χ^2 or C could be higher.

3–4. a) Expected cell frequencies are 10.00, 13.33, 16.67, 20.00, 16.67, 33.33.
b) $\chi^2 = 27.57$
c) $C = .43$

3–5. It would not be valid, because the dimensions of the two tables are not equal.

3–6. a) Rank correlation = −.132
b) Not necessarily. There may be several relationships that offset each other. For example, larger areas may be expected to have larger populations, but offsetting this is a tendency for more densely populated areas to form states with smaller areas.

3–7. a) $r_s = 1$ b) $r_s = -1$ c) $r_s = 0$

3–8. a) $r_s = .497$
b) One might be tending to cause the other, but no definite conclusion is warranted from the statistics.

3–13. $r = .804$

3–14. $r = .804$. No change occurs in the correlation coefficient when you add or subtract a constant to each of the X and Y values.

3–15. $r = .804$. No change occurs in the correlation coefficient when you multiply or divide each of the X and Y values by a constant.

3–16. $r = .804$. Transforming the X and Y values to deviations from their respective means does not change the correlation coefficient.

3–17. $r = -.785$

3–18. The relationship between age and earnings is almost certainly not linear over that range of ages. Average earnings for 15-year-olds will probably be low, they will increase through the twenties and probably reach a peak during the middle years and thereafter tend to decrease, especially after age 65. The Pearson coefficient, which measures only the *linear* relationship, will fail to describe the nonlinear pattern of earnings.

3–19. a) $\hat{Y} = -170 + 142.5X$
b) Predicted maintenance cost for 5th year = $542.50
Predicted maintenance cost for 6th year = $685.00
Therefore replace trucks at the end of the 5th year.
c) $r^2 = .781$

3–20. a) $\hat{Y} = 10 - .2X$
c) Not necessarily. A negative value of the regression coefficient could result from a variety of causes other than changes in the money supply. For example, authorities may restrict the expansion of the money supply when the outlook for inflation is strong.

3–21. a) X would be useless as a predictor, and the regression equation would be $\hat{Y} = \bar{Y}$, $r^2 = 0$.
b) A perfect predictor; $r^2 = 1$.
c) The regression equation would explain 50% of the variability of Y; $r^2 = 5$.

3–23. The statistician could compare the two values of r^2, to determine which predictor was able to explain a greater proportion of the variability in Y.

***3–25.** The equation predicts y to be -2.8 when the values X_1, X_2, and X_3 are zero.

The predicted increase in job performance is 2.04 for each increase of 1 year's experience, if the effects of years of schooling and managerial aptitude test score are held constant.

The predicted increase in job performance is 3.18 for each increase of 1 year of schooling completed, if the effects of years of experience and managerial aptitude test score are held constant.

The predicted increase in job performance is 0.93 for each increase of 1 point in managerial aptitude test score, if the effects of years of experience and years of schooling are held constant.

***3–26.** 144.74

3–27. No. Both charity giving and expenditures on liquor could be related to a third variable, such as income.

***3–28.** Not necessarily. The coefficient reflects the expected change in demand from an increase in price if all other things, including a competitor's price, are held constant. If, on the other hand, competitors are likely to react to a price change by a price change of their own, the net effect on the demand may be substantially different.

3–29. $\chi^2 = 19.200$ and $C = .3714$

3–30. a) There appears to be a slight tendency for their rankings to agree.
b) The Spearman rank correlation coefficient is +.143.

3–31. a) $r = .826$

3–33. a) $\hat{Y}_1 = -96.78 + 935.45X_1$ b) $\hat{Y}_2 = 12{,}400 + 200X_2$

***3–34.** The regression coefficient is the predicted increase in earnings associated with an increase of 1 year of schooling. The partial regression coefficient is the predicted increase in earnings associated with an increase of 1 year of schooling, for a given number of children (or with number of children held constant.

3–35. b) $r = .685$. This is a measure of the linear relationship between the two variables.

CHAPTER 4

4–1. {S = Tax passes and allocation passes; tax passes and allocation does not pass; tax does not pass and allocation passes; tax and allocation both do not pass.}

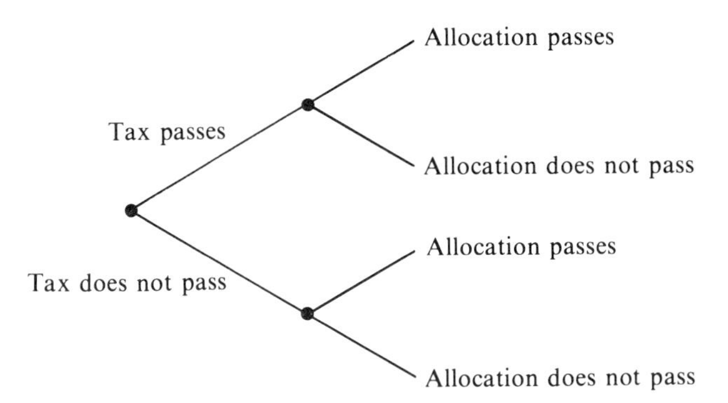

4–2. You can depict the event by an ordered triplet, the first element being the party of the President, the second element the party in control of the Senate, and the third element the party in control of the House. Then

a) (R,D,D)
b) (R,R,D)
c) (R,R,R) ∪ (D,D,D)
d) (R,D,D) ∪ (D,R,R)
e) (R,R,D) ∪ (R,D,R) ∪ (D,R,D) ∪ (D,D,R)
f) (R,R,D) ∪ (R,D,R) ∪ (R, R,R) ∪ (D,R,D) ∪ (D,D,R) ∪ (D,D,D)

4–3. Let the first element of each pair be the race of the wife and the second element the race of the husband. Then the 4 possibilities are (W,W), (W,B), (B,W), and (B,B).

4–4.

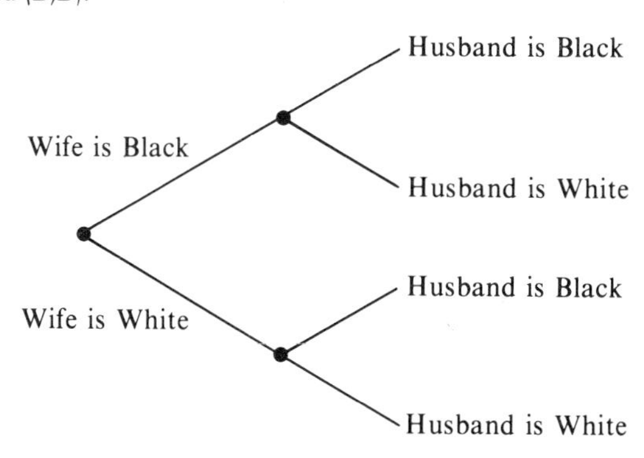

4–8. a) Either the fire alarm sounds or there is a fire (or both).
b) The fire alarm sounds and there is no fire.
c) Impossible event.
d) A false alarm.
e) No fire alarm sounds, and there is no fire.

4–9. a) It is summer and it is not summer.
b) It is summer and it is snowing.

4–10. a) A priori
b) Empirical
c) Empirical and subjective
d) Empirical
e) Empirical and subjective
f) Empirical

4–12. a) 4/15
b) You may abandon your *a priori* assignment of equal probability and use empirical probability, especially if you suspect that the drawing is not random.
c) This would depend on your subjective belief in horoscopes.

4–13. a) 1/6
b) A priori (giving each face equal probability)

4–14. If you used empirical probability you would have assigned the probability .20. If you had stuck to the *a priori* probability, ignoring sample evidence, you would have still assigned the probability 1/6. A method of combining a priori and empirical probability is discussed in Section 4.4.

4–15. a) 4/50, or .08 b) 8/50, or .16 c) 10/50, or .20

4–16. a) 13/52, or 1/4 b) 13/52, or 1/4 c) 26/52, or 1/2 d) 26/52, or 1/2

4–17. a) 78/100 b) 60/100 c) 28/78 d) 28/40 e) 90/100

4–18. a) .44
b) .20
c) Yes, if we consider the relative frequencies as probabilities.

4–20. a) $P(A \cup B) = .3 + .5 - .1 = .7$ b) $P(A' \cap B) = .4$

4–21. a) $P(A \cup B) = .7 + .8 - .6 = .9$
b) $P(B|A) = .6/.7 = 6/7$
c) $P(B'|A) = .1/.7 = 1/7$
d) $P(B|A') = .2/.3 = 2/3$

4–22. a)

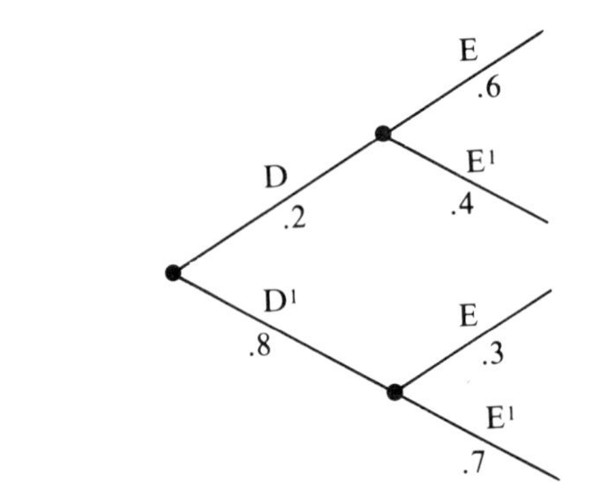

b) $P(D \cap E) = .12$ c) $P(E) = .12 + .24 = .36$

4–23. 2/3

4–24. Yes. If you draw a table of probabilities based on these frequencies, you will see that in every case the joint probabilities are the products of the marginal probabilities.

4–26. No. To be independent, all of the joint probabilities would have to be the products of the marginal probabilities.

4–27. a) It is raining, and it is Wednesday
b) It is raining, and it is not raining
c) If A and B are independent, the occurrence of one *does not change* the probability of the other. If A and B are mutually exclusive, the occurrence of one *precludes* the occurrence of the other.

***4–28.** Given: $P(A) > 0$ and $P(B) > 0$. If A and B are independent, then $P(A \cap B) = P(A)P(B)$. Since $P(A)$ and $P(B)$ are both positive then, $P(A)P(B)$ must be positive. But if A and B were mutually exclusive, then $P(A \cap B)$ would be zero, not positive. Therefore, if A and B are independent, they cannot be mutually exclusive.

***4–29.** $P(\text{Health/Woman}) = .8$

***4–30.** $P(\text{ESP}|\text{Correct Call}) = 52/61$

4–31.

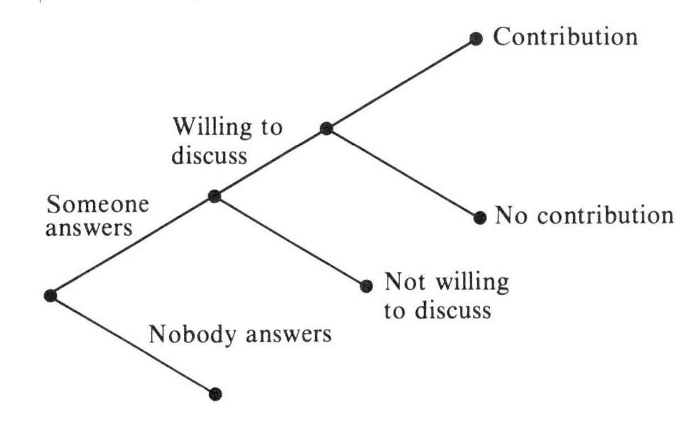

4–32. a) .3 b) .42 c) .112 d) .168

4–33. a) Persons who are not employed
b) Persons who are either employed or receiving unemployment compensation
c) Employed persons who are receiving unemployment compensation
d) Persons other than employed persons receiving unemployment compensation

4–34. a) The nominee is not black.
b) The nominee is either black or female.
c) The nominee is black female.
d) The nominee is black, female, and confirmed by the Senate.
e) The nominee is black, female, and not confirmed by the Senate.
f) The nominee is either black or female, or else confirmed by the Senate.

4–36. a) .25 b) .30 c) .70 d) .67 e) .44

4–37. a) $P(B_3)$ b) $P(A_3)$ c) $P(A_2 \cup A_3)$ d) $P(B_1|A_3)$ e) $P(A_3|B_1)$

4–38. a) 0 b) .21

4–39. a) .25 b) .33 c) No, because $(.4)(.3) \neq .10$.

4–41. .55. Use the addition theorem for two mutually exclusive events.

4–42. c. In all other cases there is probably some degree of dependence.

4–43. a) .3 b) .2 c) .06 d) .42 e) .48 f) 1/8

4–44. a) $A_1 \cap A_2 \cap A_3 \cap \ldots \cap A_n$ b) $A_1 \cup A_2 \cup A_3 \cup \ldots \cup A_n$

CHAPTER 5

5–1. a) X = the number of people who attend the rally.
b) 2000, 5000, 8000, 12000
c)

x	$f(x)$
2000	.02
5000	.18
8000	.30
12000	.50

d)

x	$f(x)$
$x < 2000$	0
$2000 \leqslant x < 5000$	.02
$5000 \leqslant x < 8000$	.20
$8000 \leqslant x < 2000$	.50
$X \geqslant 12000$	1.00

f) .5

5–2. a) OK
b) OK
c) No, because a probability cannot be negative.
d) No, because the probabilities do not sum to 1.
e) OK
f) OK
g) OK
h) No. When z is 3, then $f(2)$ becomes negative.

5–3. a)

t	10	11	12	13	14	15	16	17
$F(t)$	.05	.15	.40	.60	.75	.85	.95	1.00

b) .85
c) .60
d) $F(15) - F(11) = .70$
e) $F(14) - F(12) = .35$
f) The addition theorem for the union of two mutually exclusive events.

5–4. $f(x) = 1/61$ $x = 0, 1, \ldots, 60$, where x indicates the number of minutes past 10:00.

5–5. a)

y	2	3	4	5	6
$f(y)$	1/15	2/15	3/15	4/15	5/15
$F(y)$	1/15	3/15	6/15	10/15	1

b) 3/15
c) 6/15
d) 0
e) 1/15

5–6. The small letter x indicates a value that the random variable X takes on. The capital letter is the name of the random variable.

5–7. a) $f(x) = 1/366 \quad x = 1, 2, \ldots, 366$
b) 1/366
c) 29/366
d) $F(x) = x/366 \quad x = 1, 2 \ldots, 366$

5–8. Assume the hiring of an individual is independent of the hiring of other individuals. Then

$$f(x) = \frac{4!}{x!(4-x)!}\left(\frac{1}{4}\right)^{x}\left(\frac{3}{4}\right)^{4-x} \qquad x = 0, 1, 2, 3, 4$$

where x is the number of blacks hired.

5–9. $f(0) = 1/64 \quad f(4) = 15/64$
$f(1) = 6/64 \quad f(5) = 6/64$
$f(2) = 15/64 \quad f(6) = 1/64$
$f(3) = 20/64$

5–10. a) .4401 c) .3707
b) .8108 d) 1 − .8108 = .1892

5–11. a) .3456 b) .6826 c) .6630 d) .0778 e) .0870

5–12. a) $f(x) = \frac{15!}{x!(15-x)!}(.4)^{x}(.6)^{15-x} \qquad x = 0, 1, 2, 3, \ldots, 15$

b) $f(y) = \frac{15!}{y!(15-y)!}(.15)^{y}(.85)^{15-y} \qquad y = 0, 1, 2, \ldots, 15$

5–13. a) .0013
b) The probability is so low that it is hard to believe she could get that many wrong if she did not have the ability to discriminate. Perhaps she was deliberately pressing the "wrong" button.

5–14. a) .064 b) .288 c) .432 d) .216

5–15. $E(Y) = .6$; $\sigma^2 = 1.04$; and $\sigma = 1.02$

5–16. a) $E(X) = .5$. This is simply the proportion of women (or number 1's) in the population.
b) $\sigma^2 = .25$ and $\sigma = .5$

5–17. a) 60 cents b) 40 cents c) $200,000

5–18. a) $E(t) = 86.0$ minutes and $\sigma^2 = 20.0$

5–19. $E(X) = 3.5$ and $\sigma^2 = 2.92$

5–20. 20 weeks

5–21. $E(100X) = 100E(X) = 20{,}000$ pounds

5–22. a) 6.4 prisoners
b) $\sigma^2 = 3.84$ and $\sigma = 1.96$ prisoners
c) $\sigma_p^2 = 0.15$ and $\sigma_p = .122$

5–23. (a), (b), (d), (e), and (f) are discrete probability functions.

5–24. a) .0687 b) .5583 c) .9806 d) .0194 e) .2834

5–25. a) .1176 b) .6267 c) .0705 d) .0007

5–26.

X	1	2	3	4	5
F(x)	.2	.4	.6	.8	1.0

5–27. $E(r) = 7.30\%$ and $\sigma^2 = 20.11$

5–28. a) 3/18 e) 12/18
b) 9/18 f) 1.0
c) 0 g) 20.278 years
d) 6/18

CHAPTER 6

6–1. b) .30
c) .035
e) The smooth curve should indicate a probability somewhat less than .035.

6–2. The probability is .20

6–3. a) Discrete
b) Discrete, if the clock records time to the nearest minute
c) Continuous
d) Discrete
e) Continuous, for practical purposes

6–4. a) $P(X < 35) = P\left(X < \frac{35 - 45}{10}\right) = P(z < -1) = .5000 - .3413 = .1587$
b) .8664
c) .0062

6–5. a) .4772 b) .4772 c) .9270 d) .1012 e) .1010

6–6. b) Jenny: $\frac{71 - 65}{3} = +2.00$

Max: $\frac{72 - 69}{4} = +0.75$

6–7. $z = 1.645$

6–8. 23.435 days, or 24 days

***6–9.** $z = -2.576$ and $z = +2.576$

6–10. The probability that more than 45 people enter is approximated by the probability that a normally distributed random variable exceeds 45.5.

$$P(X > 45.5) = P\left(z > \frac{45.5 - 36}{6}\right) = P(z > 1.58) = .0571$$

6–11. a) .2894

6–12. .6856

6–13. a) .0228 b) .9772

6–14. .093

6–15. .0322

***6–16.** 24.04 cents

***6–17.** 75.32 miles per hour

6–18. .6915

6–19. a) binomial: .1792 and normal: .18

6–20. $P(X \geqslant 11) = .0039$ (using binomial)
$P(X \geqslant 10.5) = .0023$ (using normal approximation)

6–21. .896

6–22. .023

6–23. Less than .001

6–24. Using binomial, .2173
Using normal approximation, .2148

6–25. a) Less than .001 b) .984

6–26. Between 78.24 and 101.76 voters, or between 78 and 102 voters

6–27. .618

6–28. a) 1/3 b) 1/3 c) 1/12

6–29. a) .3413 b) .1587 c) .1336

6–30. a) .3413 b) .1587 c) .9282 d) .0718

***6–31.** $-1.645 \leqslant z \leqslant 1.645$

6–32. 0.5775 hours and 2.225 hours.

6–33. .2627. But binomial model is probably not appropriate because of lack of independence of children within the same family.

6–34. .7368 (using binomial table)
.7372 (using normal approximation)

6–35. a) The probability distribution of time until Senator Dome gets into the chair is normal with $\mu = 105$ minutes and $\sigma = \sqrt{300} = 17.3$ minutes.
b) The probability distribution of time until Senator Dome is finished is normal, with $\mu = 140$ minutes and $\sigma = \sqrt{400} = 20.0$ minutes.

6–36. 133.4 minutes

6–37. .6331

6–38. a) Route A b) Route B

CHAPTER 7

7–1. a) The distribution is normal, with $E(\overline{X}) = 72.0$ years and $\sigma^2_{\overline{X}} = 1$.
b) $P(\overline{X} \geqslant 75) = .0013$

7–2. a) $\mu = 40.0$ and $\sigma = 7.07$
b)

Sample	$\overline{X}$	*Sample*	$\overline{X}$
30, 35	32.5	35, 45	40.0
30, 40	35.0	35, 50	42.5
30, 45	37.5	40, 45	42.5
30, 50	40.0	40, 50	45.0
30, 40	37.5	45, 50	47.5

c) $E(\overline{X}) = 40.0$ and $\sigma_{\overline{X}} = 4.33$

d) $E(\overline{X}) = \mu = 40.0$ and $\sigma_{\overline{X}} = \dfrac{7.07}{\sqrt{2}}\sqrt{\dfrac{5-2}{5-1}} = 4.33$

7–3. a)

Sample	$\overline{X}$
30, 30	30.0
30, 35	32.5
30, 40	35.0
30, 45	37.5
30, 50	40.0
35, 35	35.0
35, 40	37.5
35, 45	40.0
35, 50	42.5
40, 40	40.0
40, 45	42.5
40, 50	45.0
45, 45	45.0
45, 50	47.5
50, 50	50.0

b) $E(\overline{X}) = 40.0$ and $\sigma_{\overline{X}} = 5.0$

c) $E(\overline{X}) = \mu = 40.0$ and $\sigma_{\overline{X}} = \dfrac{\sigma}{\sqrt{n}} = \dfrac{7.07}{\sqrt{2}} = 5.0$

d) Sampling with replacement is in effect increasing the size of the population to infinity.

7–4. a) The sampling distribution is approximately normal, with

$$E(\overline{X}) = 12.5 \qquad \text{and} \qquad \sigma_{\overline{X}} = .75$$

b) About .25

7–5. The sampling distribution is approximately normal, with $E(\overline{p}) = .70$ and $\sigma_{\overline{p}} = .05$.

7–6. a) Approximately .9 d) Approximately .975
b) Approximately .75 e) 1.000
c) Approximately .99 f) 1.000

7–7. a) 1.8 d) .083
b) 1.8 e) .8664
c) .067 f) .7698

7–8. a) Approximately normal; $E(\overline{p}) = .30$; $\sigma_{\overline{p}}{}^{1} = .10$
b) .0228
c) On the average, about one sample in 44 will show a majority smoking El Ropo.

7–9. a) Reduced by ½
b) Reduced by ⅔ (to ⅓ of what it was)
c) Reduced by ½

7–10. Approximately .843

7–11. 20.2 to 39.8 minutes

7–12. a) 7.4 ± .65 months, or 6.75 to 8.05 months
b) 7.4 ± .86 months, or 6.54 to 8.26 months

7–13. .0412 to .1588

7–14. .457 to .743

7–15. 97, in order to have 95% confidence; 166, in order to have 99% confidence

7–16. 1537 people

7–17. A census would appear to be preferable.

7–18. Only a sample would be feasible.

7–22. The interval would not be valid, since the sample was not a random sample.

7–23. a) The distribution is approximately normal, with μ equal to 11.4 days and $\sigma_{\overline{X}}$ equal to .5 days.
b) Less than .001

7–24. .102 to .298

7–25. a) One statistic V could be $\overline{X} + .1$
b) One statistic W could be $.9\overline{X}$
c) Several unbiased estimators of μ are: the sample mean $\overline{X}$; the sample median Md; and the first observation X_1.

7–26. \$1763 ± \$96.35, or \$1666.65 to \$1859.35

7–27. .618 to .782

7–29. At least 683 vaccine recipients, for 95% confidence

7–30. a) False b) True c) True

7–31. The difference is that the two fathers are selected without replacement so that the same individual cannot be repeated. On the other hand, the two dice are independent of each other, so that the same value *can* be repeated.

CHAPTER 8

8–2. a) By the nature of his statement you know that Dr. Davis's null hypothesis is that current levels of pollution are not harmful, and the alternative hypothesis is that they are harmful. Until evidence is presented that the pollution levels are harmful, he is prepared to assume (or act on the assumption) that current levels are *not* harmful.
b) He could have set the hypotheses up the other way; in such a case he would assume the levels were harmful until there was sufficient evidence that they were safe.

8–4. H_0: $\mu = 0.12$
H_1: $\mu < = 0.12$
Reject H_0 if $z < -1.64$. Since $z = -2.00$, you reject H_0. Yes, you can say that μ is below .12.

8–5. Yes. The decision rule is: Accept H_0 if $-1.96 < z < 1.96$. Since $z = -3.33$, you reject H_0.

8–6. a) One-tail b) Two-tail c) Two-tail d) One-tail

8–7. Yes. For a one-tail test the decision rule is to reject H_0 if $z < -1.645$. Since $z = -1.75$, you reject H_0.

8–8. If $\overline{X} \geq 22.33$, send out the nasty memo.

8–9. Yes. $z = 2.78$. Since this is more than 2.33, reject H_0.

8–10. No. $z = -2.00$. Since this is greater than -2.33, accept H_0.

8–11. No. $z = -1.51$. Since this is greater than -2.33, accept H_0.

8–12. According to the binomial distribution, $P(0 \text{ or } 5 \text{ heads}) = .0315 + .0315 = .0630$. Since this probability is greater than .05, you cannot reject H_0.

8–13. No. $z = -.88$. Since this is greater than -1.645, accept H_0.

8–14. No. $t = -1.75$. Since this is greater than -1.860, accept H_0.

8–15. No. $t = -1.44$. Since this is between -2.131 and 2.131, accept H_0. The orderly might take a larger sample and come up with significant results.

8–16. $t = -.75$. Since this is between -3.355 and 3.355, accept H_0.

8–17. a) Since the mean of the sample is only $2000, it suggests that μ is less than $5000.
b) No. $t_{.01}$, 2 dof (one tail) $= -6.965$. $t = -5.196$. Not significant.

8–18. Better not to pay $250. Expected Cost of Repairs is only $150.

8–19. a) Not betting. Expected payoff of betting is -32 cents.
b) Expected payoff is $2.20.

8–20. a) Yes. Expected payoff is $36,000.
b) No. The expected utility of Suing is 40; the expected utility of Not Suing is 60.

***8–21.** 40

***8–22.** a)

	Build	Do Not Build
$P = .10$	$-25,000	$0
$P = .20$	$+50,000	$0

b) Yes.
c)

p	$P(p)$
.10	.113
.20	.887

d) Expected payoff:
Build $41525
Do not build $0

Expected loss:
Build $2825
Do not build $44350

***8–23.**

e	$P(e\|x)$
e_1	.538
e_2	.462

8–24. a) Expected payoff of $a_1 = -2$; expected payoff of $a_2 = 0$; so a_2 has the higher expected payoff.
b) Expected payoff of $a_1 = +.76$; expected payoff of $a_2 = 0$; so a_2 has the higher expected payoff.

***8–25.** $P(e_1|y) = 1/33$
$P(e_2|y) = 4/33$
$P(e_3|y) = 9/33$
$P(e_4|y) = 8/33$
$P(e_5|y) = 5/33$
$P(e_6|y) = 6/33$

8–26. $z = -1.75$. Since this is less than -1.645, reject H_0.

8–27. A significant z value means only that the difference was large *in terms of its own standard error.* This could be due to a low standard error due to a large sample size rather than to a big difference between the means of boys and girls.

8–28. Yes. We can say with 100% confidence that the pupil–teacher ratio is 31,360 to 1,120, or 28.00 to 1. We have population values, so that a statistical test is not called for.

8–29. Yes. $z = 4.23$. Since this is greater than 2.33, reject H_0. (In this problem you should use the finite universe correction factor.)

8–30. No. $t = 1.44$. Since this is less than 1.753, accept H_0.

8–31. No. $t = 1.414$. Since this is less than 2.132, accept H_0. You cannot say the cars meet standards at the .05 level of significance.

8–33. a) Reject H_0.
b) Accept H_0.
c) Rejecting H_0 means the sample results are significant.

8–34. a) The Type I Error would be to say that the count was less than 100, when in fact it was at least 100.
b) The Type II Error would be to say that the count was 100, when in fact it was less than 100.

8–35. a) The result is significant at the .01 level $z = 2.67$
b) You should not come to this conclusion unless you have reason to believe that pickerel caught in the first week of August are a random sample of the lengths of pickerels.

8–36. a) No. $t = 1.333$. This is less than 2.353. Accept H_0.
b) Yes. $t = 2.444$. This is greater than 2.353. Reject H_0.
c) Yes. $t = 3.55$. This is greater than 2.353. Reject H_0.

8–37. a) The drug is not harmless.
b) Type I Error: Say the drug is harmless when in fact it is not.
Type II Error: Say the drug is not harmless when in fact it is.

8–39. a) Yes. If $p = .2$, then the probability of 2 left-handed people in a sample of 2 is .04.

8–40. b) Expected payoff of Debating is +4000 votes.
Expected payoff of Not Debating is 0 votes.
c) Debating

8–41. a) $6000. Yes.
b) No. The expected utility of suing is 41. The expected utility of not suing is 60.

***8–42.** Whatever the sum you selected, the utility of paying that amount should be equal to 99.

***8–43.** a) On the basis of prior probabilities, accepting has an expected payoff of +$100; rejecting has an expected payoff of $0. Therefore accepting has the higher expected payoff.
b) Posterior probability of e_1 is .1935.

Expected payoff of a_1 is -103.2.
Better action is a_2.

CHAPTER 9

9–1. $z = 1.60$. Since this is between -1.96 and $+1.96$, accept H_0.

9–2. No. $t_{.05}$, 14 dof $= \pm 2.145$. $t = -2.025$. Accept H_0.

9–3. Yes. $z_{.01} = \pm 2.576$. $z = 3.00$. Reject H_0.

9–4. No. $t_{.05}$, 8 dof $= -1.860$. $t = -1.00$. Not significant.

9–5. a) Yes. $t_{.05}$, 4 dof $= 2.132$. $t = \infty$. Reject H_0.

9–6. Yes. $z_{.01} = -2.33$. $z = -5.93$. Significant.

9–8. $z_{.10} = \pm 1.645$. $z = -.69$. Accept H_0.

9–9. $z_{.05} = \pm 1.96$. $z = 1.41$. Accept H_0. There is not sufficient evidence that men and women differ in their opinions.

9–10. a) Yes. You do not need to run a statistical test to answer the question.
b) Yes. $z_{.01} = \pm 2.576$. Since $z = 3.43$, the difference is significant at the .01 level.

9–11. a) You must assume that these are independent random samples of days under Republican and Democratic administrations.
b) $z = 2.66$. This is significant at the .05 and .01 levels.

9–12. $\chi^2_{.05}$, 1 dof $= 3.84$. $\chi^2 = 1.500$. Accept H_0.

9–13. No. $\chi^2_{.01}$, 2 dof $= 9.21$. $\chi^2 = 8.58$. Accept H_0.

9–14. Yes, since combining the "Jewish" and "Other" groups, we find χ^2 to be 13.84 and, $\chi^2_{.05}$, 4 dof $= 9.49$.

9–15. $\chi^2_{.01}$, 4 dof $= 13.3$. $\chi^2 = 6.26$. Accept H_0.

***9–16.** $\chi^2 = 11.01$. This is significant at the .01 level.

9–17. $\chi^2_{.01}$, 4 dof $= 13.3$. $\chi^2 = 53.51$. Reject H_0.

9–18. a) $\chi^2_{.05}$, 1 dof $= 3.84$. $\chi^2 = 0$.
b) The value of χ^2 is remarkably low, suggesting that the variables are not independent. However, in this case the dependency appears to be due to a policy of laying off equal percentages of males and females.

9–19. a) True d) Not true
b) Not true e) True
c) Not true f) True

9–20. a) Normal distribution e) t distribution
b) Normal distribution f) t distribution
c) Normal distribution g) t distribution
d) Normal distribution h) Binomial distribution

9–21. No. $z_{.05} = \pm 1.96$. $z = -1.78$. Not significant.

9–22. Find the opinions of 1000 voters, both before and after the speech.

9–23. a) No. $z_{.05} = \pm 1.96$. $z = 1.632$. Not significant.

9–24. $t_{.05}$, 10 dof $= \pm 2.228$. $t = 2.28$. Reject H_0.

9–25. a) No. $z_{.10} = \pm 1.645$. $z = -1.20$. Not significant.

c) Yes. The standard error of the difference is particularly sensitive to the smaller of the two samples.

9–26. No. $z_{.01} = \pm 2.576$. $z = 1.52$. Accept H_0.

9–27. The two teams differ by 1.5 pounds. These are population values; no statistical tests should be run.

9–28. No. $t_{.10}$, 10 dof $= \pm 1.812$. $t = -1.16$. Accept H_0.

9–29. a) Reject H_0. $t_{.05}$, 4 dof $= \pm 2.776$. $t = 2.98$.
b) One other possible source of difference is age, since husbands generally tend to be older than their wives.

9–30. No. $z_{.05} = 1.645$. $z = 1.45$. Accept H_0.

9–31. a) H_0: $p = .03$. The lot is acceptable. A low value of α (perhaps .01) is called for.
b) H_0: $\mu = 10$. The usage is at least 10 times a week. A high value of α (perhaps .10 or .20) is appropriate for this small study.
c) H_0: $p_1 = p_2$. The promotion rates for men and women (in the long run) are the same. Because of the serious consequences of rejecting H_0, perhaps a low value of α (such as .01) is appropriate.
d) In this case it does not matter which way you set up the hypothesis, if there is no prior reason for assuming that George (or Sue) is superior. The level of significance is .50.

9–32. No. $\chi^2_{.05}$, 1 dof $= 3.84$. $\chi^2 = 3.71$. We do not have significant evidence that thoroughbreds differ from mutts.

9–33. Accept H_0. $\chi^2_{.05}$, 2 dof $= 6.00$. $\chi^2 = 5.000$.

9–34. Accept H_0. $\chi^2_{.01}$, 2 dof $= 9.21$. $\chi^2 = 8.7$. Combine the teachers and principals.

9–35. a) Yes. $\chi^2_{.01}$, 4 dof $= 13.3$. $\chi^2 = 25.05$. Reject H_0.
b) Yes. $\chi^2_{.01}$, 3 dof $= 11.4$. $\chi^2 = 11.96$. Reject H_0.

CHAPTER 10

10–1. a) $\hat{Y} = 1.775 + .5083X$
c) $s_{y \cdot x} = .328$
d) $r^2 = .215$
e) $t_{.05}$, 8 dof $= \pm 2.306$ $t = 1.5$. Not significant.

10–2. $\hat{\rho}^2 = .12$

10–3. a) $\hat{Y} = 2.54X$
b) $s_{y \cdot x} = 0$
c) $r^2 = 1$
d) 2.54 ± 0

10–4. a) $r^2 = 1 - 50/120 = .583$
b) $\rho^2 = 1 - 25/40 = .375$

10–5. a) $r^2 = .388$
b) $\hat{\rho}^2 = .375$

10–7. a) $\hat{Y} = 9.2 + 2.4X$ b) 2.4 ± 1.219 c) 14 minutes

10–8. a) $r = -.857$

$t_{.05}$, 4 = ± 2.776. t = −3.3. Reject H_0.
There is evidence that $\rho \neq 0$.

10–9. a) Contagious diseases might affect more staff persons at the same time they afflict more of the general public who are admitted to the hospital.
b) $t_{.05,100}$ = ± 2.00. t = 3.14. Reject H_0.
The relationship is statistically significant. However, since r^2 is only .09, the admission rate explains only a small proportion of the sickness rate among the staff.

10–10. a) $t_{.05}$, 8 dof = ± 2.306. t = −2.4. Reject H_0.
b) If you are interested in the effect of inoculations on flu cases (and not vice versa), then a regression analysis would probably be more appropriate.

10–11. a) $s_y^2 = 7.50$
b) $s_{y \cdot x}^2 = 2.92$
c) $r^2 = .628$
d) $\hat{\rho}^2 = .611$
e) $t_{.05}$, 22 dof = 2.074. t = −6.09. Reject H_0.

10–12. a) $r = -1.00$
b) $s_x = 14.085$
$s_y = 14.085$
c) $s_{y \cdot x} = 0$
d) $s_{x \cdot y} = 0$
e) $t_{.01}$, 4 dof = ± 4.604. $t = -\infty$. Reject H_0.

***10–13.** Yes, since the value of t will be plus or minus infinity.

10–14. a) s_y^2 = 8738.83 (original units are hundreds of dollars)
$s_{y \cdot x}^2 = 8445.95$
$s_{x \cdot y}^2 = 9453.50$
b) $r^2 = .1026$
$\hat{\rho}^2 = .0335$
c) $t_{.05}$, 13 dof = ± 2.160. t = 1.22. Not significant. Accept H_0.

10–15. a) $r = .421$ and $r^2 = .177$
b) $\rho^2 = .132$
c) $t_{.05}$, 18 dof = 2.101. t = 1.97. Accept H_0.
d) $\hat{Y}^1 = 5.026 + .10526X$
e) $\hat{X}^1 = 34.136 + 1.68421Y$
f) $bb^1 = .177$

***10–16.** The statistics have the same numerator. The denominators of regression coefficients and correlation coefficients can never be negative.

10–17. No. There may be two or more cause-and-effect linkages that may offset each other, so that the statistical relationship is zero. Also, since r measures only the linear relationship, there may be a nonlinear relationship that does not show up in the linear correlation coefficient.

10–18. a) r = .258.
c) No. The set of random digits is not normally distributed.

***10–19.** a) 17.176
b) For every unit increase in class size, you can predict a decrease of .32

in math achievement score, if family income is held constant. For every unit increase in family income, you can predict an increase of .0025 in math achievement score, if class size is held constant.

***10–20.** b) $10,689
c) $6851
d) Yes, in spite of the negative coefficient of X_1. This coefficient controls for sex. But if the males in the sample tend to be much older than the females and earn more, overall you may find a positive effect on income due to age.

***10–21.** b) 12.327 units

***10–22.** b) $\hat{Y} = 168.8 + 7.68X$
c) $\hat{Y} = 160(1.048)^x$

***10–23.** a) $\hat{Y} = 1820 - 20.5X$
$s_y^2 = 555000$
$s_{y \cdot x}^2 = 179666.67$
b) $X = 76.9 - .0369Y$
$s_{x \cdot y}^2 = 323.72$
$s_x^2 = 1000$
c) $r^2 = .757$

***10–24.** $r = -.87$
$t_{.05}$, 3 dof = ± 3.182. $t = 3.06$. Accept H_0.

***10–25.** a) 56.7 years
b) Not necessarily. If there is a cumulative effect of previous cigarettes on life expectancy, then this effect will not likely be eliminated immediately if one stops smoking. And the effect of cigarette smoking on life expectancy could be partially due to other variables that are correlated with cigarette smoking that are not treated in the equation.
c) 5.27 years
d) No. Such a conclusion cannot be drawn from a variable that is not in the model.

10–26. A second degree (parabolic) equation would seem to be a good fit.

10–27. a) $\hat{Y} = 41.28 - .143X_1$
b) $\hat{Y} = 33.25 - .112X_2$

10–28. a) $\hat{Y} = 11{,}529.78 + 140.288X$
b) $t_{.05}$, 8 dof = ± 2.306. $t = .858$. Accept H_0.

10–29. a) .084 b) zero

10–30. a) $r = .94$
b) $t_{.05}$, 4 dof = ± 2.776. $t = 5.51$. Reject H_0.

10–31. a) $r = .60$
b) $t_{.10}$, 8 dof = 1.860. $t = 2.12$. Reject H_0.

10–32. a) $\hat{Y} = 84.45 + .95X$
b) $s_{y \cdot x} = 31.27$
c) $r^2 = .6045$
d) $t_{.05}$, 11 dof = 2.201. $t = 4.10$. Reject H_0.

10–33. a) $r^2 = .0625$
b) $t_{.01}$, 120 dof = 2.576. $t = 2.90$. Reject H_0.

CHAPTER 11

11–1. (1) Parametric; (2) nonparametric; (3) parametric; (4) nonparametric; (5) neither.

11–2. Lower critical value: 5 runs
Upper critical value: 15 runs
Number of runs is 9. Therefore accept H_0.
The sequence is random.

11–3. a) Lower critical value: 5 runs
Numbers of runs is 9. Accept H_0.
The sequence is random, and there is no upward or downward trend.
b) Lower critical value: 5 runs
Number of runs is 2. Reject H_0.
The sequence is not random, and there is an upward trend.

11–4. $\chi^2_{.05}$, 2 dof = 6.00. $\chi^2 = 4.00$. Not significant.

11–5. a) Use sign test. $n = 32$. $z_{.05} = \pm 1.96$. $z = 1.41$. Accept H_0.
b) Yes. There is insufficient evidence of a voter shift.

11–6. Yes. If H_0 is true, the probability that all 8 would show improvement is .004.

11–7. $\chi^2_{.01}$, 1 dof = 6.64. $\chi^2 = 6.827$. Reject H_0.

11–8. a) Only if you assume that grades can be considered interval level of measurement and that they are approximately normally distributed.
b) Median test: $\chi^2_{.05}$, 1 dof = 3.84. $\chi^2 = 2.48$. Not significant.
Mann-Whitney test: $z_{.05} = \pm 1.96$. $z = 1.36$. Not significant.

11–9. a) $t_{.05}$, 38 dof = $\pm$ 2.021. $t = 2.155$. Reject H_0.
b) Use Mann-Whitney test. $z_{.05} = \pm 1.96$. $z = 2.105$. Reject H_0.
c) The distribution of the data does not appear to support the assumption.

11–10. a) $\chi^2 = .26$ (using Yates' Correction). Not significant.
b) $t_{.10}$, 12 dof = 1.782. $t = 1.19$. Not significant.

11–11. $\chi^2_{.05}$, 2 dof = 6.00. $\chi^2 = 5.8$. Not significant.

11–12. a) Using Mann-Whitney test: $z = 1.75$. Not significant at .05 for two-tail test. Using K-S test; $\chi^2 = 3.33$. Not significant.

11–13. a) $r_s = -.60$
b) Not significant

11–14. a) $r_s = -1.00$ This is significant at the .01 level.
b) Either get 2 TV sets, give up TV, or separate from each other.

11–15. a) $r_s = +.472$
b) Significant at the .05 level

11–16. $r_s = -.47$. Not significant.

11–17. $t = -.94$. Not significant.

11–18. b) $W = .3125$
c) $\chi^2 = 10.00$. $\chi^2_{.05}$, 8 dof = 15.5. Not significant.

11–19. $W = 1.00$. Significant at .01 level.

***11–20.** One possible solution:

	I	*II*	*III*	*IV*	*V*
A	1	2	3	4	5
B	4	5	1	2	3
C	4	2	5	3	1

11–21. $n_1 = 21$; $n_2 = 9$. Number of runs is 10. Not significant. Accept H_0.

11–22. a) $z_{.05} = 1.645$. $z = 2.89$. Reject H_0.
b) You must assume that the employees who were interviewed were a random sample of the universe.

11–23. a) Using the median test: $\chi^2_{.05}$, 1 dof = 3.84. Using Yates' Correction, χ^2 is .416. Accept H_0.

11–24. a) $r_s = .456$. $t_{.05}$, 14 dof = 2.145. $t = 1.92$. Not significant.
b) $r = .1595$. $t_{.05}$, 14 dof = 2.145. $t = .604$. Not significant.
c) Although the relationship appears to be positive, one very high income was paired with a contribution of zero. This extreme pair had a much greater effect in reducing the Pearson r than the rank correlation coefficient.

11–25. a) There was 1 gain and there were 14 losses. For $n = 15$, $p = .5$, the probability of one or fewer successes is .005. Therefore we have significance at the .05 level.
b) Six people lost less than 10 pounds. Ten people lost more than 10 pounds. Under H_0, the probability that no more than 6 would lose under 10 pounds is .2272. This is not significant. Accept H_0.
c) $\chi^2_{.05}$, 1 dof = 3.84. χ^2 (using Yates' Correction, counting 13 as "above median") = .83. Not significant.
d) $r_s = -.507$. $t_{.05}$, 14 dof = $\pm$ 2.145. $t = -2.20$. Reject H_0.

11–26. $r_s = -.423$. $t_{.05}$, 11 dof = $\pm$ 2.201. $t = -1.548$. Accept H_0.

11–27. a) $r_s = -.100$. $t_{.05}$, 3 dof = $\pm$ 3.182. $t = -.17$. Not significant.
b) $r = -.705$. $t_{.05}$, 3 dof = $\pm$ 3.182. $t = -1.72$. Not significant.

11–28. $n_1 = 8$; $n_2 = 8$. 14 runs is critical at the .05 level. Reject H_0.

11–29. $SS = 254$. $SS_{.05}$ ($k = 3$, $n = 7$) = 157.3. Reject H_0. There is a tendency for students to have similar rankings.

CHAPTER 12

12–4. b) You would need to know the proportion that were cured within a week without the remedy.

12–5. b) Of the 8 white males, assign 4 to each group.
Of the 6 black males, assign 3 to each group.
Of the 10 white females, assign 5 to each group.
Of the 6 black females, assign 3 to each group.

***12–6.** $-.277$ to .277

***12–7.** 193 in each group

***12–9.** a) List the possible divisions into experimental and control groups. Unfortunately, each one leaves at least one variable uncontrolled.

Experimental Group	*Control Group*	*Variables Not Controlled*
AB	CD	College, Religion
AC	BD	Party
AD	BC	Marital Status
BC	AD	Marital Status
BD	AC	Party
CD	AB	College, Religion

12–14. Differences in abilities of students who enter
Differences in family incomes of students who enter
Differences in nationality of students who enter
Differences in ages of students who enter
Differences in sex ratios of students who enter
Differences in motivations of students who enter

12–16. a) $z_{.01} = \pm 2.58$. $z = 7.80$. Reject H_0. The groups differ.
b) No. The causation may be the other way around. Or perhaps there are common factors causing people to smoke and to be overweight.

12–17. a) No. You would need at least 8 individuals in the experimental group *and* 8 individuals in the control group, in order to have an individual in each of the 8 cells.
b) Yes.

12–20. b) $\overline{X}_1$, the mean monthly number of accidents in the 12 months *prior* to the installation of the light is 7.6666 accidents.

$\overline{X}_2$, the mean monthly number of accidents in the 12 months *after* the installation of the light is 5.00 accidents.

The difference between the means is −2.6666 accidents per month.

c) Running a test for difference between means, we have $t_{.05}$, 22 dof $= \pm 2.074$. $t = -2.242$. This is significant. It is necessary to assume that the samples are independent and come from a normally distributed population.
d) Median test: $\chi^2_{.05}$, 1 dof $= 3.84$. $\chi^2 = 2.667$. Not significant. You need to assume that the samples are independent.
e) If you can assume the corresponding months are related, then a test for paired observations is appropriate.

12–23. a) $t_{.05}$, 8 dof $= 1.860$ (one tail). $t = 2.76$. Significantly higher.
b) $t = 3.15$. Significantly higher.
c) $t_{.05}$, 8 dof $= \pm 2.306$. $t = .54$. Not significant.

12–25. a)

Group A	*Group B*	$\overline{X}_A$	$\overline{X}_B$	$\overline{X}_A \quad \overline{X}_B$
20,30	40,50	25	45	−20
20,40	30,50	30	40	−10
20,50	30,40	35	35	0
30,40	20,50	35	35	0
30,50	20,40	40	30	10
40,50	20,30	45	25	20

b) Put the 30 and 40 year old in one group, and the 20 and 50 year old in another. The mean ages in each group are the same, but you would still have a problem if you wanted to keep the variances in the two groups the same.

12–26. a) Call the four individuals M_1, M_2, F_1, F_2. Let P be the proportion of females. The six assigments are:

A	B	P_A	P_B	$P_A - P_B$
M_1M_2	F_1F_2	0	1	−1
M_1F_1	M_2F_2	.5	.5	0
M_1F_2	M_2F_1	.5	.5	0
M_2F_1	M_1F_2	.5	.5	0
M_2F_2	M_1F_1	.5	.5	0
F_1F_2	M_1M_2	1	0	1

b) Select an assignment from any of those for which $P_A - P_B = 0$. This can be done by randomly selecting a male for each group and a female for each group.

12–31. a) Maturation

b) Statistical regression, due to nonrandom assignments of the students to the two sections

Index